Official Journal of the 'Mediterranean League against Thromboembolic Diseases'

Hæmostasis

International Journal on Haemostasis and Thrombosis Research

KARGER

Printed in Switzerland on acid-free paper by Reinhardt Druck, Basel

Appears bimonthly: 1 volume per year (6 issues)

Guidelines for Authors

Hæmostasis

Submission

Only original papers written in English that communicate novel findings in the field of haemostasis and thrombosis are considered. Both contributions to the basic understanding of these processes and reports on clinical management of disorders in this field are acceptable. Manuscripts can be sent either to the managing editor or to one of the editors as listed in each issue of Haemostasis.

Manuscripts should be submitted in triplicate (with three sets of illustrations of which one is an original), typewritten double-spaced on one side of the paper, with a wide margin.

Conditions

All manuscripts are subject to editorial review. A manuscript is to be considered rejected when an author does not return a revised manuscript within 3 months of receiving the suggestions for revision, unless the author explains further delay. Manuscripts are received with the explicit understanding that they are not under simultaneous consideration by any other publication. Submission of an article for publication implies the transfer of the copyright from the author to the publisher upon acceptance. Accepted papers become the permanent property of 'Haemostasis' and may not be reproduced by any means, in whole or in part, without the written consent of both the author and the publisher. It is the author's responsibility to obtain permission to reproduce illustrations, tables, etc. from other publications.

Arrangement

Title page: The first page of each paper should indicate the title, the authors' names, and the institute where the work was conducted. A short title for use as running head as well as the full address of the author to whom correspondence should be sent are also required.

Full address: The exact postal address complete with postal code must be given at the bottom of the title page. Please also supply phone and fax numbers, as well as your e-mail address.

Key words: For indexing purposes, a list of 3–10 key words in English is essential.

Abstract: Each paper needs an abstract of up to 10 lines.

Small type: Paragraphs which can or must be set in smaller type (case histories, test methods, etc.) should be indicated with a 'p' (petit) in the margin on the left-hand side.

Footnotes: Avoid footnotes. When essential, they are numbered consecutively and typed at the foot of the appropriate page.

Tables and illustrations: Tables and illustrations (both numbered in Arabic numerals) should be prepared on separate sheets. Tables require a heading and figures a legend, also prepared on a separate sheet. For the reproduction of illustrations, only good drawings and original photographs can be accepted; negatives or photocopies cannot be used. Due to technical reasons, figures with a screen background should not be submitted. When possible, group several illustrations on one block for reproduction (max. size 144 × 196 mm) or provide crop marks. On the back of each illustration, indicate its number, the author's name, and 'top' with a soft pencil.

Color illustrations: Up to 6 color illustrations per page can be integrated within the text at the reduced price of CHF 660.– / USD 545.00 per page. Color illustrations are reproduced at the author's expense.

References: In the text identify references by Arabic numerals [in square brackets]. Material submitted for publication but not yet accepted should be noted as 'unpublished data' and not be included in the reference list. The list of references should include only those publications which are cited in the text. Do not alphabetize; number references in the order in which they are first mentioned in the text. The surnames of the authors followed by initials should be given. There should be no punctuation other than a comma to separate the authors. Cite all authors, 'et al' is not sufficient. Abbreviate journal names according to the Index Medicus system. (Also see International Committee of Medical Journal Editors: Uniform requirements for manuscripts submitted to biomedical journals. N Engl J Med 1997;336:309–315.)

(a) Papers published in periodicals: Sun J, Koto H, Chung KF: Interaction of ozone and allergen challenges on bronchial responsiveness and inflammation in sensitised guinea pigs. Int Arch Allergy Immunol 1997;112:191–195.

(b) Monographs: Matthews DE, Farewell VT: Using and Understanding Medical Statistics, ed 3, revised. Basel, Karger, 1996.

(c) Edited books: Parren PWHI, Burton DR: Antibodies against HIV-1 from phage display libraries: Mapping of an immune response and progress towards antiviral immunotherapy; in Capra JD (ed): Antibody Engineering. Chem Immunol. Basel, Karger, 1997, vol 65, pp 18–56.

Page charges

There are no page charges for papers of 6 or fewer printed pages (including tables, illustrations and references). Each additional complete or partial page is charged to the author at CHF 275.– / USD 196.00. The allotted size of a paper is equal to approx. 12 manuscript pages (including tables, illustrations and references).

Galley proofs

Unless indicated otherwise, galley proofs are sent to the first-named author and should be returned with the least possible delay. Alterations made in galley proofs, other than the correction of printer's errors, are charged to the author. No page proofs are supplied.

Reprints

Order forms and a price list are sent with the galley proofs. Orders submitted after the issue is printed are subject to considerably higher prices.

KARGER

Fax +41 61 306 12 34
E-Mail karger@karger.ch
www.karger.com

The Guidelines for Authors are available at:
www.karger.com/journals/hae/hae_gl.htm

Hæmostasis

General Information

Publication data: 'Haemostasis' is published 6 times a year. Volume 31 with 6 issues appears in 2001.

Subscription rates: Subscriptions run for a full calendar year. Prices are given per volume.
Personal subscription:
Subscription + postage and handling
CHF 322.– + CHF 27.40 Europe
CHF 322.– + CHF 40.80 Overseas
USD 247.50 + USD 27.40
+ online subscription per volume:
CHF 39.– / USD 30.00
(Must be in the name of, billed to, and paid by an individual. Order must be marked 'personal subscription'.)
Institutional subscription:
Subscription + postage and handling
CHF 644.– + CHF 34.20 Europe
CHF 644.– + CHF 46.20 Overseas
USD 495.00 + USD 34.20
+ online subscription per volume:
CHF 58.– / USD 45.00
Airmail postage: CHF 36.–, USD 30.00.
(Extra per volume.)
Discount subscription prices:
Gesellschaft für Thrombose- und Haemostase-Forschung;
International Society of Haematology.

Microforms: Please direct all enquiries and orders to: University Microfilms Inc.
300 North Zeeb Road, P.O. Box 1346
Ann Arbor, MI 48106-1346 (USA)

Subscription orders: Orders can be placed at agencies, bookstores, directly with the
Publisher S. Karger AG, P.O. Box,
CH–4009 Basel (Switzerland)
or with any of the following distributors:
Deutschland: S. Karger GmbH, Postfach
D–79095 Freiburg
France: Librairie Luginbühl
36, bd de Latour-Maubourg
F–75007 Paris
Great Britain, Ireland: S. Karger Publishers
58 Grove Hill Road
Tunbridge Wells, Kent TN1 1SP, UK
USA: S. Karger Publishers, Inc.
26 West Avon Road, P.O. Box 529
Farmington, CT 06085, USA
India, Bangladesh, Sri Lanka:
Panther Publishers Private Ltd.
39, 6th Cross
Wilson Gardens
Bangalore 560 027, India
Japan: Katakura Libri, Inc.
2F, Marushima Bldg, 4-6-17, Yushima,
Bunkyo-ku, Tokyo 113-0034, Japan

Change of address: Both old and new address should be sent to the subscription source.

Single issues and back volumes: Information can be obtained through the Publisher.

Advertising: Correspondence and rate requests should be addressed to the Publisher.

Bibliographic indices: This journal is regularly listed in bibliographic services, including *Current Contents®*.

KARGER

Fax+ 41 61 306 12 34
E-Mail karger@karger.ch
www.karger.com

The Journal Home Page is available at:
www.karger.com/journals/hae

Hæmostasis

List of Editors (continued)

KARGER

Fax+ 41 61 306 12 34
E-Mail karger@karger.ch
www.karger.com

The Journal Home Page is available at:
www.karger.com/journals/hae

Hæmostasis

EFTH

International Conference on Exogenous Factors Affecting Thrombosis and Haemostasis

Paris, July 12–13, 2001

In memoriam Stefan Niewiarowski 1926–2001

Guest Editors
Cassian Bon, Paris
Manjunatha Kini, Singapore
Francis S. Markland, Jr., Los Angeles, Calif.
Neville A. Marsh, Brisbane
Jan Rosing, Maastricht

47 figures, 2 in color, 16 tables, 2002

KARGER

Basel · Freiburg · Paris · London · New York · New Delhi · Bangkok · Singapore · Tokyo · Sydney

Vol. 31, No. 3–6, 2001

Hæmostasis

Contents

In memoriam

117 **Stefan Niewiarowski**
1926–2001
Budzynski, A.Z. (Philadelphia, Pa.)

Fibrinolysis

118 **Vampire Bat Plasminogen Activator DSPA-Alpha-1 (Desmoteplase): A Thrombolytic Drug Optimized by Natural Selection**
Schleuning, W.-D. (Berlin)

123 **Proteases from *Vipera lebetina* Venom Affecting Coagulation and Fibrinolysis**
Siigur, J.; Aaspõllu, A.; Tõnismägi, K.; Trummal, K.; Samel, M.; Vija, H.; Subbi, J.; Siigur, E. (Tallinn)

133 **Snake Venom Proteinases as Tools in Hemostasis Studies: Structure-Function Relationship of a Plasminogen Activator Purified from *Trimeresurus stejnegeri* Venom**
Wisner, A.; Braud, S.; Bon, C. (Paris)

141 **Alfimeprase: Pharmacology of a Novel Fibrinolytic Metalloproteinase for Thrombolysis**
Toombs, C.F. (Thousand Oaks, Calif.)

Substances Affecting Platelet Function

148 **Multifunctional Snake C-Type Lectins Affecting Platelets**
Clemetson, K.J.; Navdaev, A.; Dörmann, D.; Du, X.-Y.; Clemetson, J.M. (Berne)

155 **The Use of Snake Venom Toxins as Tools to Study Platelet Receptors for Collagen and von Willebrand Factor**
Andrews, R.K. (Melbourne); Kamiguti, A.S. (Liverpool); Berlanga, O. (Oxford); Leduc, M. (Paris); Theakston, R.D.G. (Liverpool); Watson, S.P. (Oxford)

KARGER

Fax+41 61 306 12 34
E-Mail karger@karger.ch
www.karger.com

Access to full text and tables of contents, including tentative ones for forthcoming issues:
www.karger.com/journals/hae/hae_bk.htm

173 **Lebecetin, a C-Lectin Protein from the Venom of *Macrovipera lebetina* That Inhibits Platelet Aggregation and Adhesion of Cancerous Cells**
Sarray, S.; Srairi, N. (Tunis); Luis, J.; Marvaldi, J. (Marseille); El Ayeb, M.; Marrakchi, N. (Tunis)

177 **New Insights on Disintegrin-Receptor Interactions: Eristostatin and Melanoma Cells**
McLane, M.A.; Kuchar, M.A.; Brando, C.; Santoli, D.; Paquette-Straub, C.A.; Miele, M.E. (Newark, Del./Philadelphia, Pa.)

183 **A Novel Snake Venom Disintegrin That Inhibits Human Ovarian Cancer Dissemination and Angiogenesis in an Orthotopic Nude Mouse Model**
Markland, F.S.; Shieh, K.; Zhou, Q.; Golubkov, V.; Sherwin, R.P.; Richters, V.; Sposto, R. (Los Angeles, Calif.)

192 **Viper Venom Components Affecting Angiogenesis**
Huang, T.F.; Yeh, C.H.; Wu, W.B. (Taipei)

207 **Lebetin Peptides: Potent Platelet Aggregation Inhibitors**
Marrakchi, N. (Tunis); Mabrouk, K. (Marseille); Regaya, I.; Sarray, S. (Tunis); Fathallah, M.; Rochat, H. (Marseille); El Ayeb, M. (Tunis)

Procoagulant Proteins and Diagnostic Agents

211 **Diagnostic Uses of Snake Venom**
Marsh, N.A. (Brisbane)

218 **Procoagulant Proteins from Snake Venoms**
Kini, R.M. (Singapore/Richmond, Va.); Rao, V.S.; Joseph, J.S. (Singapore)

225 **Snake Venom Activators of Factor X: An Overview**
Tans, G.; Rosing, J. (Maastricht)

234 **Snake Venom Prothrombin Activators Homologous to Blood Coagulation Factor Xa**
Joseph, J.S. (Singapore); Kini, R.M. (Singapore/Richmond, Va.)

241 **Factor V Activation and Inactivation by Venom Proteases**
Rosing, J.; Govers-Riemslag, J.W.P.; Yukelson, L.; Tans, G. (Maastricht)

247 **Molecular Basis for the Partition of the Essential Functions of Thrombin among Snake Venom Serine Proteinases.** The Case of Thrombin-Like Enzymes
Maroun, R.C. (Paris)

257 **Effects of Lopap on Human Endothelial Cells and Platelets**
Chudzinski-Tavassi, A.M. (São Paulo); Schattner, M. (Buenos Aires); Fritzen, M. (São Paulo); Pozner, R.G. (Buenos Aires); Reis, C.V.; Lourenço, D. (São Paulo); Lazzari, M.A. (Buenos Aires)

Anticoagulant Proteins and Diagnostic Agents

266 **Protein C Activators from Snake Venoms and Their Diagnostic Use**
Gempeler-Messina, P.M.; Volz, K.; Bühler, B.; Müller, C. (Basel)

273 **Interaction of Bothrojaracin with Prothrombin**
Zingali, R.B.; Bianconi, M.L.; Monteiro, R.Q. (Rio de Janeiro)

279 **Anticoagulant Venom and Mammalian Secreted Phospholipases A_2: Protein- versus Phospholipid-Dependent Mechanism of Action**
Mounier, C.M. (Cergy-Pontoise); Bon, C. (Paris); Kini, R.M. (Singapore/Richmond, Va.)

288 ***Lonomia* Genus Caterpillar Envenomation: Clinical and Biological Aspects**
Arocha-Piñango, C.L.; Guerrero, B. (Caracas)

294 **Identification of Anticoagulant Activities in Salivary Gland Extracts of Four Horsefly Species (Diptera, Tabanidae)**
Kazimírová, M.; Šulanová, M.; Kozánek, M.; Takáč, P.; Labuda, M. (Bratislava); Nuttall, P.A. (Oxford)

306 **Author Index Vol. 31, No. 3–6, 2001**
307 **Subject Index Vol. 31, No. 3–6, 2001**

308 **Author Index Vol. 31, 2001**
310 **Subject Index Vol. 31, 2001**
after 312 **Contents Vol. 31, 2001**

S. Karger
Medical and Scientific Publishers
Basel · Freiburg · Paris · London
New York · New Delhi · Bangkok
Singapore · Tokyo · Sydney

Drug Dosage
The authors and the publisher have exerted every effort to ensure that drug selection and dosage set forth in this text are in accord with current recommendations and practice at the time of publication. However, in view of ongoing research, changes in government regulations, and the constant flow of information relating to drug therapy and drug reactions, the reader is urged to check the package insert for each drug for any change in indications and dosage and for added warnings and precautions. This is particularly important when the recommended agent is a new and/or infrequently employed drug.

P.O. Box, CH–4009 Basel (Switzerland)
Printed in Switzerland on acid-free paper by Reinhardt Druck, Basel
ISBN 3–8055–7414–2

In memoriam

Hæmostasis

Haemostasis 2001;31:117

Stefan Niewiarowski MD, PhD

1926–2001

On August 25, 2001, the Sol Sherry Thrombosis Research Center at Temple University School of Medicine in Philadelphia lost one of its founding investigators, when Stefan Niewiarowski, MD, PhD, died after a long battle with bladder cancer. A professor at the Thrombosis Center and the Department of Physiology, he was a highly original and motivating leader in hemostasis and thrombosis research. He received his degrees from the University of Warsaw in Poland, and published his first paper in 1951. That was followed by almost 300 additional publications during his outstanding career. Dr. Niewiarowski began his research career in Warsaw, then moved on to Paris, Stockholm, Boston, Hamilton (Ontario), and finally to Philadelphia in 1972. Here, at Temple University, he was a group leader focusing his work on peptides and proteins that are secreted from blood platelets or affected platelet functions. He was a recipient of many honors and awards given in recognition of his eminent research accomplishments, but he told me that his greatest satisfaction came from opening the field of disintegrins in 1987. Notably, he studied disintegrins to the end of his productive life. Dr. Niewiarowski recognized early that agents present in snake venoms and blood-sucking animals represent precious tools for studies on the structure and function of proteins and cells involved in hemostasis. As a result, he collaborated with a number of investigators including myself. At the time of the Conference on Exogenous Factors affecting Thrombosis and Haemostasis he was very ill and unable to attend the conference in Paris, and a lecture ('New Insights on Disintegrin-Receptor Interactions') was presented for him by one of his former students. His absence as researcher, colleague and friend leaves many of us in deep sorrow and grief.

Andrei Z. Budzynski, PhD, Philadelphia, Pa.

KARGER

Fax + 41 61 306 12 34
E-Mail karger@karger.ch
www.karger.com

0301–0147/01/0316–0117$17.50/0

Accessible online at:
www.karger.com/journals/hae

Fibrinolysis

Hæmostasis

Haemostasis 2001;31:118–122

Vampire Bat Plasminogen Activator DSPA-Alpha-1 (Desmoteplase): A Thrombolytic Drug Optimized by Natural Selection

Wolf-Dieter Schleuning

Paion GmbH, Research Center Berlin, Berlin, Germany

Key Words
Plasminogen activator · DSPA · Desmoteplase · Thrombolytic · Fibrinolytic · Fibrin specificity · Stroke

Abstract
Plasminogen activators are enzymes found in all vertebrate species investigated so far. Their physiological function is the generation of localized proteolysis in the context of tissue remodeling, wound healing and neuronal plasticity. The common vampire bat *(Desmodus rotundus)* is a New World species that feeds exclusively on blood. Its saliva contains highly potent plasminogen activators, specialized in rapid lysis of fresh blood clots. Biochemical and pharmacological evidence indicates that these plasminogen activators represent a new class of thrombolytics with pharmacological and toxicological properties superior to human tissue-type plasminogen activator, the clot dissolving agent now most frequently used in medicine. A form of the enzyme produced by recombinant DNA technology is currently employed to test this hypothesis in clinical studies.

Introduction

Thrombolytic therapy with plasminogen activators (PAs) is performed routinely as a treatment of acute myocardial infarction as well as selected cases of cerebral vascular occlusion and pulmonary embolism. Thrombolytic agents such as streptokinase (SK), tissue-type PA (t-PA), and urokinase-type plasminogen activators (u-PA) are potent drugs, but they have various shortcomings related to their pharmacodynamic, pharmacokinetic

KARGER

Fax + 41 61 306 12 34
E-Mail karger@karger.ch
www.karger.com

0301–0147/01/0316–0118$17.50/0

Accessible online at:
www.karger.com/journals/hae

Wolf-Dieter Schleuning
Paion GmbH, Research Center Berlin, Tegeler Weg 33
D–10589 Berlin (Germany)
Tel. +49 30 34 35 37 12, Fax +49 30 34 35 37 33
E-Mail wd.schleuning@research.paion.de

and safety profiles. Vampire bats are the only mammalian species feeding exclusively on blood. To support this diet, their saliva contains very potent PAs, originally described by Hawkey [1]. In contrast to t-PA with its more subtle role in wound healing and neuronal plasticity, vampire bat salivary PA has been optimized by natural selection for the rapid lysis of fresh blood clots.

Biochemistry

In order to exploit this extraordinary evolutionary adaptation for medical purposes, we have cloned, expressed and characterized salivary PAs from the vampire bat *Desmodus rotundus.* Four *D. rotundus* salivary PAs (DSPAs) were identified which, like t-PA and u-PA, are composed of various conserved domains known from related proteins. DSPAα_1 and DSPAα_2 exhibit the structural formulas finger (F), EGF (E), kringle (K), protease (P), and DSPAβ and DSPAγ, the formulas EKP and KP, respectively. Subtle sequence differences and data from Southern blot hybridization indicate that the four enzymes are coded by four different genes and are not generated by differential splicing of a single primary transcript [2, 3]. A preliminary biochemical and pharmacological analysis indicated that DSPAα_1 exhibited the most favorable profile and was therefore chosen for further study. A recombinant CHO cell line for the production of DSPA and a purification protocol were established to obtain material that fulfilled the specifications for preclinical and clinical development [4]. The most important feature that distinguishes DSPA from other plasminogen activators is its extraordinary fibrin specificity. In fact, the activity of DSPAα_1 is 105,000 times higher in the presence of fibrin than in its absence. The respective factor for t-PA is only 550. Likewise, fibrinogen, a fairly potent cofactor of plasminogen activation by t-PA, has hardly any effect on DSPAα_1. Therefore the factor of fibrin selectivity expressed as the quotient of activity in the presence of fibrin versus activity in the presence of fibrinogen is 12,900 for DSPA but only 72 for t-PA [5].

In order to explore the molecular basis of this fastidious dependence on a fibrin cofactor, the recombinant catalytic domain of DSPAα_1 has been crystallized in a covalent complex with Glu-Gly-Arg-chloromethyl ketone and its structure solved at 2.9 Å resolution. The structure is similar to that of activated two-chain human t-PA. Despite its single-chain status, the activation domain is observed in an enzymatically active conformation, with a functional substrate binding site and an active site accommodating the peptidylmethylene inhibitor. The activation pocket, which normally receives the N-terminal Ile16, is occupied by the side chain of Lys156, whose distal ammonium group forms an internal salt bridge with the carboxylate group of Asp194. Lys156 is in a groove shielded from the bulk solvent by the intact 'activation loop' (Gln10-Phe21), favoring Lys156-Asp194 salt bridge formation and stabilization of a functional substrate binding site. Together with the characteristic 186 insertion loop, the activation loop could act as a switch, effecting full single-chain enzymatic activity upon binding to fibrin [6].

Neurotoxic Potential of t-PA

Kingston et al. [7] studied the effects of Alzheimer amyloid β-peptide analogues on the activity of t-PA in vitro. These authors discovered that the peptides exhibited a marked stimulatory effect upon plasminogen activation by t-PA, similar to known stimulators of t-PA such as fibrinogen. This activity appeared to increase when β-peptides formed aggregated fibrillar structures equivalent to those found in amyloid deposits. This finding may have a bearing on the pathogenesis of hereditary cerebral hemorrhage with amyloidosis Dutch type and cerebral amyloid angiopathy-related cerebral hemorrhage. Such a mechanism may contribute to the intracerebral hemorrhages that have occurred in patients undergoing t-PA treatment for acute myocardial infarction. In contrast to t-PA, DSPAα_1 is neither stimulated by native nor by aggregated amyloid β-peptide analogues [Bringmann P., Donner P. and Schleuning W.-D., unpubl. obs.]. Besides its desired thrombolytic effects, t-PA also promotes neurodegeneration after intracerebral injection of excitotoxins, such as glutamate, and contributes to the spread of neuronal damage after a cerebral vascular occlusion, which is believed to be promoted by excitotoxins. Wang et al. [8] examined the role of t-PA in experimental cerebral ischemia, independent of its effect as a thrombolytic agent. t-PA-deficient mice exhibited approximately 50% smaller cerebral infarcts than wild-type mice. Intravenous injection of t-PA into t-PA–/– or wild-type mice produced larger infarcts, indicating that t-PA increased stroke-induced injury. Since t-PA promotes desirable (thrombolytic) as well as undesirable (neurotoxic) manifestations during stroke, a thrombolytic agent that does not promote excitotoxic damage could achieve better neuroprotection after an acute cerebral infarct.

Pharmacology

On the basis of biochemical evidence, DSPAα_1 fulfills the hopes and expectations originally placed on t-PA: plasminogen activation restricted to the clot surface, without the systemic activation that leads to fibrinogen consumption, 'plasminogen steal', and degradation of clotting factors VIII and V. Pharmacological studies using a lung embolism model in rats demonstrated that recombinant DSPAα_1 is more potent and more clot specific than t-PA [9]. A higher potency, clot specificity, and prolonged half-life of DSPAα_1 over t-PA were also verified in an arterial thrombosis model in rats [10]. In a coronary thrombosis model of acute myocardial infarction in dogs, DSPAα_1 led to a faster recanalization and a lower incidence of reocclusion compared to t-PA [11]. On the other hand, a lower incidence of bleeding was observed with DSPAα_1 in a rat mesenteric vein model [12]. DSPAα_1 was characterized pharmacokinetically in rats and cynomolgus monkeys. In comparison to t-PA, a lower total clearance and a longer terminal half-life could be demonstrated [13]. These data encouraged the development of an intravenous bolus regimen in humans. A detailed toxicologic study has been performed in rats and monkeys. No direct toxic effects could be observed, and secondary effects appeared only at extremely high doses of DSPAα_1.

Clinical Studies

Myocardial Infarction

A phase II study was conducted to evaluate the efficacy, safety and tolerability of DSPAα_1 as a thrombolytic agent in the treatment of patients with acute myocardial infarction. The study was designed as a nonrandomized, open-label, prospective dose finding study in which patients received an intravenous bolus of either 0.5 mg/kg or 0.75 mg/kg of DSPAα_1 administered over 1–2 min followed by intravenous heparin. Efficacy was determined by measuring coronary angiography patency rates at 90 min after onset of thrombolysis while safety was assessed by recording the occurrence of bleedings, allergic and other early or late complications. In addition, the impact of DSPAα_1 on hemostatic parameters and the development of DSPAα_1 antibodies after treatment were determined. A total of 26 patients (19 males and 7 females) were enrolled into the study with a mean age of 61 years (range 41–75) and a mean weight of 76 kg (range 62–92). The follow-up period of observation was 6 months. 18 patients received 0.5 mg/kg and 8 patients received 0.75 mg/kg of DSPAα_1. Patency, as defined by a TIMI grade III score at the 90-min coronary angiogram, was achieved in 65% (17 of the 26 patients). Late patency (90 min–24 h) was demonstrated in 21 patients. Out of the remaining 5 patients 3 died within 8 h after inclusion and 2 refused the 24- to 36-hour angiogram. The laboratory data from the study confirmed the high fibrin specificity of DSPAα_1 demonstrated by normal levels of fibrinogen. Other laboratory parameters – plasminogen, hematocrit, hemoglobin, activated partial thromboplastin time, creatine kinase, creatine kinase-MB and α_2-antiplasmin – did not indicate any treatment-related safety issues. The safety profile of DSPAα_1 in this early study was typical for the administration of a thrombolytic in myocardial infarction. The serious adverse event reported was judged not to be DSPAα_1-related by the independent safety advisory board. These events were rather indication specific and typical for the natural course of the disease. This early study demonstrated that DSPAα_1 acted as a typical plasminogen activator with comparable activity to other thrombolytics: The patency rate was 65%. Fibrin specificity was demonstrated by the fact that no fibrinogen depletion occurred. Practically none of the bleeding episodes required any therapy. Bleeding occurred more frequently in studies where angiography was performed and when patients were treated with high doses of heparin. Conclusions with regard to efficacy and safety are hampered by the fact that the patient number was small; however, proof of concept (thrombolysis) could be shown [G. Groetzbach, D. Gulba et al., pers. commun.].

Stroke

After the favorable pharmacotoxicological and pharmacokinetic profile of DSPAα_1 had been confirmed by the results of the phase II study in acute myocardial infarction, a phase II/III study of desmoteplase in the indication acute cerebral vascular occlusion appeared promising. The Desmoteplase in Acute Ischemic Stroke study was initiated by Paion GmbH, Stolberg, Germany in January 2001. It is an international multicenter placebo-controlled study in the time window 3–6 h after onset of stroke using MRI-measured mismatch as a key inclusion criterion next to standard inclusion and exclusion criteria typical for the use of a thrombolytic. Centers have to be qualified by a central imaging unit for their MRI capabilities. First results of the study will become available in 2002.

References

1 Hawkey C: Plasminogen activator in the saliva of the vampire bat *Desmodus rotundus.* Nature 1966;211:434–435.

2 Krätzschmar J, Haendler B, Langer G, Boidol W, Bringmann P, Alagon A, Donner P, Schleuning W-D: The plasminogen activator family from the salivary gland of the vampire bat *Desmodus rotundus*: Cloning and expression. Gene 1991;105:229–237.

3 Schleuning W-D, Alagon A, Boidol W, Bringmann P, Petri T, Krätzschmar J, Haendler B, Baldus B, Witt W, Donner P: Plasminogen activators from the saliva of *Desmodus rotundus* (common vampire bat): Unique fibrin specificity. Ann NY Acad Sci 1992;667:395–403.

4 Petri T, Langer G, Bringmann P, Cashion L, Shallow S, Schleuning W-D: Production of vampire bat plasminogen activator DSPAα_1 in CHO and insect cells. J Biotechnol 1995;39:75–83.

5 Bringmann P, Gruber D, Liese A, Toschi J, Kraetzschmar J, Schleuning W-D, Donner P: Structural features mediating fibrin selectivity of vampire bat plasminogen activators. J Biol Chem 1995;270:25596–25603.

6 Renatus M, Stubbs MT, Huber R, Bringmann P, Donner P, Schleuning W-D, Bode W: Catalytic domain structure of vampire bat plasminogen activator: A molecular paradigm for proteolysis without activation cleavage. Biochemistry 1997;36:13483–13493.

7 Kingston IB, Castro MJ, Anderson S: In vitro stimulation of tissue-type plasminogen activator by Alzheimer amyloid beta-peptide analogues. Nat Med 1995;1:138–142.

8 Wang YF, Tsirka SE, Strickland S, Stieg PE, Soriano SG, Lipton SA: Tissue plasminogen activator (tPA) increases neuronal damage after focal cerebral ischemia in wild-type and tPA-deficient mice. Nat Med 1998;4:228–231.

9 Witt W, Baldus B, Bringmann P, Cashion L, Donner P, Schleuning W-D: Thrombolytic properties of *Desmodus rotundus* (vampire bat) salivary plasminogen activator in experimental pulmonary embolism in rats. Blood 1992;79:1213–1217.

10 Muschick P, Zeggert D, Donner P, Witt W: Thrombolytic properties of *Desmodus* (vampire bat) salivary plasminogen activator DSPα_1, Alteplase and streptokinase following intravenous bolus injection in a rabbit model of carotid artery thrombosis. Fibrinolysis 1993;7:284–290.

11 Witt W, Maass B, Baldus B, Hildebrand M, Donner P, Schleuning W-D: Coronary thrombolysis with *Desmodus* salivary plasminogen activator in dogs. Circulation 1994;90:421–426.

12 Gulba DC, Praus M, Witt W: DSPA α_1-properties of the plasminogen activators of the vampire bat *Desmodus rotundus.* Fibrinolysis 1995;9:91–96.

13 Hildebrand M, Bunte T, Bringmann P, Schutt A: Development of an ELISA for the measurement of DSPAα_1 (*Desmodus rotundus* salivary plasminogen activator) in plasma and its application to investigate pharmacokinetics in monkeys. Fibrinolysis 1995;9:107–112.

Fibrinolysis

Hæmostasis

Haemostasis 2001;31:123–132

Proteases from *Vipera lebetina* Venom Affecting Coagulation and Fibrinolysis

J. Siigur A. Aaspõllu K. Tõnismägi K. Trummal M. Samel
H. Vija J. Subbi E. Siigur

National Institute of Chemical Physics and Biophysics, Tallinn, Estonia

Key Words

Fibrino(geno)lytic enzymes · Lebetase · Factor X activator · Factor V activator · Proteases · Snake venom · *Vipera lebetina* · Glycoprotein · MALDI-TOF MS · Substrate specificity

Abstract

Our studies of the venom from the Levantine viper *Vipera lebetina* have demonstrated the existence of both coagulants and anticoagulants in the same venom. We showed that *V. lebetina* venom contains: (1) proteases that degrade fibrinogen, but not fibrin; (2) fibrinolytic enzyme (lebetase); (3) factor X activator (VLFXA); (4) factor V activator (VLFVA). Fibrinolytic enzyme and VLFXA are metalloproteases; the other studied enzymes are serine proteases. α-Fibrinogenase has no homolog among known serine proteases. β-Fibrinogenase is a typical thermostable arginine esterase that hydrolyzes esters and amides of arginine and attacks the β-chain of fibrinogen. Lebetase is a direct-acting fibrinolytic zinc metalloendopeptidase related in amino acid sequence to reprolysins. We used the matrix-assisted laser desorption/ionization time-of-flight mass spectrometry technique for the recovery and identification of peptides released by protease hydrolysis and for the detection of human factor X cleavage products after VLFXA hydrolysis. VLFXA cleaves the Arg^{52}-Ile^{53} bond in the heavy chain of human factor X and the Arg^{226}-Val^{227} bond in human factor IX precursor; VLFVA cleaves Arg^{1545}-Ser^{1546} in factor V.

Introduction

Snake venoms, particularly those belonging to the Crotalidae and Viperidae families, contain a large variety of proteases that induce alterations in the blood coagulation

KARGER

Fax + 41 61 306 12 34
E-Mail karger@karger.ch
www.karger.com

0301–0147/01/0316–0123$17.50/0

Accessible online at:
www.karger.com/journals/hae

Jüri Siigur
National Institute of Chemical Physics and Biophysics
Akadeemia tee 23
12618 Tallinn (Estonia)
Tel. +372 6 398 360, Fax +372 6 398 313, E-Mail siigur@kbfi.ee

cascade [1–8]. These highly specific proteases which cleave limited bond(s) in the blood coagulation factors are usually divided into two groups: (1) metalloproteases which need Ca^{2+} or Zn^{2+} (or both) for their hydrolytic activity and are inhibited by metal-chelating agents (factor X activator, prothrombin activator, α(β)-fibrinogenases), (2) serine proteases (factor V activator, protein C activator, plasminogen activator, kinin-releasing and thrombin-like enzymes, β-fibrinogenases). Recent studies revealed that some snake venoms contain factor X activators and prothrombin activators that are serine proteases [8].

Zinc metalloproteinases occur widely and participate in a number of important biological, physiological and pathophysiological processes (hemorrhage, fertilization, thrombolysis, cancer metastasis). Because of the possible therapeutic and diagnostic role of snake venom proteases, these enzymes merit further investigation. *Vipera lebetina* (Levantine viper) is a snake found in the South-East parts of Europe, in South-West Asia and in North-West Africa. Its venom comprises various proteases, which act on coagulation through both pro- and anticoagulant mechanisms.

V. lebetina venom contains different metalloproteases – hemorrhagic and nonhemorrhagic (including fibrinolytic enzyme, lebetase [9–15]) and factor X activator (VLFXA) [16]. Snake venom metalloproteases have homologous sequences and a typical active site structure for reprolysins HEXXHXXGXXH [17]. The amino acid sequence of lebetase was deduced from its cDNA. The primary structure of lebetase shows extensive sequence homology with fibrolase and several other small snake venom metalloproteinases [11].

However, the substrate specificity of snake venom metalloproteases is rather different. Lebetase, a metalloprotease with thrombolytic activity, is a direct-acting fibrinolytic agent, as it acts *via* direct cleavage of fibrin, not by plasminogen activation. In general, the fibrin(ogen)olytic enzymes cleave off fragments from the C-terminals of α- and β-chains of fibrinogen, rendering it unclottable by thrombin. Fibrin is dissolved by the same mechanism of cleavage. Lebetase has the typical active site for reprolysins. The enzyme readily hydrolyzes the Aα chain and more slowly the Bβ chain of fibrinogen. Previous studies performed on the B chain of oxidized insulin showed that lebetase has primary specificity for hydrophobic P1′ residues, such as Leu (Ala^{14}-Leu^{15}, Tyr^{16}-Leu^{17}) [10]. It was demonstrated that lebetase cleaves the 'bait' region in α_2-macroglobulin and hydrolyzes pregnancy zone protein (PZP) [14]. In order to define the substrate specificity of the lebetase proteolytic activity more clearly, different peptides were used as substrates.

Furthermore, lebetase is an anticoagulant that acts by inhibiting platelet aggregation. It is a zinc-metalloproteinase with three disulfide bridges [13].

V. lebetina venom is able to activate factor X [16]. Factor X (substrate for factor X activator) is a vitamin-K-dependent protein that occupies a pivotal position in the coagulation process. Factor X is essential for hemostasis since the reduction in its functional activity results in a bleeding disorder known as Stuart-Prower factor deficiency. During physiological hemostasis, factor X can be activated by factor IXa, requiring Ca^{2+}, phospholipid, and factor VIIIa, or by factor VIIa, requiring Ca^{2+} and tissue factor. Factor X is activated to the serine protease factor Xa. The activation results from the cleavage of the Arg^{52}-Ile^{53} bond in the heavy chain of human factor X and release of a 52-residue activation peptide [8].

The enzymes degrading the β-chain of fibrinogen without fibrinolysis belong to serine proteinases. *V. lebetina* β-fibrinogenase is a typical representative of arginine esterases without caseinolytic activity [18]. The alkaline serine protease – a nonfibrinolytic α-fibrinogenase – has a unique specificity hydrolyzing casein but not arginine esters [19]. Factor V activator from *V. lebetina* venom (VLFVA) resembles the activator of *Vipera russelli* venom [20, 21].

Due to the potential use of snake venom proteases as diagnostic and thrombolytic agents, it is important to know their specificity against biologically active proteins and peptides. In this report we studied the specificity of *V. lebetina* proteinases. The substrates used were biologically active peptides and 6–10 amino acid residues containing peptides synthesized according to the literature provided protease cleavage regions in proteins, such as human factor X, factor IX, factor V, fibrinogen, α_2-macroglobulin and PZP. Peptides synthesized according to the amino acid sequence of lebetase proprotein and mature protein region were also used as substrates for metalloproteases.

Sequence analysis of peptide fragments after enzyme hydrolysis has been the traditional method of determining the cleavage sites in peptides and proteins, but it is a time-consuming and rather expensive method. The matrix-assisted laser desorption/ionization time-of-flight (MALDI-TOF MS) methodology is sensitive and quick, which has led us to use this technology to obtain the primary information on the enzyme cleavage sites in peptides.

Materials and Methods

Materials

Bradykinin and substance P were purchased from Serva (Heidelberg, Germany). Ferulic acid, 2,5-dihydroxybenzoic acid (DHB) and human factor X (lot 27H93059) were from Sigma (St. Louis, Mo., USA). All other reagents used were of analytical grade.

Methods

Purification of Proteases. The purification of proteases has been described in our previous studies [10, 12, 18–20]. Enzymes were analyzed by SDS-PAGE, isoelectric focusing, HPLC and MALDI TOF MS.

Carbohydrate Test. Neutral sugars were determined by the phenol-sulfuric acid method [22], using a standard solution containing *D*-glucose. Hexosamines were assayed by the procedure of Winzler [23] after acid hydrolysis in 3 *N* HCl for 4 h at 100 °C. Glucose amine was used as a standard. Neuraminic acid was measured by the thiobarbituric acid method of Warren [24] using N-acetylneuraminic acid as a standard.

Fibrinogen Degradation. Specific cleavage of fibrinogen was shown on 5–15% polyacrylamide gels. 0.5 ml of 1% fibrinogen solution was incubated with 25 μg of enzyme at 37 °C in 0.05 *M* Tris-saline buffer (pH 7.4). At various time intervals, 50-μl aliquots were withdrawn and added to 50 μl of denaturing solution (10 *M* urea, 4% SDS, 4% 2-mercaptoethanol). The samples were reduced and denatured overnight at 37 °C before being electrophoresed.

Enzyme Activities. Caseinolytic activity was assayed by the method of Kunitz as modified by Mebs [25]. Azocaseinolytic activity was estimated according to the method of Charney and Tomarelli [26]. Fibrinolytic activity was estimated by the fibrin-plate method of Astrup and Müllertz [27].

Kinin-releasing activity was determined using heated human plasma (56 °C, 3 h) as substrate. Plasma was centrifuged and dialysed against 0.05 *M* ammonium acetate. 5–10 μl of enzyme solution (1 mg/ml) was added to 200 μl of treated plasma containing 1 m*M* *o*-phenanthroline and the mixture was incubated at 37 °C for 20 min. The mixture was ultrafiltrated (Ultra spin ultrafilters molecular weight cut-off 10,000). The kinin was detected in filtrate by MALDI-TOF MS. Arginine esterase activity was detected by the method of Schwert and Takenaka [28], using BAEE as substrate.

The activation effect of factor X by VLFXA was measured by the amidolytic activity of the factor Xa that was formed according to the method described by Hofmann and Bon [29]. Bovine or human factor X was used as substrate for VLFXA. Factor Xa activity was determined with S-2222 or S-2337 by recording the liberation of *p*-nitroaniline at 405 nm. The degradation products of human factor X treated with VLFXA were detected by MALDI TOF MS.

Molecular Mass of Enzymes

SDS-PAGE was carried out in 10 and 12.5% gels and gradient gels. The following molecular mass indicators were used: bovine serum albumin (67 kD), ovalbumin (43 kD), bovine carbonic anhydrase (29 kD), soybean trypsin inhibitor (20.1 kD), cytochrome C (12.3 kD). Staining was performed with Coomassie Brilliant Blue R250.

MALDI-TOF MS

The MALDI mass spectra were measured with a home-built gridless TOF MALDI MS designed for maximum flexibility in use (National Institute of Chemical Physics and Biophysics) [13]. Cytochrome C and bovine carbonic anhydrase were used for mass calibration. Ferulic acid was used as matrix.

Peptide Synthesis. All peptides were synthesized at the 100-μmol scale on Applied Biosystems 431A Peptide Synthesizer using BOC (t-butyl-oxycarbonyl) chemistry, as suggested by the manufacturer.

The purity of peptides was assessed by analytical reverse-phase-high-performance liquid chromatography (HPLC) and MALDI-TOF MS.

Monitoring of the Enzymatic Reaction: MALDI-TOF MS of Peptides. Peptide and peptide cleavage analyses by MALDI-TOF MS have been previously reported [15]. All peptide solutions were directly prepared in 0.1 *M* NH_4HCO_3, at concentrations of about 1–5 mg/ml, and kept frozen at –20 °C until use. The enzymatic hydrolysis of peptides was carried out in 0.1 *M* NH_4HCO_3, at 37 °C in an Eppendorf tube. In a typical experiment, 100 μl of 0.1 *M* NH_4HCO_3 solution of substrate (1 mg/ml) in an Eppendorf tube was thermally equilibrated to 37 °C in a thermostated rack. The reaction was started by addition of 15 μl of enzyme solution (1 mg/ml in 0.1 *M* NH_4HCO_3). At predetermined time intervals (5 min, 0.5 h, 20 h), the aliquot (10 μl) was diluted with 100 μl of H_2O and 2 μl of 6 *N* HCl was added to stop the reaction. One half microliter of diluted mixture was used for MALDI-TOF MS. The matrix used for peptide analyses was 2,5-dihydroxybenzoic acid (DHB). 10 mg of DHB was dissolved in 1 ml of a 1:1 mixture of 0.1% trifluoroacetic acid and acetonitrile for sample preparation. 0.5 μl of this mixture was deposited on a stainless steel probe tip, mixed there with 0.5 μl of reaction mixture of peptides and enzyme and allowed to dry at room temperature. External mass calibration was accomplished by using as peptide standards, substance P and Lys-bradykinin.

For peptide analysis, the mass spectrometer was operated in reflectron mode with 2.6-kV pulsed extraction, 14.6 kV total acceleration voltage and 500 ns delay between laser and extraction pulse. A double multichannel plate detector was used for ion detection.

The cleavage products of human factor X after treating with VLFXA were detected by MALDI-TOF MS. The enzymatic hydrolysis of human factor X (75 μg/ml, 25 μl) was carried out in 0.1 *M* NH_4CH_3COO containing 0.025 m*M* $CaCl_2$ (5 μl) pH 8.0, 37 °C. We applied a high enzyme-substrate ratio (1:20 on a weight basis) to determine the cleavage. At predetermined time intervals (5 min, 0.5 h, 20 h), the reaction mixture (25 μl) was frozen to stop the reaction. Before MALDI-TOF analysis, samples were purified from salts using ZipTip C_4 or ZipTip C_{18} according to Millipore instructions. The matrix used for human factor X, VLFXA and cleavage products of factor X was ferulic acid, cytochrome C was used for mass calibration.

Results and Discussion

V. lebetina venom contains serine (α-fibrinogenase, β-fibrinogenase, factor V activator) and metalloproteinases (fibrinolytic enzyme lebetase, factor X activator) that affect coagulation and fibrinolysis. Molecular masses, detected by MALDI TOF MS and SDS-PAGE, are provided in table 1. Most *V. lebetina* proteases (except lebetase) are glycosylated (table 2). Two glycosylated serine fibrinogenases and VLFVA have homologous N-terminal sequences:

α-fibrinogenase	VIGGRPCNINQHRSLALLY,
β-fibrinogenase	VVGGDECNKNEHRSLVFLY,
VLFVA	VVGGDECDINEHPFLVALY,

but can be clearly differentiated according to substrate specificity, glycosylation levels, molecular mass and fibrinogen degradation. α-Fibrinogenase has no homolog among known serine proteinases. It has N-terminal similarity with snake venom arginine esterases, but does not hydrolyze the esters of arginine, lysine and tyrosine. The enzyme has strong caseinolytic activity and degrades α-chain of fibrinogen [31] altering its clottability by thrombin. β-Fibrinogenase is a typical arginine esterase which hydrolyzes esters and amides of arginine and attacks the β-chain of fibrinogen [18]. VLFVA hydrolyzed human

Table 1. Molecular masses of *V. lebetina* venom proteases

Enzyme	Nonreduced (MALDI) D	Reduced (SDS) D
Lebetase	22,719; 22,912 [13]	23,700 [10]
VLFXA	86,500–91,000	57,500; 17,400; 14,500
VLFVA	28,400 [20]	30,000 [20]
β-Fibrinogenase	42,200 [31]	52,500 [18]
α-Fibrinogenase	31,100 [31]	37,000 [19]

Table 2. Carbohydrate composition of *V. lebetina* venom proteases

Enzyme	Hexoses, %	Hexosamines, %	Sialic acid, %
VLFXA	8.0	2.5	2.5
VLFVA	4.4	4.4	3.1
α-Fibrinogenase	8.5	5.8	0.8
β-Fibrinogenase	23.0	8.3	1.0

factor V peptide fragments, containing 7–9 amino acids (YLRSNNG, WYLRSNNG and AWYLRSNNG), synthesized according to *V. russelli* factor V activator cleavage region in human factor V. VLFVA cleaves Arg-Ser bond in these peptides (fig. 1). We may suppose that VLFVA cleaves Arg^{1545}-Ser^{1546} bond in human factor V. We have shown earlier that *V. lebetina* venom contains two different bradykinin-releasing serine enzymes [30]. The released bradykinin from plasma was detected in this work by MALDI TOF MS (fig. 2). This method successfully replaces the rat uterus test formerly used for kinin detection.

Specificity of metalloproteinases was studied against different proteins and peptides. Lebetase specificity was characterized in our previous publications [10, 13–15]. Lebetase hydrolyzed peptides synthesized according to the proprotein and mature protein region of Le-3 (table 3). Le-3 designates the deduced amino acid sequence of the precursor for lebetase [11]. Lebetase is synthesized in the venomous gland in a latent zymogen form containing prosequences. The mature protease may be released autocatalytically. The Le-3 fragment is also hydrolyzed by some other *V. lebetina* proteases from I–II fraction of Sephadex G-100 gel filtration.

Comparison of amino acid sequences for a variety of substrates shows that lebetase has no strict P1-P1′ specificity requirement (table 3). Hydrolysis at sites with a Pro residue at P1 was observed with bradykinin, substance P, peptide fragments of PZP and fibrinogen α-chain as substrates.

The purified VLFXA had no effect on fibrinogen, prothrombin, plasminogen, indicating that it might activate factor X specifically. The effect of VLFXA on human factor X was studied by measuring the amidolytic and the coagulant activities of the activated factor X (factor Xa). The factor-X-activating enzyme VLFXA has no amidolytic activity against factor Xa substrates S-2337 or S-2222, while *V. lebetina* venom itself hydrolyzes

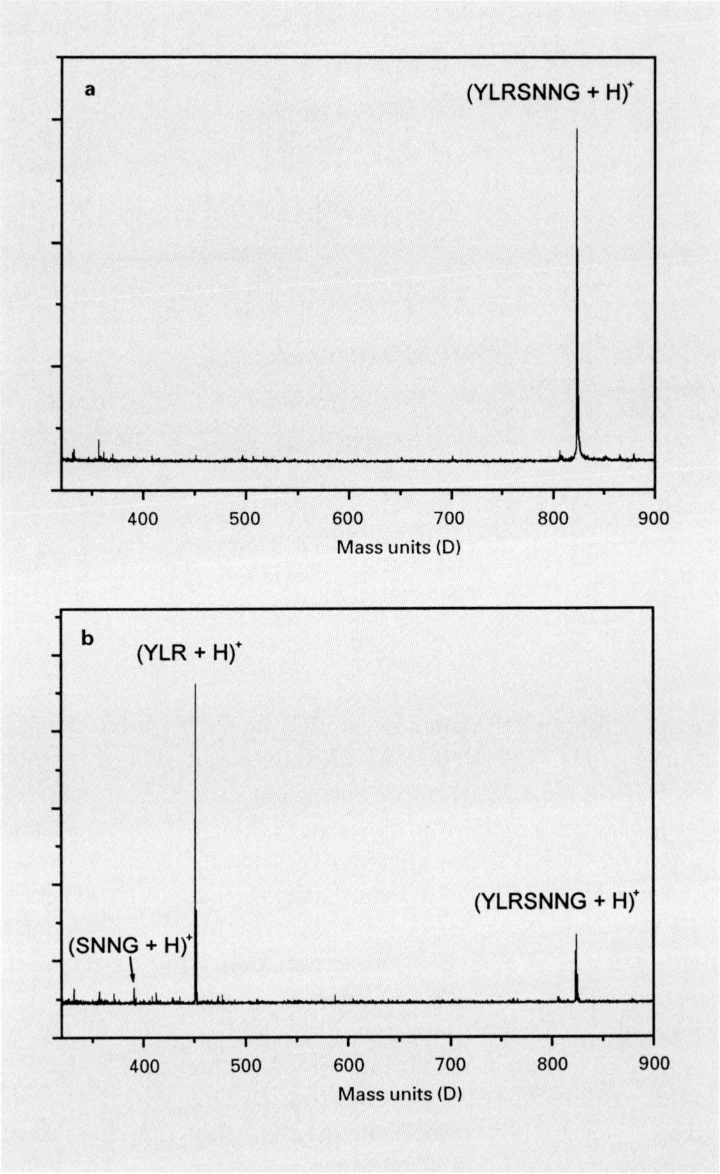

Fig. 1. a MALDI TOF mass spectrum of human factor V peptide fragment 1543–1549 YLRSNNG (823.4 D). **b** Mass spectrum of cleavage products (YLR 451.3 D, SNNG 391.2 D) of human factor V peptide fragment 1543–1549 after treatment with VLFVA.

these substrates. The activator converts the inactive factor X to the active form Xa in the presence of Ca^{2+} ions. The activation without Ca^{2+} was about 5%. The enzyme has maximum activity on factor X in the interval of pH 7.5–8.0.

VLFXA hydrolyzed 6–9 amino acid residues containing peptide fragments (TRIVGG, LTRIVGG and NNLTRIVGG), synthesized according to the physiological cleavage region of human factor X. VLFXA cleaved Arg-Ile bond in peptides. NNLTRI-

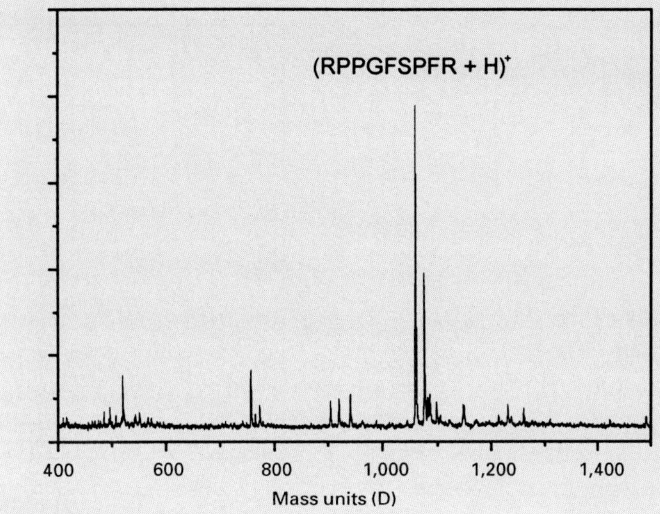

Fig. 2. MALDI TOF mass spectrum of bradykinin (1,060.6 D) liberated after 0.5 h of treatment of heated human plasma at 37 °C with bradykinin-releasing enzyme fraction. A contaminating component with a mass of 1,076.6 D was found in nontreated plasma. Details are provided in Methods.

Table 3. Hydrolysis of peptide substrates by lebetase

Name of substrate	Cleavage P1↓P1′
Substance P	RP↓KP↓QQFFG↓LM* [15]
Bradykinin	RPPGFSP↓FR [15]
Le-3 fragment 192–198	TPE↓Q↓QRF
Le-3 fragment 191–198	LTPE↓Q↓QRF
PZP fragment 686–693	PYVP↓QLGT [15]
α_2-M fragment 676–684	GPEG↓LRVGF [15]
α_2-M fragment 693–700	GHAR↓LVHV [15]
Fibrinogen fragment 513–520	FFSP↓MLGE [15]
Fibrinogen fragment 408–417	EYHTEK↓LVTS [15]

↓ = Cleavage site of peptide by lebetase; *P↓Q is the main cleavage site (after 5 min hydrolysis) of substance P by lebetase.

VGG was cleaved most effectively. Activator hydrolyzed 9 amino acid residues containing peptide fragment-NDFTRVVGG, synthesized according to the physiological cleavage region of human factor IX, in position Arg-Val (fig. 3). These peptides were used for localization of VLFXA in different steps of purification and also in inhibitory tests.

The catalytic cleavage of human factor X by the activators from *V. lebetina* (VLFXA) and *V. russelli* (RVVXAE, Sigma) venoms was examined by MALDI-TOF MS. The

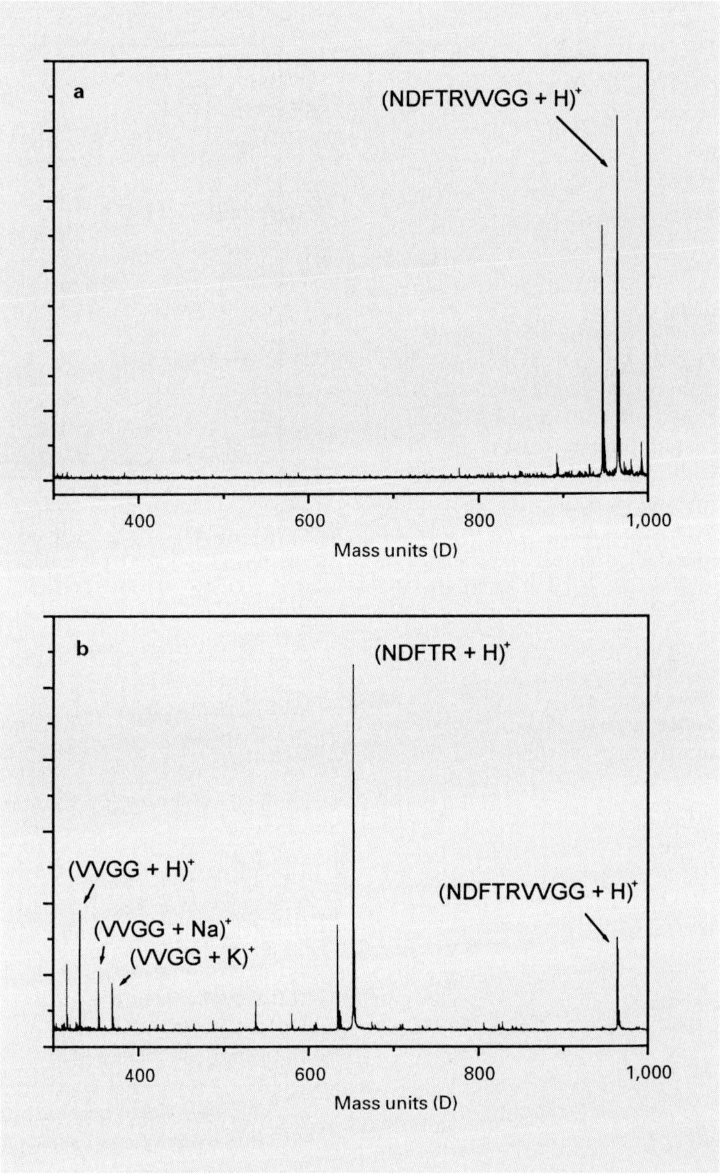

Fig. 3. a MALDI TOF mass spectrum of human factor IX peptide fragment 222–230 NDFTRVVGG (964.5 D), contaminating component – N-terminus cyclization product of peptide. **b** Mass spectrum of cleavage products (NDFTR 651.7 D, VVGG 331.2 D) of human factor IX peptide fragment after treating with VLFXA.

masses of cleavage products of human factor X were 43.1 and 11.6 kD after treating with VLFXA and RVV-X. The activation was already detectable after 5 min of hydrolysis in the case of VLFXA. On the basis of cleavage of peptides and human factor X, we can conclude that VLFXA hydrolyses human factor X at the positions Arg^{52}-Ile^{53} and human factor IX precursor at the positions Arg^{226}-Val^{227}. Factor X can be completely converted to factor Xa by VLFXA and the specific activity of factor Xa is the same as that of factor Xa obtained after activation by the activator from Russell's viper venom [32].

The use of MALDI-TOF MS has several advantages over traditional methods for the elucidation of cleavage sites by proteases in peptides and proteins. MALDI-TOF MS has high sensitivity that allows the analysis of small aliquots removed from reaction mixture. Our results show that MALDI-TOF MS is a very informative tool and can easily characterize the composition of cleaved peptides and proteins. However, as it stands now, MALDI is not quantitative enough for detecting kinetic parameters of enzyme reactions.

Acknowledgements

We are grateful to Dr. Nisse Kalkkinen (Institute of Biotechnology, University of Helsinki) for performing the N-terminal analysis. The work was financially supported by Estonian Science Foundation Grants No. 4228 and No. 3842.

References

1 Kornalik F: The influence of snake venom enzymes on blood coagulation. Pharmacol Ther 1985;29:353–405.

2 Kornalik F: Toxins affecting blood coagulation and fibrinolysis; in Shier WT, Mebs D (eds): Handbook of Toxinology. New York, Marcel Dekker, 1990, pp 683–759.

3 Seegers WH, Ouyang C: Snake venoms and blood coagulation; in Lee CY (ed): Snake Venoms, Handbook of Experimental Pharmacology. Berlin, Springer, 1979, vol 52, pp 684–750.

4 Stocker K: Snake venom proteins affecting hemostasis and fibrinolysis; in Stocker K (ed): Medical Use of Snake Venom Proteins. Boca Raton, CRC Press, 1990, pp 97–160.

5 Ouyang C, Teng CMT, Huang F: Characterization of snake venom components acting on blood coagulation and platelet function. Toxicon 1992;30:945–966.

6 Braud S, Bon C, Wisner A: Snake venom proteins acting on hemostasis. Biochimie 2000;82:851–859.

7 Matsui T, Fujimura Y, Titani K: Snake venom proteases affecting hemostasis and thrombosis. Biochim Biophys Acta 2000;1477:146–156.

8 Morita T: Proteases which activate factor X; in Bailey GS (ed): Enzymes from Snake Venom. Alaken, Fort Collins, 1998, pp 179–208.

9 Siigur E, Siigur J: The direct-acting α-fibrin(ogen)olytic enzymes from snake venoms. J.Toxicol Toxin Rev 1992;11:91–113.

10 Siigur E, Siigur J: Purification and characterization of lebetase, a fibrinolytic enzyme from *Vipera lebetina* (snake) venom. Biochim Biophys Acta 1991;1074:223–229.

11 Siigur E, Aaspôllu A, Tu AT, Siigur J: cDNA cloning and deduced amino acid sequence of fibrinolytic enzyme (lebetase) from *Vipera lebetina* snake venom. Biochem Biophys Res Commun 1996;224:229–236.

12 Siigur J, Tõnismägi K, Tu AT, Siigur E: Cross-reactivities of polyclonal antibodies against lebetase, fibrinolytic enzyme of Levantine viper *(Vipera lebetina)* venom. Toxicon 1996;34:608–613.

13 Siigur J, Samel M, Tõnismägi K, Subbi J, Siigur E, Tu AT: Biochemical characterization of lebetase, a direct-acting fibrinolytic enzyme from *Vipera lebetina* snake venom. Thromb Res 1998;90:39–49.

14 Saidi N, Samel M, Siigur J, Jensen PEH: Lebetase, an α(β)-fibrin(ogen)olytic metalloproteinase of *Vipera lebetina* snake venom, is inhibited by human α-macroglobulins. Biochim Biophys Acta 1999;1434: 94–102.

15 Trummal K, Vija H, Subbi J, Siigur J: MALDI-TOF mass spectrometry analysis of substrate specificity of lebetase, a direct-acting fibrinolytic metalloproteinase from *Vipera lebetina* snake venom. Biochim Biophys Acta 2000;1476:331–336.

16 Siigur J, Samel M, Tõnismägi K, Siigur E: Factor X activators from Viperidae venoms. Toxicon 1995;33: 258.

17 Bjarnason JB, Fox JW: Snake venom metalloendopeptidases: Reprolysins. Methods Enzymol 1995;248: 345–368.

18 Siigur E, Mähar A, Siigur J: β-Fibrinogenase from the venom of *Vipera lebetina.* Toxicon 1991;29: 107–118.

19 Mähar A, Siigur E, Siigur J: Purification and properties of a proteinase from *Vipera lebetina* (snake) venom. Biochim Biophys Acta 1987;925: 272–281.

20 Siigur E, Samel M, Tõnismägi K, Subbi J, Reintamm T, Siigur J: Isolation, properties and N-terminal amino acid sequence of a factor V activator from *Vipera lebetina* (Levantine viper) snake venom. Biochim Biophys Acta 1998;1429:239–248.

21 Siigur E, Aaspõllu A, Siigur J: Molecular cloning and sequence analysis of a cDNA for factor V activating enzyme, a coagulant protein from *Vipera lebetina* snake venom. Biochem Biophys Res Commun 1999; 262:328–332.

22 Dubois M, Gilles KA, Hamilton KK, Rebers PA, Smith F: Colorimetric method for determination of sugars and related substances. Anal Chem 1956;28:350–356.

23 Winzler RJ: Determination of serum glycoproteins; in Glick D (ed): Methods of Biochemical Analysis. New York, Interscience, 1958, vol 2, pp 279–311.

24 Warren OL: The thiobarbituric acid assay of sialic acids. J Biol Chem 1959;234:1971–1975.
25 Mebs D: A comparative study of enzyme activities in snake venoms, Int J Biochem 1970;1:335–342.
26 Charney J, Tomarelli RM: A colorimetric method for the determination of the proteolytic activity of duodenal juice. J Biol Chem 1947;131: 501–505.
27 Astrup T, Müllertz S: The fibrin plate method for estimating of fibrinolytic activity. Arch Biochem Biophys 1952;40:346–351.
28 Schwert GW, Takenaka Y: A spectrophotometric determination of trypsin and chymotrypsin. Biochim Biophys Acta 1955;16:570–575.
29 Hofmann H, Bon C: Blood coagulation induced by the venom of *Bothrops atrox*. 2. Identification, purification and properties of two factor X activators. Biochemistry 1987;26: 780–787.
30 Siigur EP, Siigur JR, Aaviksaar AA, Kibirev VK, Fedoryak DM: Separation of a bradykinin-releasing enzyme from the proteolytic complex of Levantine viper (in Russian). Biokhimiya 1982;47:1730–1737.
31 Samel M, Subbi J, Siigur J, Siigur E: Biochemical characterization of fibrinogenolytic serine proteinases from *Vipera lebetina* snake venom. Toxicon 2002;40:51–54.
32 Takeya H, Nishida S, Miyata T, Kawada S, Saisaka Y, Morita T, Iwanaga S: Coagulation factor X activating enzyme from Russell's viper venom (RVV-X). A novel metalloproteinase with disintegrin (platelet aggregation inhibitor)-like and C-type lectin-like domains. J Biol Chem 1992;267:14109–14117.

Fibrinolysis

Hæmostasis

Haemostasis 2001;31:133–140

Snake Venom Proteinases as Tools in Hemostasis Studies: Structure-Function Relationship of a Plasminogen Activator Purified from *Trimeresurus stejnegeri* Venom

A. Wisner S. Braud C. Bon

Unité des Venins, Institut Pasteur, Paris, France

Key Words

Snake venom · Proteinase · Plasminogen activator · Structure-function · Haemostasis

Abstract

Snake venom serine proteinases affect many steps of the blood coagulation cascade. Each of them usually acts selectively on one coagulation factor. They are therefore potentially useful components to study the mechanisms of action, the regulation and the structure-function relationships of human serine proteinase coagulation factors. This strategy is illustrated for a plasminogen activator purified from *Trimeresurus stejnegeri* venom.

Serine Proteinase Family

Serine proteinases are found in microorganisms [1], plants [2] and numerous animals [3, 4]. These enzymes share many biochemical and structural properties, such as a catalytic triad (Ser^{195}, His^{57}, Asp^{102}), the main criterion for classification of a protein as a serine proteinase. They exhibit structural differences which can be used to classify this family into three subgroups: trypsin-like, subtilisin-like and carboxypeptidase-like serine proteinase [3].

Human Serine Proteinases Involved in Hemostasis

Human serine proteinases have diverse functions and may be involved in digestion, complement activation, cellular differentiation and hemostasis [3]. In particular, human trypsin-like serine proteinases are involved in

KARGER

Fax + 41 61 306 12 34
E-Mail karger@karger.ch
www.karger.com

0301–0147/01/0316–0133$17.50/0

Accessible online at:
www.karger.com/journals/hae

Dr. Cassian Bon
Head of the Venoms Unit, Institut Pasteur
25–28, rue du Docteur-Roux
F–75724 Paris Cedex 15 (France)
Tel. +33 1 45 68 86 85, Fax +33 1 40 61 30 57, E-Mail cbon@pasteur.fr

hemostasis, with roles in platelet aggregation, coagulation and fibrinolysis (fig. 1). Their activity is regulated by various families of inhibitors, notably the serine proteases inhibitors (serpins), which slowly form irreversible complexes with the protease.

Human serine proteinases active in hemostasis generally contain at least a catalytic domain structurally similar to trypsin. Some also possess, at their N-terminal end, one or several additional domains (kringle, apple, finger or epidermal growth factor domains) that are involved in protease interactions with cofactors and cellular receptors. Alignments of the amino acid sequences of their catalytic domain show insertions at the variable regions, which are involved in the interaction of the serine proteinases with their macromolecular substrates, and/or specific inhibitors, as demonstrated by site-directed mutagenesis studies [5–8].

The crystal structures of more than twenty trypsin-like serine proteinases have been determined. They include those of bovine chymotrypsin [9], bovine trypsin [10], porcine kallikrein [11], human thrombin [12] and the catalytic domains of human factor X [13] and human tissue-type plasminogen activator (t-PA) [14]. The overall structure of serine proteinases is highly conserved in all members of the family [15, 16], since they are all folded into two six-stranded β-barrels, separated by the catalytic residues, and a C-terminal helical segment. In trypsin-like serine proteinases, the S_1 pocket is highly conserved, particularly the aspartate residue at position 189, which forms a canonical ion pair interaction with the positively charged side chain of the P_1 residue of the substrate molecule [17].

However, there are still major unanswered questions concerning the structure of the human serine proteinases involved in hemostasis. These questions mainly concern the dynamics of activation of the zymogen in potent enzymes, inhibition by physiological inhibitors and the molecular effects of these enzymes on their protein targets. To characterize the structural elements involved in target molecule recognition, snake venom trypsin-like serine proteases provide excellent models.

Snake Venom Serine Proteinases Affecting Hemostasis

Serine proteinases are very abundant in Viperidae and Crotalidae venoms in which they may account for 20% of the total protein content. They are not lethal by themselves, but they contribute to the toxic effect of the venom when associated with other venom proteins. They selectively affect many steps in the blood coagulation cascade by the activation or the inhibition of specific blood factors involved in platelet aggregation, coagulation or fibrinolysis (fig. 1) [18–20].

Various snake venom serine proteinases have a platelet-aggregating activity. The mechanism of action of MSP1 *(Bothrops moojeni)*, cerastobin *(Cerastes vipera)*, cerastocytin and cerastotin *(Cerastes cerastes)* on platelet activation is unknown [21, 22], but it has been reported that thrombocytin *(Bothrops atrox)* and PA-Bj *(Bothrops jararaca)* activate proteinase-activated receptors 1 and 4 of thrombin (PAR-1 and PAR-4) in human blood platelets [23, 24]. Other snake venom serine proteinases affect coagulation factors. A coagulant factor from Russell's viper venom (RVV-V), selectively activates factor V, converting it to FVa by cleaving a single peptide bond [25]. An anticoagulant enzyme from *Agkistrodon contortrix contortrix* venom activates protein C (ACC-C) [26]. The resulting activated protein C then degrades factor Va and factor VIIIa, resulting in an anticoagulant effect. Many other Viperidae serine proteinases convert fibrinogen into fibrin by cleaving fibrinopeptides A and/or B [27–29].

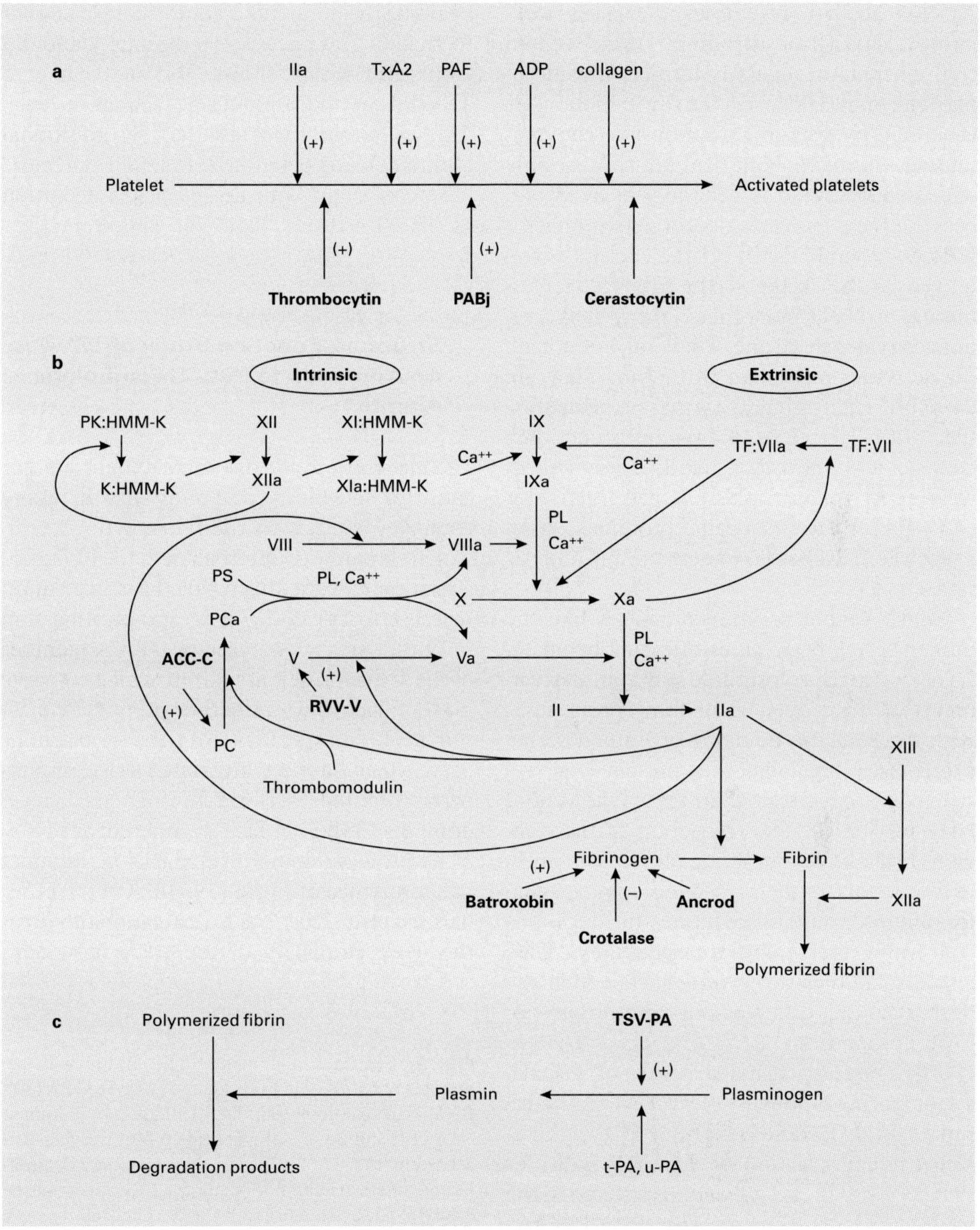

Fig. 1. Physiological processes of hemostasis. Action of various snake venom serine proteinases on the three pathways of hemostasis: platelet aggregation (**a**), coagulation (**b**) and fibrinolysis (**c**). Snake venom serine proteinases are in bold.

As this activity resembles the most well-known activity of thrombin, these venom components are generally termed 'thrombin-like' enzymes. However, the fibrin monomers generated by venom thrombin-like enzymes undergo limited polymerization, because only one fibrinopeptide is generally cleaved and because these enzymes do not activate factor XIII, to generate factor XIIIa. The clot is stabilized as the factor XIIIa catalyzed cross-linking of fibrin monomers. Many snake venom enzymes share one or several, but not all, the activities of human thrombin. They are therefore interesting tools for investigating the various activities of this multifunctional enzyme. On the other hand, *Trimeresurus stejnegeri* venom plasminogen activator (TSV-PA) is the first serine proteinase from snake venom reported to activate fibrinolysis [30].

Snake venom serine proteinases like the human serine proteinases affecting hemostasis belong to the trypsin-like subfamily. Alignments of the sequences of these proteinases with the catalytic domains of human serine proteinases highlight the presence of restricted variable regions in the snake venom enzymes (fig. 2). The restriction of these regions in snake venom serine proteinases probably accounts for their resistance to physiological inhibitors and raises interesting questions concerning their substrate specificity. They also display high levels of sequence identity, despite having different but specific physiological properties.

The crystallographic structure of a snake venom serine proteinase (TSV-PA) was determined [31]. Detailed analysis of the three-dimensional structure of TSV-PA indicates that access to the active site is restricted to the west by a loop segment. This loop has one more amino acid in TSV-PA than in trypsin, at position 218, and a proline is present in a *cis* conformation at position 219. The main chain of this loop has a conformation similar to that of the ancestral molecule, glandular kallikrein, which shows 36.5% identity to TSV-PA at the amino acid sequence level. The kallikrein loop also has an additional amino acid, at position 218, and a *cis* Pro^{219} [11], consistent with the proposed evolution of TSV-PA from glandular kallikrein [32–34].

Structure-Function Study of TSV-PA: An Approach to New Thrombolytic Agents

Since thrombotic disorders remain a major cause of morbidity and mortality in many countries, studies on the fibrinolytic system, which uniquely counterbalances the blood coagulation cascade, have called for intense research efforts [35–37] The rate-limiting step in fibrinolysis is catalyzed by t-PA, a member of the serine proteinase family which converts plasminogen into an active proteinase plasmin [37]. The specificity of t-PA for plasminogen, which has been attributed to its simultaneous binding to fibrin through its kringle domains [38], is in fact an inherent property of its protease domain since it is maintained in the absence of fibrin [39, 40]. On the other hand, recent structural investigations confirm the close similarity of the t-PA proteinase

Fig. 2. Sequence alignments. The amino acid sequences of batroxobin [43], Ancrod [44], TSV-PA [30], RVV-V [45], porcine glandular kallikrein [11], human t-PA [14] and bovine chymotrypsin [46] were aligned according to their topological equivalence. The numbering refers to chymotrypsinogen. Residues of the catalytic triad are marked by stars. Particularly close sequence homologies between all seven proteins are given in pale gray boxes, where identities are typed in dark gray. The figure was prepared using the Clustal W program.

```
             16  20        30                 40         50         60
Batroxobin   VIGGDECDINEHPFLAFMYYSP------RYFCGMTLINQEWVLTAAHCNRR-----FMRI
Ancrod       VIGGDECNINEHRFLALVYAN-------GSLCGGTLINQEWVLTARHCDRG-----NMRI
TSV-PA       VFGGDECNINEHRSLVVLFNSN------GFLCGGTLINQDWVVTAAHCDSN-----NFQL
RVV-V        VVGGDECNINEHPFLVALYTSTS----STIHCGGALINREWVLTAAHCDRR-----NIRI
Kallikrein   IIGGRECEKNSHPWQVAIYHYS------SFQCGGVLVNPKWVLTAAHCKND-----NYEV
Human t-PA   IKGGLFADIASHPWQAAIFAKHRRSPGERFLCGGILISSCWILSAAHCFQERFPPHHLTV
Chymotrypsin IVNGEEAVPGSWPWQVSLQDKTG-----FHFCGGSLINENWVVTAAHCGVT----TSDVV
                                                           *

                70         80       90              100        110
Batroxobin   HLGKHAGSVANYDEVVRYPKEKFICP--NKKKN-VITDKDIMLIRLDRPVK----NSEHI
Ancrod       YLGMHNLKVLNKDALRRFPKEKYFCL--NTRND-TIWDKDIMLIRLNRPVR----NSAHI
TSV-PA       LFGVHSKKILNEDEQTRDPKEKFFCP--NRKKD-DEVDKDIMLIKLDSSVS----NSEHI
RVV-V        KLGMHSKNIRNEDEQIRVPRGKYFCL--NTKFP-NGLDKDIMLIRLRRPVT----YSTHI
Kallikrein   WLGRHNLFENENTAQFFGVTADFPHPGFNLSADGKDYSHDLMLLRLQSPAK----ITDAV
Human t-PA   ILGRTYRVVPGEEEQKFEVEKYIVHK----EFDDDTYDNDIALLQLKSDSSRCAQESSVV
Chymotrypsin VAGEFDQGSSSEKIQKLKIAKVFKNS----KYNSLTINNDITLLKLSTAAS----FSQTV
                                                     *

               120        130      140         150        160        170
Batroxobin   APLSLPSNPPSVG--SVCRIMGWGAITTSED--TYPDVPHCANINLFNNTVCRE-AYNGL
Ancrod       APLSLPSNPPSVG--SVCRIMGWGTITSPNA--TLPDVPHCANINILDYAVCQA-AYKG-
TSV-PA       APLSLPSSPPSVG--SVCRIMGWGKTIPTKE--IYPDVPHCANINILDHAVCRT-AYSWR
RVV-V        APVSLPSRSRGVG--SRCRIMGWGKISTTED--TYPDVPHCTNIFIVKHKWCEP-LYPWV
Kallikrein   KVLELPTQEPELG--STCEASGWGSIEPGPDBFEFPDEIQCVQLTLLQNTFCAB-AHPB-
Human t-PA   RTVCLPPADLQLPDWTECELSGYGKHEALSP—FYSERLKEAHVRLYPSSRCTSQHLLNR
Chymotrypsin SAVCLPSASDDFAAGTTCVTTGWGLTRYTNA--NTPDRLQQASLPLLSNTNCK--KYWGT

                  180              190       200        210       220
Batroxobin   --PAKTLCAGVLQGG------IDTCGGDSGGPLICNG----QFQGILSWGSDPCAEPRKP
Ancrod       --AATTLCAGILEGG------KDTCKGDSGGPLICNG----QFQGILSVGGNPCAQPRKP
TSV-PA       QVANTTLCAGILQGG------RDTCHFDSGGPLICNG----IFQGIVSWGGHPCGQPGEP
RVV-V        PADSRTLCAGILKGG------RDTCHGDSGGPLICNG----QIQGIVAGGSEPCGQHLKP
Kallikrein   KVTESMLCAGYLPGG------KDTCMGDSGGPLICNG----MWQGITSWGHTPCGSANKP
Human t-PA   TVTDNMLCAGDTRSGGPQANLHDACQGDSGGPLVCLNDGRMTLVGIISWGLG-CGQKDVP
Chymotrypsin KIKDAMICAGASG--------VSSCMGDSGGPLVCKKNGAWTLVGIVSWGSSTCS-TSTP
                                         *

                 230       240
Batroxobin   AFYTKVFDYLPWIQSIIAGNKTATCP-
Ancrod       GIYTKVFDYTDWIQSIISGNTDATCPP
TSV-PA       GVYTKVFDYLDWIKSIIAGNKDATCPP
RVV-V        AVYTKVFDYNNWIQNIIAGNRTVTCPP
Kallikrein   SIYTKLIFYLDWINDTITENP------
Human t-PA   GVYTKVTNYLDWIRDNMRP--------
Chymotrypsin GVYARVTALVNWVQQTLAAN-------
```

domain to that of nonspecific proteinases such as trypsin [14, 16]. Nevertheless, the molecular basis for the specificity of t-PA for plasminogen remains poorly understood. TSV-PA, which like t-PA, selectively converts plasminogen into plasmin by cleavage of the peptide bond Arg^{561}-Val^{562}, is a useful structural template for the identification of the site of interactions of these proteins with plasminogen and the plasminogen activator inhibitor-1 (PAI-1).

Although TSV-PA shares only 23% sequence identity with the catalytic domain of t-PA, its tridimensional structure demonstrates a close structural similarity to t-PA [31]. Close to its active site, TSV-PA possesses the 99-loop which projects from the molecular surface and forms a highly charged rim (Lys^{94}-Lys^{95}-Asp^{96}-Asp^{97}-Glu^{98}-Val^{99}). A salt bridge between Asp^{97} and Arg^{174} stabilizes this loop and occludes the S_4 binding pocket. The replacement of the sequence Asp^{96}-Asp^{97}-Glu^{98} to Asn-Val-Ile resulted in a loss of plasminogenolytic activity, indicating the key role this sequence plays in substrate recognition by the enzyme. Separate replacement of each of these negatively charged residues at the site indicated that Asp^{97} was the most critical residue. The Asp^{97}-Val point mutation resulted in a 125-fold decrease in TSV-PA activity for plasminogen activation [41]. Mutational studies around the segment 172–176, however, have shown little or no effect on the plasminogenolytic activity of TSV-PA. It is therefore reasonable to assume that Asp^{97} changes its conformation upon complex formation and participates in secondary contacts with plasminogen. The conservation of Asp^{97} in urokinase-type PA (u-PA) and t-PA is noteworthy. It is likely that Asp^{97} of t-PA and u-PA have a similar role in plasminogen recognition.

The ability of a serpin to neutralize a given serine proteinase appears to depend both upon the sequence of its reactive center loop and the contribution of one (or more) secondary binding site(s). TSV-PA escapes some serpins by the use of at least two molecular mechanisms: the first involves Phe^{193}, a residue of the catalytic groove, the second originates from the absence of secondary binding sites as positively charged residues in the variable region-1 (VR1) of TSV-PA. While all other trypsin-like proteinases have a glycine at position 193 in the chymotrypsin numbering system, TSV-PA has a phenylalanine, and this phenylalanine restricts access to the S′2 pocket. The TSV-PA mutant (F193G), in which Phe^{193} had been replaced by a glycine, was inhibited by PAI-1, α_2-antiplasmin and α_1-antitrypsin with association rate constant (kon) values at least twenty-fold higher than for the wild-type enzyme. On the other hand, substitution of the VR1 of TSV-PA for either that of t-PA or u-PA also increased the kon value for the inhibition by PAI-1 over hundred times. Mutations were additive, for the inhibition by PAI-1: the double TSV-PA variant F193G/t-PA (VR1) was inhibited with a kon value 4 orders of magnitude higher than the wild-type enzyme. We conclude that these two motifs of TSV-PA fully account for its resistance to PAI-1. This opens the doors to therapeutic improvement of presently used fibrinolytic agents, since construction of a t-PA variant containing a Phe at position 193 could give a new thrombolytic agent with a longer half-life since it resists longer to PAI-1 not only in systemic circulation but also at the clot site [42]. Moreover, the TSV-PA variant F193G is a good candidate for the crystallographic resolution of a complex between a proteinase and a serpin since it produces SDS-stable complexes with PAI-1, α_2-AP and α_1-AT. Indeed, the accession of this kind of complex could improve our understanding of the inhibitory mechanism of serpins.

References

1 Siezen RJ: Multi-domain, cell-envelope proteinases of lactic acid bacteria. Antonie Van Leeuwenhoek 1999;76:139–155.

2 Rudenskaya GN, Bogdanova EA, Revina LP, Golovkin BN, Stepanov VM: Macluralisin – a serine proteinase from fruits of *Maclura pomifera* (Raf.). Schneid Planta 1995;196: 174–179.

3 Neurath H: Evolution of proteolytic enzymes. Science 1984;224:350–357.

4 Neurath H: Proteolytic enzymes, past and present. Fed Proc 1985;44: 2907–2913.

5 DiBella EE, Scheraga HA: Thrombin specificity: Further evidence for the importance of the beta-insertion loop and Trp96. Implications of the hydrophobic interaction between Trp96 and Pro60B Pro60C for the activity of thrombin. J Protein Chem 1998;17:197–208.

6 Hedstrom L, Szilagyi L, Rutter WJ: Converting trypsin to chymotrypsin: The role of surface loops. Science 1992;255:1249–1253.

7 Le Bonniec BF, Guinto ER, MacGillivray RT, Stone SR, Esmon CT: The role of thrombin's Tyr-Pro-Pro-Trp motif in the interaction with fibrinogen, thrombomodulin, protein C, antithrombin III, and the Kunitz inhibitors. J Biol Chem 1993;268:19055–19061.

8 Madison EL, Goldsmith EJ, Gerard RD, Gething MJ, Sambrook JF: Serpin-resistant mutants of human tissue-type plasminogen activator. Nature 1989;339:721–724.

9 Matthews BW, Sigler PB, Henderson R, Blow DM: Three-dimensional structure of tosyl-alpha-chymotrypsin. Nature 1967;214:652–656.

10 Ruhlmann A, Kukla D, Schwager P, Bartels K, Huber R: Structure of the complex formed by bovine trypsin and bovine pancreatic trypsin inhibitor. Crystal structure determination and stereochemistry of the contact region. J Mol Biol 1973;77:417–436.

11 Bode W, Chen Z: The X-ray structures of porcine pancreatic kallikrein and of its complex with bovine pancreatic trypsin inhibitor. Adv Exp Med Biol 1983;156:289–308.

12 Bode W, Mayr I, Baumann U, Huber R, Stone SR, Hofsteenge J: The refined 1.9 Å crystal structure of human alpha-thrombin: Interaction with *D*-Phe-Pro-Arg chloromethylketone and significance of the Tyr-Pro-Pro-Trp insertion segment. EMBO J 1989;8:3467–3475.

13 Padmanabhan K, Padmanabhan KP, Tulinsky A, Park CH, Bode W, Huber R, Blankenship DT, Cardin AD, Kisiel W: Structure of human des(1–45) factor Xa at 2.2 Å resolution. J Mol Biol 1993;232:947–966.

14 Lamba D, Bauer M, Huber R, Fischer S, Rudolph R, Kohnert U, Bode W: The 2.3 angstrom crystal structure of the catalytic domain of recombinant two-chain human tissue-type plasminogen activator. J Mol Biol 1996:258:117–135.

15 Greer J: Comparative modeling methods: Application to the family of the mammalian serine proteases. Proteins 1990;7:317–334.

16 Perona JJ, Craik CS: Structural basis of substrate specificity in the serine proteases (Review). Protein Sci 1995:4:337–360.

17 Perona JJ, Craik CS: Evolutionary divergence of substrate specificity within the chymotrypsin-like serine protease fold. J Biol Chem 1997; 272:29987–29990.

18 Kini RM, Evans HJ: Effects of snake venom proteins on blood platelets. Toxicon 1990;28:1387–1422.

19 Markland FS Jr: Snake venoms. Drugs 1997;3:1–10.

20 Mitrakul C: Effect of five Thai snake venoms on coagulation, fibrinolysis and platelet aggregation. Southeast Asian J Trop Med Public Health 1979;10:266–275.

21 Marrakchi N, Zingali RB, Karoui H, Bon C, el Ayeb M: Cerastocytin, a new thrombin-like platelet activator from the venom of the Tunisian viper *Cerastes cerastes.* Biochim Biophys Acta 1995;1244:147–156.

22 Serrano SM, Mentele R, Sampaio CA, Fink E: Purification, characterization, and amino acid sequence of a serine proteinase, PA-BJ, with platelet-aggregating activity from the venom of *Bothrops jararaca.* Biochemistry 1995;34:7186–7193.

23 Kirby EP, Niewiarowski S, Stocker K, Kettner C, Shaw E, Brudzynski TM: Thrombocytin, a serine protease from *Bothrops atrox* venom. 1. Purification and characterization of the enzyme. Biochemistry 1979;18: 3564–3570.

24 Santos ABF, Serrano SMT, Kuliopulos A, Niewiarouski S: Interaction of viper venom serine peptidases with thrombin receptors on human platelets. FEBS Lett 2000;477:199–202.

25 Kisiel W: Molecular properties of the factor V-activating enzyme from Russell's viper venom. J Biol Chem 1979;254:12230–12234.

26 Stocker K, Fischer H, Meier J, Brogli M, Svendsen L: Characterization of the protein C activator Protac from the venom of the southern copperhead *(Agkistrodon contortrix)* snake. Toxicon 1987;25:239–252.

27 Pirkle H, Theodor I: Thrombin-like venom enzymes: Structure and function. Adv Exp Med Biol 1990; 281:165–175.

28 Pirkle H: Thrombin-like enzymes from snake venoms: An updated inventory. Scientific and Standardization Committee's Registry of Exogenous Hemostatic Factors. Thromb Haemost 1998:79:675–683.

29 Stocker K, Fischer H, Meier J: Thrombin-like snake venom proteinases. Toxicon 1982;20:265–273.

30 Zhang Y, Wisner A, Xiong Y, Bon C: A novel plasminogen activator from snake venom. Purification, characterization, and molecular cloning. J Biol Chem 1995;270: 10246–10255.

31 Parry MA, Jacob U, Huber R, Wisner A, Bon C, Bode W: The crystal structure of the novel snake venom plasminogen activator TSV-PA: A prototype structure for snake venom serine proteinases. Structure 1998;6: 1195–1206.

32 Nakashima K, Ogawa T, Oda N, Hattori M, Sakaki Y, Kihara H, Ohno M: Accelerated evolution of *Trimeresurus flavoviridis* venom gland phospholipase A2 isozymes. Proc Natl Acad Sci USA 1993;90: 5964–5968.

33 Nakashima K, Nobuhisa I, Deshimaru M, Nakai M, Ogawa T, Shimohigashi Y, Fukumaki Y, Hattori M, Sakaki Y, Hattori S: Accelerated evolution in the protein-coding regions is universal in Crotalinae snake venom gland phospholipase A2 isozyme genes. Proc Natl Acad Sci USA 1995;92:5605–5609.
34 Ogawa T, Oda N, Nakashima K, Sasaki H, Hattori M, Sakaki Y, Kihara H, Ohno M: Unusually high conservation of untranslated sequences in cDNAs for *Trimeresurus flavoviridis* phospholipase A2 isozymes. Proc Natl Acad Sci USA 1992;89:8557–8561.
35 Collen D: On the regulation and control of fibrinolysis. Edward Kowalski Memorial Lecture. Thromb Haemost 1980;43:77–89.
36 Collen D, Lijnen HR, Todd PA, Goa KL: Tissue-type plasminogen activator. A review of its pharmacology and therapeutic use as a thrombolytic agent. Drugs 1989;38:346–388.
37 Collen D, Lijnen HR: Basic and clinical aspects of fibrinolysis and thrombolysis. Blood 1991;78:3114–3124.
38 Lijnen HR, Collen D: Molecular interactions between tissue-type plasminogen activator and plasminogen. Methods Enzymol 1993;223:197–206.
39 Kohnert U, Hellerbrand K, Martin U, Stern A, Popp F, Fischer S: The recombinant *Escherichia coli*-derived protease-domain of tissue-type plasminogen activator is a potent and fibrin specific fibrinolytic agent. Fibrinolysis 1996;10:93–102.
40 Martin U, Kohnert U, Stern A, Popp F, Fischer S: Comparison of the recombinant *Escherichia coli*-produced protease domain of tissue-type plasminogen activator with alteplase, reteplase and streptokinase in a canine model of coronary artery thrombolysis. Thromb Haemost 1996;76:1096–1101.
41 Zhang Y, Wisner A, Maroun RC, Choumet V, Xiong Y, Bon C: *Trimeresurus stejnegeri* snake venom plasminogen activator. Site-directed mutagenesis and molecular modeling. J Biol Chem 1997;272:20531–20537.
42 Shohet RV, Spitzer S, Madison EL, Bassel-Duby R, Gething MJ, Sambrook JF: Inhibitor-resistant tissue-type plasminogen activator: An improved thrombolytic agent in vitro. Thromb Haemost 1994;71:124–128.
43 Itoh N, Tanaka N, Mihashi S, Yamashina I: Molecular cloning and sequence analysis of cDNA for batroxobin, a thrombin-like snake venom enzyme. J Biol Chem 1987;262:3132–3135.
44 Au LC, Lin SB, Chou JS, Teh GW, Chang KJ, Shih CM: Molecular cloning and sequence analysis of the cDNA for ancrod, a thrombin-like enzyme from the venom of *Calloselasma rhodostoma.* Biochem J 1993;294(pt 2):387–390.
45 Tokunaga F, Nagasawa K, Tamura S, Miyata T, Iwanaga S, Kisiel W: The factor V-activating enzyme (RVV-V) from Russell's viper venom. Identification of isoproteins RVV-V alpha, -V beta, and -V gamma and their complete amino acid sequences. J Biol Chem 1988;263:17471–17481.
46 Wang D, Bode W, Huber R: Bovine chymotrypsinogen A X-ray crystal structure analysis and refinement of a new crystal form at 1.8 Å resolution. J Mol Biol 1985;185:595–624.

Fibrinolysis

Hæmostasis

Haemostasis 2001;31:141–147

Alfimeprase: Pharmacology of a Novel Fibrinolytic Metalloproteinase for Thrombolysis

Christopher F. Toombs

Product Development, Amgen Inc., Thousand Oaks, Calif., USA

Key Words

Fibrolase · Alfimeprase · Thrombosis · Fibrinolysis · Metalloprotease · Thrombolysis

Abstract

Alfimeprase is a recombinantly produced, truncated form of fibrolase, a known directly fibrinolytic zinc metalloproteinase that was first isolated from the venom of the southern copperhead snake *(Agkistrodon contortrix contortrix).* Both fibrolase and alfimeprase have been shown to have direct proteolytic activity against the fibrinogen Aα chain. In vivo pharmacology studies have shown that thrombolysis with alfimeprase is up to 6 times more rapid than with plasminogen activators. Alfimeprase can be bound and neutralized by serum α_2-macroglobulin, a prevalent mammalian protease inhibitor which is capable of forming a macromolecular complex with alfimeprase. As a result, systemic bleeding complications have been greatly reduced due to the inhibitory effects of α_2-macroglobulin. This article reviews the biochemical in vitro and in vivo characteristics of this novel acting thrombolytic.

Introduction

Alfimeprase is a recombinantly produced, truncated form of fibrolase, a known fibrinolytic zinc metalloproteinase that was first isolated from the venom of *Agkistrodon contortrix contortrix,* the southern copperhead snake [1]. Fibrolase is a member of clan MB of metallopeptidases, family M12, subfamily B (the reprolysins), a grouping of proteolytic enzymes that are comprised of many enzymes originally characterized from snake venoms.

The alfimeprase molecule differs slightly from fibrolase at the amino terminus where alfimeprase contains 201 amino acids with an

KARGER

Fax + 41 61 306 12 34
E-Mail karger@karger.ch
www.karger.com

0301–0147/01/0316–0141$17.50/0

Accessible online at:
www.karger.com/journals/hae

Christopher F. Toombs, PhD
Product Development/Mailstop 27-5-A, Amgen Inc.
1, Amgen Center Drive, Thousand Oaks, CA 91320 (USA)
Tel. +1 805 447 8121, Fax +1 805 375 6524
E-Mail ctoombs@amgen.com

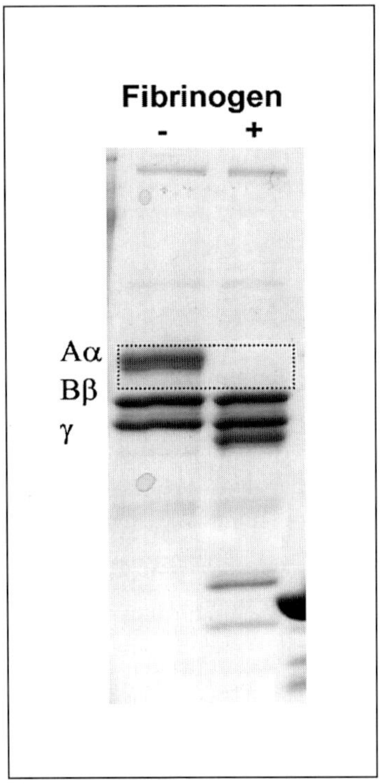

Fig. 1. Coomassie-based visualization of proteolytic cleavage of fibrinogen Aα chain. Samples of human fibrinogen (741 μg) were incubated with 1 μg alfimeprase (+ lane) or buffer (– lane) at 37 °C for 30 min. The microgram quantity of fibrinogen was selected such that alfimeprase was combined with each substrate in a 1:50 molar ratio (alfimeprase:substrate). Separation of the reaction mixture by SDS-PAGE and visualization of proteins with Coomassie staining reveals proteolytic degradation of the fibrinogen Aα chain in the presence of alfimeprase (dashed box).

N-terminal sequence of SFPQR---, in contrast to the N-terminal sequence of fibrolase, which begins with EQRFPQR--- and is 203 amino acids in length [2]. While alfimeprase is a modest 2-amino-acid truncation of fibrolase, the two enzymes are similar with respect to enzymatic activity and the ability to be bound and inhibited by α_2-macroglobulin. This is consistent with the data of Manning [3] indicating that the active site of the fibrolase molecule spans amino acids 139–159, and its location, in three-dimensional space, is distant from the amino terminus where the truncation was made.

Fibrinolytic Activity of Alfimeprase

Both fibrolase and alfimeprase have been shown to be fibrinolytic, and fibrolase has been previously documented to have proteolytic activity against the fibrinogen Aα chain, with reduced proteolytic cleavage of the Bβ chain and no activity against the γ chain of fibrinogen [4]. In contrast to the plasminogen activator class of thrombolytic drugs, fibrolase and alfimeprase do not rely on the endogenous fibrinolytic system (conversion of plasminogen to plasmin). Hence, fibrolase and alfimeprase can be distinguished from the plasminogen activators by their unique mode of action and are defined as directly fibrinolytic agents.

Substrates of Alfimeprase and Fibrolase

Published literature on venom fibrolase has demonstrated the proteolytic activity against fibrinogen at the Lys^{413}-Leu^{414} site [5] and against the oxidized β-chain of insulin at the Ala^{14}-Leu^{15} site [6, 7]. Alfimeprase has also been determined to have proteolytic activity on these substrates. The degradation of the fibrinogen Aα chain can easily be demonstrated in a gel-based method, as first described with fibrolase by Retzios and Markland [8] and replicated in figure 1 using alfimeprase.

α_2-Macroglobulin

α_2-Macroglobulin is a prevalent protease inhibitor present in mammalian serum and is one of the largest of the serum proteins (725 kD). The properties of α_2-macroglobulin that are relevant to alfimeprase are described in this section, and for a more thorough review, see Barrett [9]. The specificity of α_2-macroglobulin for proteases is broad, encompassing serine, cysteine, aspartic and metalloprotease classes. The α_2-macroglobulin molecule is a tetramer of identical subunits that are disulfide bonded in pairs with a noncovalent association of the half molecules. Thus, under reducing conditions, α_2-macroglobulin can be dissociated into its four monomeric subunits.

Each subunit of α_2-macroglobulin possesses a region that is very susceptible to proteolytic cleavage (the 'bait' region). Proteolysis of the bait region induces a conformational change in α_2-macroglobulin, which entraps the protease within the α_2-macroglobulin molecule. This process is described in the literature as a 'Venus-flytrap' interaction. Once the protease is entrapped, the protease is sterically hindered and therefore cannot access macromolecular substrate.

In addition, a covalent bond can form between α_2-macroglobulin and many of the proteases that it entraps. Thus, within the general circulation, α_2-macroglobulin can effectively neutralize a variety of proteases by entrapment of the protease which, in some instances, also becomes covalently bonded to α_2-macroglobulin.

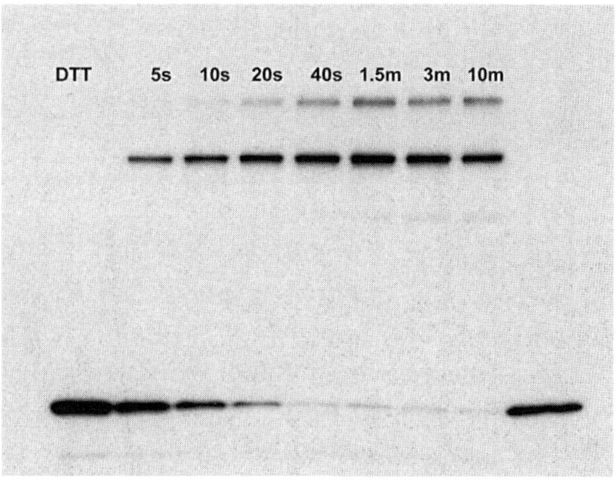

Fig. 2. Rapidity of alfimeprase: α_2-macroglobulin complex formation. Visualization of the rate at which alfimeprase can form complex with purified human α_2-macroglobulin. Reaction products were detected using a SDS-PAGE separation and Western blotting with a polyclonal antibody that recognizes alfimeprase. Alfimeprase and purified human α_2-macroglobulin are incubated together at 37 °C with α_2-macroglobulin present in a 3-fold molar excess of alfimeprase. At various time points, the formation of complex is stopped by the addition of a buffer containing the reducing agent dithiothreitol (DTT). If α_2-macroglobulin is treated with DTT prior to the addition of alfimeprase (DTT lane), formation of the complex is completely prevented, indicating that DTT is an effective stopping agent. Upon the addition of alfimeprase, complex formation begins as early as 5 s (5s lane) and is nearly complete within 1.5 min (1.5m lane). Purified alfimeprase appears alone in the last lane.

Interaction of Alfimeprase with Purified Human α_2-Macroglobulin

As previously mentioned, α_2-macroglobulin is a prevalent protease inhibitor present in mammalian serum, which is capable of forming a macromolecular complex with alfimeprase. Unlike some proteases that can form a dissociable complex with α_2-macroglobulin, alfimeprase forms a complex and cannot be dissociated from α_2-macroglobulin under physiologic conditions.

As shown in figure 2, alfimeprase remains bound to α_2-macroglobulin even under the harsh laboratory conditions used to process samples for Western blotting (reducing agent, detergent, and boiling), confirming the non-dissociable nature of the alfimeprase-α_2-macroglobulin complex.

α_2-Macroglobulin Binding in Human Plasma Is Saturable

Although α_2-macroglobulin is one of the major plasma proteins, there is nonetheless a finite quantity of α_2-macroglobulin available to bind and neutralize alfimeprase, and therefore the α_2-macroglobulin binding capacity is saturable. Using Western blotting techniques a titration experiment can be performed where a range of alfimeprase concentrations is added to a plasma sample from an individual human donor. When low concentrations of alfimeprase are added to plasma, only the high molecular weight alfimeprase-α_2-macroglobulin complex form is detectable. At concentrations that approach and then slightly exceed the α_2-macroglobulin binding capacity, the unbound alfimeprase begins to become detectable. Once the α_2-macroglobulin binding capacity has been exceeded, the concentration of unbound alfimeprase rises proportionally as additional alfimeprase is added to the sample. Using this methodology, the alfimeprase binding capacity of human plasma was estimated to be ~ 40 μg/ml plasma (data not shown).

In vitro Lysis of Clotted Human Whole Blood

The activity of alfimeprase has been demonstrated in an in vitro model of clot lysis. Human blood can be clotted by the addition of a high calcium concentration. Once formed, these clots can be solubilized with thrombolytic agents. The rate of clot lysis can be assessed by visual inspection of the clot mass.

As can be visually observed in the data of figure 3, when a fixed amount of alfimeprase is used to degrade clots of varying sizes, the time required for clot dissolution increases proportionally with the mass of the clot.

Studies Which Demonstrate Biologic Activity in Animals: Rat Carotid

Pharmacology studies were conducted in carotid arteries that have been acutely thrombosed by locally injuring the vessel with anodal current. Once an occlusive thrombus forms, the thrombosed carotid artery is observed for a period of 30 min to assure that the carotid occlusion is stable. Heparin and aspirin are then administered intravenously to prevent further propagation of the thrombus. In these studies, a small needle (30 gauge) that was introduced into the artery in the vicinity of the occlusive thrombus and then cemented into place such that test agents could be administered locally.

Thrombolysis was attempted by intra-arterial infusion of test agents in a blinded fashion using numbered, coded syringes. While study drugs were being administered, flow through the carotid artery was monitored with a perivascular flow probe to determine the time at which reflow occurs. The time to lysis was recorded and group means were calculated for those experiments where clot lysis was successful. Blood pressure and heart rate were observed during the study to monitor any untoward effects of the test agents. As a measure of the hemorrhagic potential of the test agents, any blood that was shed from the surgical site was collected with gauze swabs. The swabs were placed in a detergent solution to solubilize red blood cells and release hemoglobin, which was then quantified spectrophotometrically. Shed hemoglobin was mathematically converted into a volume of blood

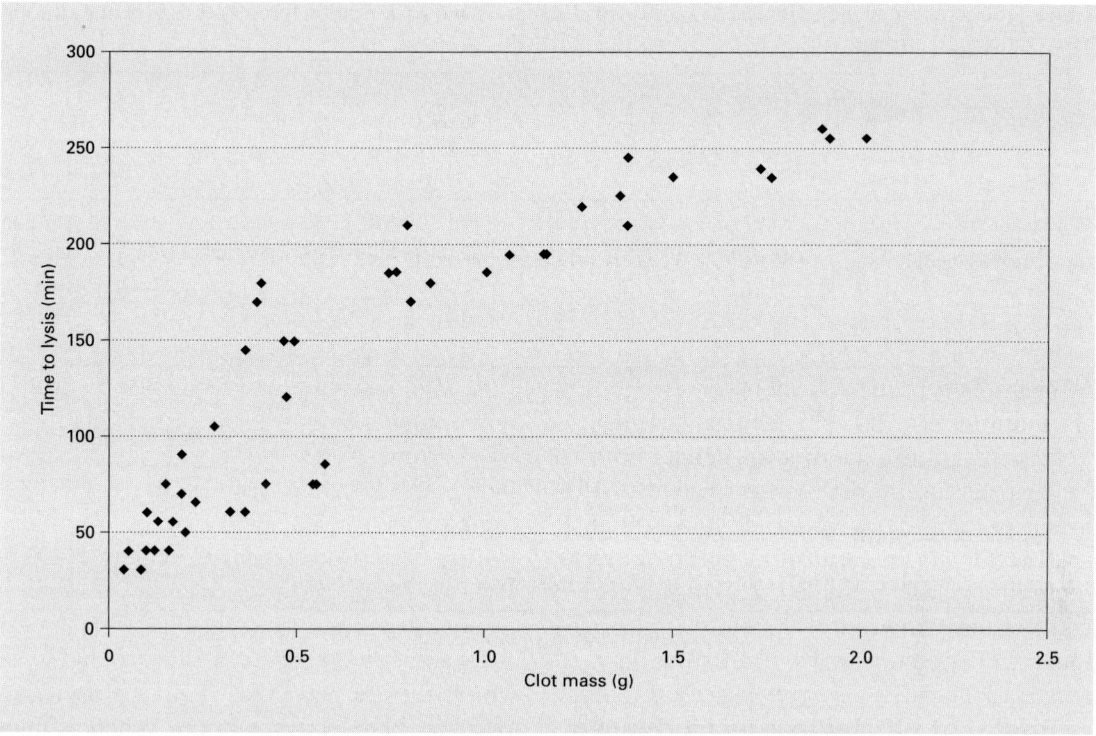

Fig. 3. In vitro time to clot lysis with alfimeprase is directly proportional to clot mass. Scatter plot of individual clot lysis experiments. Retracted whole human blood clots were formed from citrated whole blood by the addition of excessive calcium and incubation for 90 min. The clotted blood was cut into various sizes, weighed and transferred to new tubes containing 2 mg of alfimeprase (0.4 ml of a 5 mg/ml solution). The tubes were inspected visually over a period of 5 h and time to clot lysis was determined as the time point at which no visible clot mass remained.

loss by using the hemoglobin concentration of whole blood (in g/dl), which was measured for each animal.

As shown in table 1, clot lysis in the 2-mg alfimeprase group was more rapid than in all other groups. In addition, urokinase (at 250 U/min) was associated with a significant amount of blood loss from the surgical site. While mean blood loss in the 2-mg alfimeprase and 25 U/min urokinase groups was higher than in the saline group; these changes did not reach statistical significance. While some changes in mean arterial pressure reached statistical significance in the saline and urokinase 250 U/min groups, no significant changes were noted in the alfimeprase treatment group.

These results indicate that alfimeprase has the potential to be more effective than a reference plasminogen activator in terms of both efficacy (speed of lysis) and safety (hemorrhagic complication).

Table 1. Selected parameters from a comparison of alfimeprase and urokinase on thrombolysis in acutely thrombosed rat carotid arteries

	Incidence of lysis %	Time to lysis min	Carotid flow ml/min		Blood volume loss at 90′, ml	Mean arterial pressure, mm Hg		Heart rate min^{-1}	
			base	90′ Rx		base	15′ Rx	base	15′ Rx
Saline (n = 6)	0 (0 of 6)	N/A	4.3±1.5	0.0±0.0	0.10±0.10	85.2±3.6	76.5±3.2[1]	315±23	311±20
Urokinase 25 U/min (n = 15)	47% (7 of 15)	55.3±4.3	3.2±0.4	0.7±0.3	1.06±0.41	87.7±3.6	87.1±3.4	306±9	313±8
Urokinase 250 U/min (n = 15)	87% (13 of 15)	33.5±4.1	4.1±0.7	1.8±0.5	1.43±0.38[2]	89.8±5.6	82.5±4.1[1]	320±8	317±8
Alfimeprase 2 mg (n = 14)	71% (10 of 14)	6.3±1.6[3]	3.8±0.5	2.1±0.7	0.97±0.21	87.1±4.6	84.5±3.6	305±9	310±9

All data are presented as mean ± standard error.

Base: baseline time point following surgical instrumentation, but prior to the formation of thrombus. 15′ Rx and 90′ Rx, time points at 15 and 90 min following the administration of test agents.

1 Significant difference ($p < 0.05$) from baseline time point by paired t test.

2 Significant difference ($p < 0.05$) from saline group by ANOVA.

3 Significant difference ($p < 0.05$) vs. all other groups by Kaplan Meier estimates.

Effects of Alfimeprase on Fibrinogen and Platelet Aggregation in vitro

Alfimeprase has been evaluated for potential effects on human plasma fibrinogen and platelet aggregation using citrated whole blood.

In general, we found that plasma fibrinogen was not affected by incubation with alfimeprase at concentrations of 10 and 20 μg/ml. Presumably, the alfimeprase that was added to the sample was completely inhibited by the α_2-macroglobulin that was also present in the sample. In contrast, incubation of plasma with alfimeprase at 100 μg/ml resulted in complete degradation of fibrinogen in all plasma samples within the first 30 min of the incubation (data not shown), suggesting that at an alfimeprase concentration of 100 μg/ml, the capacity for α_2-macroglobulin to bind and neutralize alfimeprase has been exceeded.

Incubation of platelet-rich plasma with alfimeprase up to and including 40 μg/ml did not affect platelet aggregation. When alfimeprase was incubated with platelet-rich plasma at 100 μg/ml, a concentration which completely degraded plasma fibrinogen in all subjects studied (as noted above), a platelet aggregation response was still present, although it was diminished compared to untreated controls.

The retention of a platelet aggregation response when alfimeprase is incubated at 100 μg/ml appears to be consistent with the ability of activated platelets to release the fibrinogen that is stored in platelet α-granules. This release of fibrinogen appears sufficient to support some, but not fully normal, platelet aggregation.

Conclusion

In conclusion, the pharmacology of alfimeprase appears to demonstrate a high degree of novelty in relation to the plasminogen activator class of thrombolytic agents. Specifically, the animal data thus far indicate the speed of lysis is greatly accelerated and the potential risk of hemorrhagic complications greatly diminished. Despite the potential drawback in the requirement for local delivery of alfimeprase, the advantage gained in local delivery is more rapid thrombolysis combined with a reduced risk of bleeding. If these attributes are maintained as clinical safety and efficacy are assessed, alfimeprase will represent the first of a new therapeutic class (fibrinolytic metalloproteases) for use in thrombolysis.

References

1 Guan AL, Retzios AD, Henderson GN, Markland FS: Purification and characterization of a fibrinolytic enzyme from venom of the southern copperhead snake *(Agkistrodon contortrix contortrix)*. Arch Biochem Biophys 1991:289;197–207.

2 Randolph A, Chamberlain SH, Chu HL, Retzios AD, Markland FS Jr, Masiarz FR: Amino acid sequence of fibrolase, a direct-acting fibrinolytic enzyme from *Agkistrodon contortrix contortrix* venom. Protein Sci 1992;1:590–600.

3 Manning MC: Sequence analysis of fibrolase, a fibrinolytic metalloproteinase from *Agkistrodon contortrix contortrix*. Toxicon 1995;33:1189–1200.

4 Ahmed NK, Tennant KD, Markland FS, Lacz JP: Biochemical characteristics of fibrolase, a fibrinolytic protease from snake venom. Haemostasis 1990;20:147–154.

5 Retzios AD, Markland FS: Fibrinolytic enzymes from the venoms of *Agkistrodon contortrix contortrix* and *Crotalus basiliscus basiliscus:* cleavage site specificity towards the alpha-chain of fibrin. Thromb Res 1994;74:355–367.

6 Pretzer D, Schulteis BS, Smith CD, Vander VD, Mitchell JW, Manning MC: Stability of the thrombolytic protein fibrolase: Effect of temperature and pH on activity and conformation. Pharmacol Res 1991;8: 1103–1112.

7 Pretzer D, Schulteis B, Vander VD, Smith CD, Mitchell JW, Manning MC: Effect of zinc binding on the structure and stability of fibrolase, a fibrinolytic protein from snake venom. Pharmacol Res 1992;9:870–877.

8 Retzios AD, Markland FS: A direct-acting fibrinolytic enzyme from the venom of *Agkistrodon contortrix contortrix:* Effects on various components of the human blood coagulation and fibrinolysis systems. Thromb Res 1988:52;541–552.

9 Barrett AJ: Alpha-2 macroglobulin; in Barrett AJ (ed): Methods in Enzymology. Philadelphia, Academic Press, 1981, pp 737–754.

Substances Affecting Platelet Function

Hæmostasis

Haemostasis 2001;31:148–154

Multifunctional Snake C-Type Lectins Affecting Platelets

Kenneth J. Clemetson   Alexei Navdaev   Dagmar Dörmann
Xiao-Yan Du   Jeannine M. Clemetson

Theodor Kocher Institute, University of Berne, Switzerland

Key Words
Snake venom · C-type lectin · Platelet receptor · Binding sites · Mechanism

Abstract
Snake venoms contain a wide range of components, many of which affect haemostasis by activation or inhibition of platelets or coagulation factors. They can be classified into groups based on structure and mode of action. One group is the snake C-type lectins, so called because of the typical folding which closely resembles that found in classical C-type lectins, such as selectins and mannose-binding proteins. Unlike the classic C-type lectins, those from snakes are generally heterodimeric with two subunits, α and β. Some are multimeric heterodimers. The subunits have homologous sequences and are generally linked by a disulphide bond as well as by swapping loops. One of the first C-type lectins with a defined function was echicetin which was demonstrated to bind to platelet GPIb and block several functions of this receptor. Since then, many proteins with similar structure have been reported to act on platelet receptors or coagulation factors and several have been crystallized. These proteins were thought to be specific for a single platelet receptor or coagulation factor, i.e. they had only one receptor per heterodimer. Recent studies show that most of these C-type lectins have binding sites for more than one ligand and have complex mechanisms of action.

Introduction

Venoms produced by snakes have a wide variety of components which kill or weaken their preys. Venoms from some snake families contain mostly neurotoxic proteins, while others such as those of the Viperidae and Cro-

KARGER

Fax + 41 61 306 12 34
E-Mail karger@karger.ch
www.karger.com

0301–0147/01/0316–0148$17.50/0

Accessible online at:
www.karger.com/journals/hae

Dr. K.J. Clemetson
Theodor Kocher Institute, University of Berne
Freiestrasse 1, CH–3012 Berne (Switzerland)
Tel. +41 31 631 41 48, Fax +41 31 921 54 43
E-Mail clemetson@tki.unibe.ch

talidae are mainly haemorrhagic. Snake C-type lectins are an important class among the protein family that has been shown to have haemorrhagic effects. This family is named after the type of folding that occurs in classic C-type lectins, such as mannose-binding protein [1, 2] and the selectins [3]. The 'C' is from calcium-dependent and the 'lectin' from the sugar-binding function but, in fact, most of the snake C-type lectins do not show these properties and the calcium-binding loop is deleted in their structure. We have nevertheless retained this nomenclature since it is that of the primary literature. C-type lectins were first identified as part of metalloprotease/disintegrin/C-type lectin structures without any specific function. A large number of snake C-type lectins have now been characterized with effects on either coagulation factors or platelets. Those affecting platelets either inhibit or activate them by binding to specific receptors like GPIb, $\alpha_2\beta_1$, and GPVI.

Snake C-Type Lectin Targets

C-type lectins that act via GPIb to agglutinate platelets include alboaggregins [4–6], flavocetins A and B [7] and mamushigin [8]. Most of the inhibitory C-type lectins described so far bind to GPIb. These include echicetin [9], jararaca GPIb binding protein [10, 11], tokaricetin [12], CHH-A and B [4], and agkicetin [13]. Echicetin, a heterodimeric snake C-type lectin from *Echis carinatus*, has been shown by several authors to bind specifically to platelet GPIb and to block platelet interactions with von Willebrand factor (vWf) [9] and thrombin [14]. There has been considerable interest in using C-type lectins such as echicetin as antithrombotics, blocking the interaction between vWf and platelets. However, when echicetin or similar snake C-type lectins were injected into small animals to study their effects in vivo, induction of thrombocytopenia has often been reported [9, 15]. Generally, the platelet count dropped to 20–30% of the control value and then gradually recovered over several hours. This phenomenon has remained unexplained. Other snake C-type lectins bind to different platelet receptors, for example, convulxin, a trimeric heterodimer, has been shown to be specific for GPVI [16]. Several other snake C-type lectins were reported to bind to other platelet receptors, aggretin (or rhodocytin) [17] and rhodocetin [18], from *Calloselasma rhodostoma* venom and EMS16, a C-lectin type protein from *Echis multisquamatus* venom [19] were shown to interact with $\alpha_2\beta_1$, and, in the case of aggretin (or rhodocytin) with GPIb and possibly with (an)other receptor(s), which has not yet been identified. Rhodocetin was also the first example described of a snake C-type lectin where the subunits associate noncovalently and there is no disulphide bond between them. This might allow more rapid evolution by enabling subunits to exchange more rapidly. A wide range of other snake C-type lectins has been described that activate coagulation factors or affect other plasma components involved in hemostasis. These include factor IX/factor X binding proteins from *Trimeresurus flavoviridis* [20] and *Echis carinatus leucogaster* [21] venom and coagulation factor X binding protein from *Deinagkistrodon acutus* venom [22].

Bothrojaracin, is a 27 kD C-type lectin from *Bothrops jararaca* venom which has been characterized as a potent thrombin inhibitor [23]. Bothrojaracin binds to exosites I and II, with a dissociation constant of 0.7 n*M*, and influences but does not block the proteinase catalytic site. Bothrojaracin also binds the partially exposed anion-binding exosite (proexosite I) of prothrombin and blocks its activation to thrombin [24].

Botrocetin forms a bimolecular complex with vWF that is capable of binding specifically to GPIbα, a component of the GPIb-IX-V complex on the platelet membrane [25]. The venoms of several snake species in the genus *Bothrops* and at least one species in the genus *Bitis* contain an activity with the properties ascribed to botrocetin, but detailed studies have been conducted mostly with the protein isolated from *B. jararaca.*

Bitiscetin from *Bitis arietans* venom [26] forms a bimolecular complex with vWF, like botrocetin; however the binding site of bitiscetin is within the A3 domain [27]. Binding also involves the GPIb-binding site of vWF located on the A1 domain because it was inhibited by a monoclonal antibody to vWF whose epitopes are within this domain and that block binding of vWF to platelets induced by ristocetin or botrocetin.

Ligand Binding Sites

After the characterization of the first few snake C-type lectins it was generally assumed that the α- and β-subunits had separate functions, and several studies supported this conclusion. It is difficult to compare the positive and negative results from such studies because the individual subunits were prepared under a wide variety of conditions. It nevertheless seems plausible that the subunits prepared, after separation under reducing conditions, can refold if the thiol groups are not blocked, to give at least partially active structures. Under these conditions, several authors were able to show that α- and β-subunits had different binding characteristics [28]. However, the first determination of the X-ray crystal structure of a snake C-type lectin by Mizuno et al. [29] indicated that the two subunits were more tightly associated than had previously been suspected and demonstrated the formation of a concave surface formed by loop-swapping between the two subunits as most likely the binding site for the ligand. This was substantiated by a recent study from the same group showing that the binding site for the Gla domain of factor X does indeed lie in this region between the two subunits of the factor X binding protein [30]. Thus, the consensus of opinion moved to a model where snake C-type lectins have essentially one binding site made up of parts of both subunits.

While this model could be valid for some snake C-type lectin interactions with single ligands, it soon became clear that this is not always the case and, in fact, more complex situations may be commoner. One of the first examples of a C-type lectin interacting with two different ligands was echicetin [31]. As mentioned above, echicetin was one of the first snake C-type lectins interacting with platelet receptors to be identified and characterized. It was shown to interact with GPIb and to inhibit GPIb-related platelet functions [9]. Since then C-type lectins with similar properties were identified in a variety of species of snakes [4, 10, 13, 32]. The first indication that echicetin binds to components other than GPIb came from the observation that echicetin induced platelet agglutination in platelet-rich plasma but not with washed platelets. This suggested that there is a component in plasma that is recognized by echicetin and can cross-link echicetin bound to GPIb to link platelets together. Indeed, this component was identified as a subgroup of IgMκ and can be used to affinity-purify echicetin [31]. The results from earlier studies with isolated subunits suggest that the β-subunit is probably involved in binding to GPIb which might indicate (from steric considerations alone) that the α-subunit is mainly involved in binding IgMκ.

Another example of a snake C-type lectin that binds to more than one ligand is alboaggregin A [4, 33–35]. The situation is slightly more complicated here because this is a dimeric heterodimer or tetramer and probably binding sites are present on both heterodimers. Alboaggregin A was first identified as a GPIb-binding protein from *Trimeresurus albolabris* venom as well as alboaggregins B and C [33]. However, unlike the two others it showed much stronger activating effects on platelets. At first this was ascribed to the tetrameric structure, whereas alboaggregin B is simply heterodimeric, leading to more clustering of receptors and hence a stronger signal [4, 34]. However, the fact that the tyrosine phosphorylation pattern in platelets activated by alboaggregin A was different from that in platelets activated by alboaggregin B pointed to the involvement of other receptors. Investigation of the platelet receptors which alboaggregin A binds to clarified this issue by showing that not only GPIb but also GPVI is involved [35, 36]. Activation of platelets by clustering GPVI is well known as the mechanism used by collagen, collagen-related peptides and convulxin, also a snake C-type lectin. It is intriguing that of the four chains making up alboaggregin A, presumably two α (AlboA chains 1 and 2)- and two β (AlboA chains 3 and 4)-chains, two have sequences that resemble the α- and β-subunits of convulxin (2 and 4) whereas the other two (1 and 3) more closely resemble α- and β-subunits of C-type lectins that interact with GPIb [6].

Another example of a C-type lectin with multiple ligands is aggretin (or rhodocytin) [17, 37]. This was originally reported to be $\alpha_2\beta_1$-specific and not to interact with GPIb [17, 37]. Later studies indicated that both $\alpha_2\beta_1$ and GPIb but not GPVI were involved [38, 39]. On the other hand, platelets genetically lacking $\alpha_2\beta_1$, where GPIb was also removed proteolytically, still showed the same reactivity with this C-type lectin, suggesting that yet another receptor played a critical role [40].

There are other examples of C-type lectins which were thought to have a simple monovalent activity, but which now appear more complex. A classic example of these is botrocetin, mentioned above. This was thought to cause vWf to bind to GPIb by binding to the A1 domain and changing its configuration to an 'activated' state. More recent work based upon X-ray crystallographic structure determination [41] indicates that this may not be the whole story and that following initial binding between botrocetin/A1 domain and GPIb, botrocetin also engages a second binding site with GPIb and so may link vWf and GPIb together physically.

In the case of bothrojaracin, it is also not clear whether only one binding site is adequate to inhibit both the exosites on thrombin or, again, if two independent bind sites on the C-type lectin are necessary [42]. In order to elucidate this and other problems of C-type lectin-binding sites it will be necessary to have more structural information based on X-ray crystallographic data.

Structure-Function Relationships

Modelling the structure of snake C-type lectins has been a fairly useful approach to establishing putative binding sites on their surfaces based on the premise that interactions with ligands involve electrostatic as well as shape-matching interactions. Comparison of linear sequence data is not always a very helpful indication of similarities in binding ligands for the simple reason that even where ligands are shared between two C-type lectins, for example, from different snake species, it does not mean that the binding site on the ligand is identical. Species differences between the sequence and structure of the target ligand also need to be considered. If the two snake species have a different distribution of prey species then evolution will optimize C-type lectin/prey ligand interactions for those sequences which are most encountered or which offer the snake the best prey mass/venom protein relationship. Analysis of this situation is complicated by the fact that it is the whole venom that is optimized by evolution and not just the single molecular species but the principle holds nevertheless. Why does one venom contain a number of components, including C-type lectins which can bind to the same ligand and may indeed compete for binding sites? A plausible answer is that again we are dealing with a kind of combinatory library that is being screened constantly by another library, namely the molecules present in the various prey species. Venom components interact with a wide variety of other cells, endothelial cells, smooth muscle cells, nerve cells, and not only with plasma proteins or platelets in blood, despite these being prime targets. Even if specific components may appear to compete when tested against platelets, they may have individual advantages against other cells in particular prey species.

There are a variety of C-type lectins which are generally considered to interact with a single platelet ligand. A typical example is convulxin, thought to be specific for GPVI. However, even in this case a number of questions can be asked. One of these is whether convulxin has two different binding sites for GPVI with different affinities [43]. An additional possibility is that convulxin has (a) binding site(s) for other convulxin molecules [44]. Whether these are always present or are exposed after binding to GPVI are additional questions that have not been clearly resolved. It may be optimal in terms of amount of venom to attain maximum platelet activation to deliver (or assemble) a few large multimeric structures to the platelet surface rather than a larger number of smaller multimers. These questions can only be addressed by a physicochemical approach when recombinant C-type lectin molecules with systematically mutated binding sites become available.

Acknowledgement

Work at the Theodor Kocher Institute was supported by grants from the Swiss National Science Foundation No. 31-52396.97 and 31-063868.00 to K.J.C.

References

1 Weis WI, Kahn R, Fourme R, Drickamer K, Hendrickson WA: Structure of the calcium-dependent lectin domain from a rat mannose-binding protein determined by MAD phasing. Science 1991;254:1608–1615.

2 Weis WI, Drickamer K, Hendrickson WA: Structure of a C-type mannose-binding protein complexed with an oligosaccharide. Nature 1992;360:127–134.

3 Chou KC: Knowledge-based model building of the tertiary structures for lectin domains of the selectin family. J Protein Chem 1996;15:161–168.

4 Andrews RK, Kroll MH, Ward CM, Rose JW, Scarborough RM, Smith AI, Lopez JA, Berndt MC: Binding of a novel 50-kilodalton alboaggregin from *Trimeresurus albolabris* and related viper venom proteins to the platelet membrane glycoprotein Ib-IX-V complex. Effect on platelet aggregation and glycoprotein Ib-mediated platelet activation. Biochemistry 1996;35:12629–12639.

5 Peng M, Lu W, Kirby EP: Alboaggregin-B: A new platelet agonist that binds to platelet membrane glycoprotein Ib. Biochemistry 1991;30:11529–11536.

6 Kowalska MA, Tan L, Holt JC, Peng M, Karczewski J, Calvete JJ, Niewiarowski S: Alboaggregins A and B. Structure and interaction with human platelets. Thromb Haemost 1998;79:609–613.

7 Taniuchi Y, Kawasaki T, Fujimura Y, Suzuki M, Titani K, Sakai Y, Kaku S, Hisamichi N, Satoh N, Takenaka T, et al: Flavocetin-A and -B, two high molecular mass glycoprotein Ib binding proteins with high affinity purified from *Trimeresurus flavoviridis* venom, inhibit platelet aggregation at high shear stress. Biochim Biophys Acta 1995;1244:331–338.
8 Sakurai Y, Fujimura Y, Kokubo T, Imamura K, Kawasaki T, Handa M, Suzuki M, Matsui T, Titani K, Yoshioka A: The cDNA cloning and molecular characterization of a snake venom platelet glycoprotein Ib-binding protein, mamushigin, from *Agkistrodon halys blomhoffii* venom. Thromb Haemost 1998;79: 1199–1207.
9 Peng M, Lu W, Beviglia L, Niewiarowski S, Kirby EP: Echicetin: A snake venom protein that inhibits binding of von Willebrand factor and alboaggregins to platelet glycoprotein Ib. Blood 1993;81:2321–2328.
10 Fujimura Y, Ikeda Y, Miura S, Yoshida E, Shima H, Nishida S, Suzuki M, Titani K, Taniuchi Y, Kawasaki T: Isolation and characterization of jararaca GPIb-BP, a snake venom antagonist specific to platelet glycoprotein Ib. Thromb Haemost 1995; 74:743–750.
11 Kawasaki T, Fujimura Y, Usami Y, Suzuki M, Miura S, Sakurai Y, Makita K, Taniuchi Y, Hirano K, Titani K: Complete amino acid sequence and identification of the platelet glycoprotein Ib-binding site of jararaca GPIb-BP, a snake venom protein isolated from *Bothrops jararaca*. J Biol Chem 1996;271:10635–10639.
12 Kawasaki T, Taniuchi Y, Hisamichi N, Fujimura Y, Suzuki M, Titani K, Sakai Y, Kaku S, Satoh N, Takenaka T, et al: Tokaracetin, a new platelet antagonist that binds to platelet glycoprotein ib and inhibits von Willebrand factor-dependent shear-induced platelet aggregation. Biochem J 1995;308:947–953.
13 Chen YL, Tsai IH: Functional and sequence characterization of agkicetin, a new glycoprotein Ib antagonist isolated from *Agkistrodon acutus* venom. Biochem Biophys Res Commun 1995;210:472–477.
14 Peng M, Emig FA, Mao A, Lu W, Kirby EP, Niewiarowski S, Kowalska MA: Interaction of echicetin with a high affinity thrombin binding site on platelet glycoprotein GPIb. Thromb Haemost 1995;74:954–957.
15 Fujimura Y, Kawasaki T, Titani K: Snake venom proteins modulating the interaction between von Willebrand factor and platelet glycoprotein Ib. Thromb Haemost 1996;76: 633–639.
16 Polgar J, Clemetson JM, Kehrel BE, Wiedemann M, Magnenat EM, Wells TNC, Clemetson KJ: Platelet activation and signal transduction by convulxin, a C-type lectin from *Crotalus durissus terrificus* (tropical rattlesnake) venom via the p62/GPVI collagen receptor. J Biol Chem 1997;272:13576–13583.
17 Huang TF, Liu CZ, Yang SH: Aggretin, a novel platelet-aggregation inducer from snake *(Calloselasma rhodostoma)* venom, activates phospholipase C by acting as a glycoprotein Ia/IIa agonist. Biochem J 1995; 309:1021–1027.
18 Wang R, Kini RM, Chung MC: Rhodocetin, a novel platelet aggregation inhibitor from the venom of *Calloselasma rhodostoma* (Malayan pit viper): Synergistic and noncovalent interaction between its subunits. Biochemistry 1999;38:7584–7593.
19 Marcinkiewicz C, Lobb RR, Marcinkiewicz MM, Daniel JL, Smith JB, Dangelmaier C, Weinreb PH, Beacham DA, Niewiarowski S: Isolation and characterization of EMS16, a C-lectin type protein from *Echis multisquamatus* venom, a potent and selective inhibitor of the alpha2beta1 integrin. Biochemistry 2000;39:9859–9867.
20 Matsuzaki R, Yoshiara E, Yamada M, Shima K, Atoda H, Morita T: cDNA cloning of IX/X-Bp, a heterogeneous two-chain anticoagulant protein from snake venom. Biochem Biophys Res Commun 1996;220: 382–387.
21 Chen YL, Tsai IH: Functional and sequence characterization of coagulation factor IX/factor X-binding protein from the venom of *Echis carinatus leucogaster*. Biochemistry 1996;35:5264–5271.
22 Atoda H, Ishikawa M, Mizuno H, Morita T: Coagulation factor X-binding protein from *Deinagkistrodon acutus* venom is a Gla domain-binding protein. Biochemistry 1998; 37:17361–17370.
23 Zingali RB, Jandrot-Perrus M, Guillin M-C, Bon C: Bothrojaracin, a new thrombin inhibitor isolated from *Bothrops jararaca* venom: Characterization and mechanism of thrombin inhibition. Biochemistry 1993;32:10794–10802.
24 Monteiro RQ, Zingali RB: Inhibition of prothrombin activation by bothrojaracin, a C-type lectin from *Bothrops jararaca* venom. Arch Biochem Biophys 2000;382:123–128.
25 Read MS, Smith SV, Lamb MA, Brinkhous KM: Role of botrocetin in platelet agglutination: Formation of an activated complex of botrocetin and von Willebrand factor. Blood 1989;74:1031–1035.
26 Hamako J, Matsui T, Suzuki M, Ito M, Makita K, Fujimura Y, Ozeki Y, Titani K: Purification and characterization of bitiscetin, a novel von Willebrand factor modulator protein from *Bitis arietans* snake venom. Biochem Biophys Res Commun 1996;226:273–279.
27 Obert B, Houllier A, Meyer D, Girma JP: Conformational changes in the A3 domain of von Willebrand factor modulate the interaction of the A1 domain with platelet glycoprotein Ib. Blood 1999;93:1959–1968.
28 Peng M, Holt JC, Niewiarowski S: Isolation, characterization and amino acid sequence of echicetin beta subunit, a specific inhibitor of von Willebrand factor and thrombin interaction with glycoprotein Ib. Biochem Biophys Res Commun 1994; 205:68–72.
29 Mizuno H, Fujimoto Z, Koizumi M, Kano H, Atoda H, Morita T: Crystal structure of coagulation factor IX-binding protein from habu snake venom at 2.6 Å: Implication of central loop swapping based on deletion in the linker region. J Mol Biol 1999; 289:103–112.
30 Mizuno H, Fujimoto Z, Atoda H, Morita T: Crystal structure of an anticoagulant protein in complex with the Gla domain of factor X. Proc Natl Acad Sci USA 2001;98:7230–7234.

31 Navdaev A, Dormann D, Clemetson JM, Clemetson KJ: Echicetin, a GPIb-binding snake C-type lectin from *Echis carinatus,* also contains a binding site for IgMkappa responsible for platelet agglutination in plasma and inducing signal transduction. Blood 2001;97:2333–2341.
32 Chen YL, Tsai KW, Chang T, Hong TM, Tsai IH: Glycoprotein Ib-binding protein from the venom of *Deinagkistrodon acutus*–cDNA sequence, functional characterization, and three-dimensional modeling. Thromb Haemost 2000;83:119–126.
33 Peng M, Lu W, Kirby EP: Characterization of three alboaggregins purified from *Trimeresurus albolabris* venom. Thromb Haemost 1992;67:702–707.
34 Falati S, Edmead CE, Poole AW: Glycoprotein Ib-V-IX, a receptor for von Willebrand factor, couples physically and functionally to the Fc receptor gamma-chain, Fyn, and Lyn to activate human platelets. Blood 1999;94:1648–1656.
35 Dormann D, Clemetson JM, Navdaev A, Kehrel BE, Clemetson KJ: Alboaggregin A activates platelets by a mechanism involving glycoprotein VI as well as glycoprotein Ib. Blood 2001;97:929–936.
36 Asazuma N, Marshall SJ, Berlanga O, Snell D, Poole AW, Berndt MC, Andrews RK, Watson SP: The snake venom toxin alboaggregin-A activates glycoprotein VI. Blood 2001;97:3989–3991.
37 Shin Y, Morita T: Rhodocytin, a functional novel platelet agonist belonging to the heterodimeric C-type lectin family, induces platelet aggregation independently of glycoprotein Ib. Biochem Biophys Res Commun 1998;245:741–745.
38 Navdaev A, Clemetson JM, Polgar J, Kehrel BE, Glauner M, Magnenat E, Wells TN, Clemetson KJ: Aggretin, a heterodimeric C-type lectin from *Calloselasma rhodostoma* (Malayan pit viper) stimulates platelets by binding to $\alpha 2\beta 1$ integrin and GPIb, activating Syk and PLCγ2, but does not involve the GPVI/Fcγ collagen receptor. J Biol Chem 2001;276:20882–20889.
39 Chung CH, Peng HC, Huang TF: Aggretin, a C-type lectin protein, induces platelet aggregation via integrin alpha(2)beta(1) and GPIb in a phosphatidylinositol 3-kinase independent pathway. Biochem Biophys Res Commun 2001;285:689–695.
40 Bergmeier W, Bouvard D, Eble JA, Mokhtari-Nejad R, Schulte V, Zirngibl H, Brakebusch C, Fassler R, Nieswandt B: Rhodocytin (aggretin) activates platelets lacking $\alpha 2\beta 1$ integrin, GPVI, and the ligand binding domain of GPIbα. J Biol Chem 2001;276:25121–25126.
41 Sen U, Vasudevan S, Subbarao G, McClintock RA, Celikel R, Ruggeri ZM, Varughese KI: Crystal structure of the von Willebrand factor modulator botrocetin. Biochemistry 2001;40:345–352.
42 Arocas V, Zingali RB, Guillin MC, Bon C, Jandrot-Perrus M: Bothrojaracin: A potent two-site-directed thrombin inhibitor. Biochemistry 1996;35:9083–9089.
43 Niedergang F, Alcover A, Knight CG, Farndale RW, Barnes MJ, Francischetti IM, Bon C, Leduc M: Convulxin binding to platelet receptor GPVI: Competition with collagen related peptides. Biochem Biophys Res Commun 2000;273:246–250.
44 Marlas G: Purification and preliminary structure of a potent platelet aggregating glycoprotein isolated from the venom of *Crotalus durissus cascavella.* Toxicon 1982;20:289–290.

Substances Affecting Platelet Function

Hæmostasis

Haemostasis 2001;31:155–172

The Use of Snake Venom Toxins as Tools to Study Platelet Receptors for Collagen and von Willebrand Factor

Robert K. Andrews[e] Aura S. Kamiguti[b] Oscar Berlanga[a] Mireille Leduc[c]
R. David G. Theakston[d] Steve P. Watson[a]

[a]Departments of Pharmacology, University of Oxford, Oxford, [b]Department of Haematology, Royal University Hospital, University of Liverpool, Liverpool, UK; [c]L'Unité des Venins, Institut Pasteur, Paris, France; [d]Alistair Reid Venom Research Unit, Liverpool School of Tropical Medicine, Liverpool, UK; [e]Baker Medical Research Institute, Melbourne, Vic., Australia

Key Words

Collagen · Convulxin · von Willebrand factor · GPVI · GPIb-IX-V · Snake toxin · C-type lectin · Metalloproteinase · Botrocetin

Abstract

A large proportion of the biologically active proteins and peptides present within snake venoms interact with components of the haemostatic system to promote or inhibit the normal sequence of events that lead to clot formation. The venom proteins achieve their effects through interaction with various components of the coagulation cascade, endothelial matrix and platelets. Within the latter group, a number of venom proteins target the interaction of platelets with the major adhesive proteins, von Willebrand factor and collagen. The venom proteins bind either the adhesive protein itself or their receptors on the platelet surface, notably GP-Ib-IX-V and GPVI. This review discusses the substantial contribution that venom proteins have made to our understanding of the role of these two adhesive proteins and their receptors (excluding GPIIb-IIIa) in platelet regulation.

Introduction

Snake venoms contain a wide variety of biologically active components that enable the snake to paralyse, kill and digest its prey. Generally, venomous snakes are divided into two major categories depending on whether their venom has predominantly neurotoxic or

KARGER

Fax + 41 61 306 12 34
E-Mail karger@karger.ch
www.karger.com

0301–0147/01/0316–0155$17.50/0

Accessible online at:
www.karger.com/journals/hae

Steve P. Watson
Department of Pharmacology, University of Oxford
Mansfield Road
Oxford OX1 3QT (UK)
E-Mail steve.watson@pharm.ox.ac.uk

vascular effects. Among those with vascular effects, the result can be either prolonged bleeding or accelerated clotting, the latter resulting in fibrinogen consumption and often incoagulable blood. Although the former is more common, both effects frequently occur concurrently following a bite. The active components react with coagulation factors, platelets, endothelial cells and the extracellular matrix. The major classes of venom proteins which interact with platelets include metalloproteinases, phospholipases, disintegrins and C-type lectins.

Venomous Snakes

The prime aim of snake venom is the rapid immobilisation of the prey, whilst a secondary purpose is defence. Snake bite in humans is mainly a rural and occupational hazard and, as such, farmers and landworkers are at greatest risk. Although death frequently results from snake bite, more commonly (and fortunately) little or no envenoming occurs because only a small amount or no venom is injected. On the other hand, mortality can be high when adequate medical treatment is not given for severe systemic envenoming; it can be as high as 50% in seasnake envenoming and 10–15% in envenoming by the viper *Echis ocellatus* in West Africa.

Venomous snakes are divided into three major groups: elapids (F. Elapidae; e.g. cobras, mambas, kraits, coral snakes), seasnakes (F. Hydrophiidae) and vipers (F. Viperidae). The Viperidae is subdivided into the true vipers (SF. Viperinae; e.g. Russell's viper, carpet viper, puff adder) and the pit vipers (SF. Crotalinae; rattlesnakes, green tree vipers, Malayan pit viper, jararaca). There is also a small group of venomous colubrids (F. Colubridae; back-fanged snakes), which includes the Boomslang *(Dispholidus typus)* of Africa.

Venoms

Venoms from snakes contain a complex mixture of toxins, mainly proteins, which can be experimentally fractionated into enzymes and polypeptides with widely varying biological, pharmacological and autopharmacological properties; they also contain low molecular weight components which have not yet been thoroughly investigated. The toxic effects depend on both the type and amount of venom injected. Although viper venoms mainly cause haemorrhage and coagulation disorders, and elapid venoms cause neurotoxicity, there are exceptions to this rule and one should be wary of an excessively dogmatic classification.

Procoagulant enzymes, present mainly in venoms of vipers but also in that of a few elapids, stimulate blood coagulation by activating different steps of the coagulation cascade. The result in envenomed humans is fibrinogen consumption and therefore incoagulable blood (consumption coagulopathy), whereas in a small prey animal, massive intravascular coagulation causes rapid immobilisation and death. Zinc metalloproteinase haemorrhagins, also present mainly in viper venoms, damage the endothelium and subendothelium of the blood vessel walls, causing both local and systemic haemorrhage. Many venoms possess cytolytic or necrotising toxins; these are digestive hydrolases (proteolytic enzymes and phospholipases A_2), polypeptide toxins and other factors which increase permeability, resulting in local swelling, as well as causing tissue breakdown usually in the region of the bite site. Haemolytic and myolytic phospholipases A_2 are often also present which result in damage to cell membranes, endothelium, skeletal muscle, nerves and erythrocytes. Postsynaptic neurotoxins, present in the venom of many elapids, such as cobras, compete with acetylcholine for receptors in the neuromuscular junction and lead to cu-

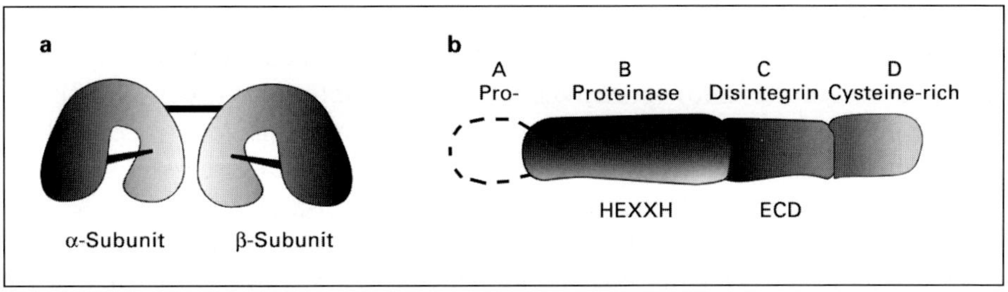

Fig. 1. a Structural representation of the 25-kD form of the αβ heterodimeric proteins of the snake venom C-type lectin-like family. Only major inter- and intra-subunit disulphide bonds common to several forms are shown (solid lines). **b** Metalloproteinase-disintegrin family proteins. The domain structure of a form lacking the pro-peptide domain (dashed line) is shown. HEXXH, His-Glu-X-X-His and ECD, Glu-Cys-Asp are sequences common to many forms.

rare-like paralysis. Presynaptic neurotoxins, present mainly in elapids but also in a few viperids, damage nerve endings, releasing acetylcholine transmitter or interfering with its release.

Over the years, snake venoms, especially those from certain viperine snakes, have been shown to contain a large number of toxins that exert powerful inhibitory and stimulatory effects on platelets. Several of these have made significant contributions to the understanding of the events underlying platelet activation. The purpose of this article is to consider the use of snake venom toxins as tools to study receptors for two of the major platelet adhesive proteins, collagen and von Willebrand factor, where extensive advances have been made within the last few years. For a more widespread discussion of the action of snake venoms on the haemostatic system, the reader is referred to a recent, extensive review [1].

Structural Classes of Snake Venom Toxins

The snake venom proteins that interfere with the action of adhesion proteins on platelets fall into three distinct structural classes, namely snake C-type lectins, disintegrins and metalloproteinases. The metalloproteinases and C-type lectins may have evolved from a common precursor [1]. The disintegrins target the major platelet aggregatory receptor, GPIIb-IIIa ($\alpha_{IIb}\beta_3$). The disintegrin domain is also found in a number of metalloproteinases where it has a distinct recognition site of ECD rather than RGD and is unable to bind GPIIb-IIIa. The structural organization of snake C-type lectins and a metalloproteinase-disintegrin is shown in figure 1.

C-Type Lectins. C-type lectins were first described as carbohydrate-binding (hence lectin) and calcium-dependent (hence C-type). They contain a carbohydrate recognition domain (CRD) of 130 amino acids, which reveals a series of invariant amino acid residues in a characteristic pattern [2, 3]. In this pattern, four cysteines form intracatenary disulphide bridges.

[C^{*1}xxxWxxxxxxC^{*1}]--(W/F/Y)------C^{*2}---(W/Y/F)xG------W----P-----C^{*2}-----WxxxxC^{*3}--------C^{*3} (the cysteines are disulphide linked to the cysteine with the same number).

Recent crystallographic analyses of C-type lectins show: (1) amino acid residues involved in the carbohydrate binding are located in the C-terminal hydrophilic region which forms a peripheral loop; (2) critical amino acid residues define carbohydrate-binding specificity of C-type lectins: EPN for the mannose-specific motif and QPD for the galactose-specific motif; Ca^{2+} is bound to the motif Q-D in galactose-specific lectins [4, 5]. The snake venom class of C-type lectins, however, do not bind sugars nor require calcium, but have other strong similarities with typical lectin domains.

During the last decade, several C-type snake proteins have been cloned and sequenced. Snake venom C-type lectins are found in true vipers and pit vipers but not in elapids and seasnakes [6]. They are made up of homodimers and heterodimers, often present as multiples. Amino acid sequences of each subunit contain between 125 and 135 residues, which corresponds to the CRD length proposed by Drickamer [2]. The amino acid identity between the snake C-type lectins is 50% or higher [7]. In all of these venoms, the sequences were found to contain the invariant amino acids of CRD as proposed by Spiess [3], with cysteine residues forming intra-disulphide bridges as chemically confirmed in botrocetin [8], isolated from the venom of *Bothrops jararaca*. Only the sequences of the homodimer of the rattlesnake lectin from *Crotalus atrox* and LmsL from *Lachesis muta stenophyrs* contain the amino acids in their CRD for binding galactose and Ca^{2+}. The subunits of the heterodimeric proteins do not contain these amino acids and do not bind mannose or galactose [9]. However, in some cases, these heterodimeric proteins are able to bind Ca^{2+}, and this has been shown to be necessary for their biological activity.

On the basis of the crystal structure of mannose-binding protein, the crystal structure of several snake venom C-type lectins have been reported including factor IX/factor X-binding protein [10], factor IX-binding protein [11], botrocetin [12] and flavocetin A [12]. These crystal structures show that dimerisation of their CRDs is an example of loop-swapping. It has been proposed that this is responsible for their loss of carbohydrate binding activity. The crystal structures also confirmed the absence of a canonical Ca^{2+} binding site, but revealed the presence of a new one localized in the C-type lectin fold. Such dimerisation of the CRD in the snake venom toxin was proposed as a strategy for survival against animal predators and exemplifies an evolutionary gain of function for the C-type lectin domain.

Disintegrins. These are cysteine-rich venom components (5–10 kD) with potent inhibitory activity on integrin GPIIb-IIIa-dependent platelet aggregation. A vast number of disintegrins have been described since trigramin from *Trimeresurus gramineus* [13] and echistatin from *Echis* species [14] venoms were first characterised. All disintegrins contain an RGD or KGD sequence that binds with high affinity to GPIIb-IIIa and therefore competes with adhesive macromolecule ligands, capable of sustaining platelet aggregation (i.e., fibrinogen and von Willebrand factor [15, 16]). It is interesting that, unlike fibrinogen or von Willebrand factor, venom disintegrins bind to GPIIb-IIIa without requiring prior activation of the integrin. This is because of their unique structure with the RGD sequence located at the tip of a Cys-Cys loop, facilitating binding to the receptor and expression of a ligand-induced binding site on the β_3 subunit of the integrin [17]. The RGD-containing venom disintegrins inhibit platelet aggregation induced by the majority of surface receptors consistent with the central role of GPIIb-IIIa in platelet function. Disintegrins are not discussed further, and the reader is referred to a recent review for additional information [1].

Metalloproteinases. Venom metalloproteinases from true vipers and pit vipers are responsible for local and systemic haemorrhage due to proteolysis of the major components of the extracellular matrix [18]. They range from 20 to 100 kD, with several sizes often being found in the same venom. In general, the larger metalloproteinases are more active in inducing haemorrhage.

The complete primary structure of a high molecular weight metalloproteinase, HR1b from *Trimeresurus flavoviridis* venom was shown for the first time by Takeya et al. [19] in 1990. A unique structure was unravelled comprising an N-terminus with the catalytic site, followed by a sequence with high homology to the RGD-disintegrins except that, instead of an RGD, an ECD sequence was found together with a cysteine-rich C-terminal domain. The sequences of the cDNA clones of high molecular weight metalloproteinases from *Bothrops jararaca* venom, jararhagin [20], and from *C. atrox* venom, Ht-e [21], also have a similar structural organisation (fig. 1). In addition, a pro-enzyme peptide domain at the N-terminus was found, which must be cleaved from the mature protein to allow enzyme activity. This class of metalloproteinases is named MDC (metalloproteinase/disintegrin/cysteine-rich) enzymes and shares homology to the cell membrane-bound family of metalloproteinases known as ADAM (a disintegrin and metalloproteinase) [22]. ADAMs are responsible for cell surface protein shedding or activation, one classical example being TACE (TNF-alpha converting enzyme) [23].

Based on their sizes, the venom metalloproteinases are classified as P-I (metalloproteinase domain only), P-II (metalloproteinase and disintegrin-like domains), P-III (MDC) and P-IV (MDC and C-type lectin domains) [18]. Because disintegrin-like structures containing the cysteine-rich region (approximately 30 kD) without the metalloproteinase domain can be isolated from the whole venom, it is believed that they are products of autoproteolysis. Examples are jararhagin C [24] and catrocollastatin C [25], derived respectively from the 52-kD jararhagin from *B. jararaca* venom and 50-kD catrocollastatin from *C. atrox* venom [26]. Disintegrin-like components from metalloproteinases have also been produced by recombinant expression [27, 28].

Snake Venom Proteins That Target GPIb-IX-V

GPIb-IX-V is one of the two major receptors for von Willebrand factor on the platelet surface. The second major receptor, the integrin GPIIb-IIIa, is not discussed further in this review. GPIb-IX-V is composed of two, disulphide-linked GPIbα and β chains, two GPIX chains and one GPV chain. There is no similar receptor complex in the genome. The interaction of von Willebrand factor with GPIb-IX-V is indispensable for platelet adhesion to the damaged subendothelium at high shear. Von Willebrand factor promotes the initial adhesion of platelets to the exposed subendothelium due to the fast on-rate of association with GPIb-IX-V. Because of a fast off-rate of association of von Willebrand factor from the receptor complex, however, a second interaction is required for stable adhesion, such as that provided through the interaction of collagen with platelet surface receptors. Von Willebrand factor is not required for adhesion under low shear conditions.

There is also overwhelming evidence in support of a role of GPIb-IX-V in the generation of intracellular signals that lead, among other responses, to activation of GPIIb-IIIa [29]. Despite a remarkable repertoire of intracellular signals that have been reported to lie downstream of the GPIb-IX-V complex, the sequence of events which underlies activation of the integrin, however, remains unclear; this situation is at least partly due to the relatively weak nature of the signal from the receptor complex.

Over the years, snake venom toxins have provided valuable information on the sites of interaction of von Willebrand factor with the GPIbα subunit and have also been routinely used to promote the interaction of the adhesion protein with the receptor complex, an interaction that does not normally take place in the absence of high shear because of conformational constraints on von Willebrand factor. Snake venom toxins have also provided important new insights into the signalling pathways linked to the GPIb-IX-V receptor complex.

C-Type Lectins. A variety of viper venom proteins of the C-type lectin family target the platelet GPIb-IX-V complex. These venom proteins bind within the N-terminal 282 residues of GPIbα [30], the same region of the receptor that contains the binding site for von Willebrand factor and other ligands such as Mac-1, P-selectin and α-thrombin [29]. One of the reasons for the current interest in GPIbα-targeting viper venom proteins is the extent to which these reagents mimic von Willebrand factor. In this regard, the mechanism by which von Willebrand factor binding to GPIb-IX-V induces signal transduction and platelet activation is contentious. For instance, is platelet activation that follows engagement of GPIb-IX-V by von Willebrand factor (or other ligands) initiated by cross-linking of one or more GP Ib-IX-V complexes or through the involvement of other receptors that may be associated with GPIb-IX-V? Recent studies of GPIb-IX-V-targeting venom proteins have provided fresh insights into potential functional consequences of ligand binding to this receptor.

Platelets are activated in response to binding of native multimeric von Willebrand factor or asialo-von Willebrand factor [29, 31] or recombinant von Willebrand factor A1 domain, multimerised by expression on the surface of COS cells [32]. In contrast, a monomeric, soluble fragment of von Willebrand factor encompassing the A1 domain (Glu480-Gly718) binds to GPIbα and inhibits binding of native von Willebrand factor, but does not induce platelet aggregation [29].

The von Willebrand factor A1 domain fragment induced phosphorylation of Syk in platelets, and its association with Src, but only native multimeric von Willebrand factor induced activation of αIIbβ3 [33]. Like monomeric von Willebrand factor fragment, a number of approximately 25-kD GPIbα-binding venom proteins of the C-type lectin family also block von Willebrand factor binding, but do not activate platelets [30, 34]. The C-type lectin echicetin, from the saw-scaled viper *Echis carinatus*, is one such lectin which binds to GPIbα and inhibits binding of both von Willebrand factor and thrombin [34, 35]. Echicitin prevents agglutination of washed platelets by von Willebrand factor [34, 35]. However, in plasma, echicetin becomes cross-linked via IgMκ, leading to agglutination, increased expression of P-selectin and activation of GPIIb-IIIa [36]. These responses can also be induced by echicetin in washed platelets by addition of IgMκ or via cross-linking of biotinylated echicetin with avidin, and are accompanied by measurable tyrosine phosphorylation of proteins including Lyn and Syk. Thus, the major physiological action of echicitin in plasma is to promote platelet agglutination rather than inhibition, as previously thought. This may account for the thrombocytopenia seen in vivo in response to echicitin and underscores the need to test venom proteins in plasma and whole blood, where they are able to interact with other endogenous constituents. These results add further support to the model in which cross-linking is required for GPIb-IX-V to induce intracellular signalling events.

A higher molecular weight form of a C-type lectin-like GPIbα-binding venom protein, 50-kD alboaggregin A from the white lipped tree viper *Trimeresurus albolabris,* is also a potent platelet agonist [37–39]. However, recent reports suggest that the platelet collagen receptor, GPVI, makes a significant contribution to alboaggregin A-dependent platelet activation [40, 41]. Treatment of platelets with alboaggregin A resulted in phosphorylation of Fc receptor γ-chain (FcR γ-chain), Syk and phospholipase Cγ2 (PLCγ2), and alboaggregin A-dependent platelet aggregation and tyrosine phosphorylation were inhibitable by the Src kinase family inhibitor PP1, consistent with a signalling pathway primarily involving GPVI [40]. Alboaggregin A also bound to recombinant GPVI, expressed on K562 cells. However, selectively masking GPIbα by anti-GPIbα antibodies, inhibitory GPIbα-binding venom proteins or proteolysis with mocarhagin partially impairs alboaggregin A-dependent platelet activation [37, 38, 40, 41]. This suggests binding to GPIbα may support the interaction of alboaggregin A with GPVI. The observation, however, that alboaggregin A induces aggregation of Bernard Soulier syndrome platelets demonstrates that GPIbα is not essential for activation of GPVI [40].

A related C-type lectin-like protein, aggretin/rhodocytin, from the Malayan pit viper *Calloselasma rhodostoma* has also been proposed to bind GPIbα, with signalling responses dependent on an additional interaction with the collagen receptor GPIa-IIa [42]. The requirement for GPIbα and GPIa-IIa in the response to aggretin/rhodocytin has, however, been recently challenged, as responses are maintained in platelets deficient in these two surface glycoproteins [43]. The mechanism of action of aggretin/rhodocytin is discussed below.

The involvement of receptors other than GPIb-IX-V, such as GPVI or possibly GPIa-IIa, complicates interpretation of signalling responses induced by GPIbα-binding toxins. However, such mechanisms may parallel platelet stimulation by von Willebrand factor, where initial binding of von Willebrand factor to GPIb-IX-V may similarly lead to activation of additional signalling receptors. For example, there is evidence supporting a linkage of GPIb-IX-V with the low-affinity immune receptor, FcγRIIa [44] and the FcR γ-chain [45]. Cross-linking of these receptors leads to tyrosine phosphorylation of the immunoreceptor tyrosine-based activation motif (ITAM) in FcγRIIa and FcR γ-chain and the binding and phosphorylation of Syk, thereby initiating a signalling pathway that activates PLCγ2 and other effector enzymes (see signalling by the collagen receptor GPVI, below). Tyrosine phosphorylation of FcR γ-chain by von Willebrand factor has been shown to contribute to platelet aggregation [45]. Interestingly, FcR γ-chain has been shown to be associated with GPIbα in platelets after stimulation by alboaggregin A [39], a reagent that binds GPVI in addition to GPIbα (referred to above). These findings raise the intriguing possibility of a potential functional association between GPIbα, FcR γ-chain and/or GPVI.

The family of venom C-type lectin-like proteins, therefore, includes forms which target GPIb-IX-V and inhibit ligand binding or activate platelets by mechanisms that may involve either receptor cross-linking or enlistment of other receptors, such as GPVI or possibly GPIa-IIa. Whereas alboaggregin A has the capacity to recognise more than one receptor, other C-type lectin family members, such as convulxin (see below), appear to selectively target GPVI and are not known to bind GPIbα.

Metalloproteinases. Mocarhagin is a metalloproteinase from the venom of the Mozambique spitting cobra, *Naja mossambica mossambica.* Originally described as a lectin, mocarhagin was affinity-purified on heparin and found to have a profound effect on platelet reactivity [46]. Mocarhagin treatment of platelets irreversibly blocked the interaction of platelet receptor glycoprotein GPIb-IX-V with its ligand, von Willebrand factor, which initiates (patho)physiological thrombus formation at high shear stress. It was subsequently shown that mocarhagin was an approximately 55-kD metalloproteinase, and had N-terminal sequence homology with members of the venom metalloproteinase-disintegrin family. Mocarhagin cleaved the von Willebrand factor-binding GPIbα subunit of the GPIb-IX-V complex as its sole detectable substrate on surface-labelled platelets [46]. The cleavage site on GPIbα was determined to be between residues Glu-282/Asp-283, releasing a soluble, approximately 40-kD N-terminal fragment (His1-Glu282) and producing an extracellular, sialomucin-like domain that remained membrane-associated (and disulphide-linked to GPIbα).

Mocarhagin has proven a useful probe of the structure and function of GPIbα. Firstly, analysis of mocarhagin-treated platelets and mocarhagin-derived fragments of purified receptor confirmed that the von Willebrand factor recognition site on GPIb-IX-V was contained within His1-Glu282 [29]. The same domain on GPIbα also binds other physiological ligands of GPIb-IX-V, which include Mac-1, P-selectin and α-thrombin. Secondly, a comparison of the mocarhagin (His1-Glu282), trypsin (His1-Arg293) and cathepsin G (His1-Leu275)-generated proteolytic fragments together showed that three tyrosine residues at Tyr276, Tyr278 and Tyr279 adjacent to the mocarhagin and cathepsin G cleavage sites were sulphated in the native receptor [46], consistent with the sulphation state of recombinant receptor [47, 48]. Thirdly, the anionic sulphated sequence between Asp269-Glu282 was shown to mediate von Willebrand factor binding to GPIbα particularly in the presence of the von Willebrand factor modulator botrocetin [46–48], and the same sequence was also confirmed as a critical element in regulating thrombin binding to GPIbα [46, 48, 49].

Interestingly, while GPIbα was the only detectable mocarhagin substrate on platelets, mocarhagin treatment of neutrophils abolished binding of P-selectin as a result of the specific cleavage of P-selectin glycoprotein ligand-1 (PSGL-1). The mocarhagin cleavage site on PSGL-1 was determined to be within an N-terminal anionic sequence containing sulphated tyrosine residues and analogous to that found in GPIbα [50]. Like in GPIbα, the PSGL-1 cleavage site was also adjacent to a sialomucin-like, membrane-associated domain. It has subsequently been shown that, like PSGL-1, GPIbα also specifically binds P-selectin [51], and its interaction with both PSGL-1 and GPIbα involves, at least in part, their sulphated tyrosine motifs [50–52]. Mocarhagin, therefore, which selectively cleaves both PSGL-1 and GPIbα, potentially has both 'anti-thrombotic' and 'anti-inflammatory' activity, by specifically targeting these functionally analogous receptors involved in initial contact of circulating neutrophils (PSGL-1) or platelets (GPIbα) with the blood vessel wall.

Snake Venom Proteins That Target Collagen Receptors

Subendothelial collagen fibres, being the most thrombogenic component of the vessel wall, support adhesion of platelets to the damaged subendothelium and promote activation. Collagen fibres support adhesion indirectly through von Willebrand factor, which binds to GPIb-IX-V, and directly through platelet surface receptors, of which the integrin GPIa-IIa ($\alpha_2\beta_1$) and GPVI are considered to be the most important. Collagen induces platelet activation through GPVI, which is present in a complex with the FcR γ-chain. Activation of GPVI leads to tyrosine phosphorylation of FcR γ-chain on a conserved motif known as an ITAM by an Src family kinase. A second tyrosine kinase, Syk, binds to the phosphorylated ITAM, initiating a downstream signalling cascade that culminates in activation of downstream effectors including PLCγ2, phosphatidylinositol 3-kinase and low molecular weight GTP binding proteins. The effector pathways elicit powerful platelet activation leading to GPIIb-IIIa-dependent aggregation, secretion and procoagulant activity. It is unclear whether the integrin GPIa-IIa also generates intracellular signals which contribute to functional responses. The interaction of platelets with collagen has been the subject of a number of recent reviews [53–56].

It is only within the last few years that snake venom toxins have begun to have a major impact on our understanding of platelet collagen interactions. The realization in 1997 that the C-type snake major venom convulxin induced platelet activation through GPVI helped to bring this receptor to the forefront of the platelet-collagen field and was pivotal in the eventual cloning of the glycoprotein. Several snake venom toxins are now recognised to target GPVI, and it seems likely that many more will be found in the coming years.

Snake Venom Toxins Acting on GPVI

C-Type Lectin. Convulxin is a C-type lectin from the venom of the South American tropical rattlesnake *Crotalus durissus terrificus*. Convulxin was cloned in 1998 [7]. It consists of two subunits, α and β, joined by disulphide bridges in a trimer-trimer organisation. Convulxin was initially described in 1970 [57] and shown to be a powerful platelet agonist in 1980 [58]. Within 2 years of this report, convulxin was proposed to activate platelets through a pathway that was also used by collagen on the basis of cross-desensitisation between the two stimuli [59]. However, it was a further 15 years before it was realised to mediate activation through the major activatory collagen receptor, GPVI [9, 60]. Several lines of evidence were presented in support of this conclusion including: (1) the ability of a polyclonal antiserum to GPVI to block platelet activation by convulxin; (2) the ability of lower molecular weight fragments of convulxin to block platelet activation by collagen; (3) the similar pattern of signal transduction events mediated by convulxin and collagen including marked tyrosine phosphorylation of FcR γ-chain, Syk and PLCγ2, and (4) a ligand blot assay demonstrated that convulxin bound to a protein of 62 kD which comigrated with GPVI under non-reducing conditions [60, 61]. The activity of convulxin on GPVI has since been confirmed on cells transfected with the glycoprotein receptor [62].

There are also reports that the action of convulxin is partially inhibited by antibodies to GPIa-IIa, raising the possibility that the integrin is, in some unknown way, linked to the action of GPVI in platelets [9, 63]. This observation should be treated with caution, however, in that antibodies to the integrin also have weak inhibitory effects on platelet activation by other agonists which signal independence of the integrin including thrombin [64] and aggretin/rhodocytin (see 'The Mechanism of Action of Aggretin/Rhodocytin'). The observation that iodinated-convulxin binds to one [65] or possibly two sites on the platelet surface [9] should also be treated with caution, bearing in mind the multisubunit composition of the toxin and that full iodination destroys activity. The two binding sites might reflect heterogeneity within the iodinated-convulxin causing affinity differences. Interestingly, collagen-related peptides, which are specific GPVI, only partially (approximately 30%) inhibit the binding of convulxin demonstrating binding to distinct sites on GPVI [66].

Convulxin has subnanomolar affinity for GPVI and, therefore, functions as a pseudo-irreversible ligand [65]. This makes it an ideal reagent as an affinity ligand, a property that was exploited by Clemetson et al. [62] in 1999 in the cloning of GPVI. GPVI is a novel member of the Ig family of surface glycoproteins with strong sequence homology to the Fcα receptor and natural killer inhibitory receptors. GPVI has recently been extensively reviewed by Clemetson et al. [55].

Convulxin is now widely used as a specific agonist in the dissection of GPVI-regulated signal transduction events because of its selective action on GPVI [67, 68]. However, it should be borne in mind that convulxin is a much more powerful agonist than collagen and, therefore, much less reliant on release of secondary mediators to bring about platelet activation [69]. Convulxin is, therefore, not an accurate mimetic of collagen-based platelet interactions.

A number of other C-type lectins which mediate their actions through GPVI have been described. Trimucytin from the venom of *Trimeresurus mucrosquamatus* was originally described as a ligand for GPIa-IIa on account of the weak inhibitory action of the GPIa-IIa mAb 6F1 against platelet activation [70]. However, Hers et al. [64] reported a similar weak inhibitory effect of mAb 6F1 against a number of other platelet agonists and further demonstrated that the monoclonal antibody was unable to block responses to higher concentrations of the snake venom toxin. Moreover, in common with other GPVI-selective agonists, trimucytin stimulated marked tyrosine phosphorylation of the FcR γ-chain, Syk and PLCγ2, suggesting it may be an agonist at GPVI [64]. This hypothesis requires confirmation of the activity of trimucytin on GPVI-transfected cells. It is also possible that trimucytin binds to both GPVI and GPIa-IIa.

A further example of a C-type lectin that binds to GPVI is alboaggregin A which, as discussed above, also binds to GPIbα [40, 41]. A second C-type lectin with dual affinity for GPVI and GPIbα, alboluxin, has also been identified in the venom from *Trimeresurus albolabris* [71]. It is interesting to speculate whether activity at more than one platelet surface glycoprotein will be common to many venom proteins.

Metalloproteinases. The 60-kD metalloproteinase-disintegrin alborhagin from the venom of the white-lipped tree viper *Trimeresurus albolabris* has recently been shown to be a potent agonist at GPVI, demonstrating the versatility of snake venom toxins of distinct structures in targeting a common substrate [72]. Interestingly, alborhagin was isolated from the same venom as for alboaggregin A, demonstrating that a single venom can contain proteins of distinct structural classes with common functions.

The evidence demonstrating that alborhagin mediates its activation through GPVI includes the similar pattern of protein tyrosine phosphorylation to that induced by convulxin in platelets, including tyrosine phosphorylation of PLCγ2 and the ability of the toxin to activate GPVI expressed in Jurkat T cells and K562 cells [72]. Alborhagin was unable to displace the binding of convulxin to GPVI in platelets or in GPVI-transfected cells, using a flow cytometry-based assay demonstrating that the two toxins bind to the glycoprotein receptor at distinct sites. On the other hand, the rat anti-GPVI monoclonal antibody, JAQ1, which has been previously shown to block platelet activation by collagen-related peptides [73], also blocked activation by alborhagin, suggesting a common site of interaction on the glycoprotein receptor. This makes alborhagin a novel probe to study the function of GPVI.

Snake Venom Toxins Acting on GPIa-IIa ($\alpha_2\beta_1$ Integrin)

C-Type Lectin. The snake C-type lectins trimucytin and aggretin/rhodocytin were originally proposed to interact with GPIa-IIa [70, 74], although this has been brought into question as discussed in 'C-type lectins' and 'The Mechanism of Action of Aggretin/Rhodocytin'. There is therefore no definitive evidence, at present, for a role of GPIa-IIa in the action of platelet-activatory snake venom toxins. Whether this reflects a minimal role of the integrin in the generation of intracellular signalling events in platelets is a subject for speculation. On the other hand, an inhibitory C-type lectin whose binding prevents binding of collagen to GPIa-IIa has been described [75, 76]. Rhodocetin is a venom component of the Malayan pit viper *Calloselasma rhodostoma,* which is also the source of aggretin/rhodocytin. Rhodocetin is a heterodimer of two subunits of 16 and 14 kD. Neither subunit has a collagenous triple helical domain or an RGD-binding sequence, and binding to the integrin is independent of divalent cations. The structural basis of the way in which rhodocetin binds to GPIa-IIa is, therefore, of considerable interest as this may pave the way for development of non-helical ligands for the integrin.

It is interesting to consider why a single snake venom might express a toxin which stimulates platelet activation and one that inhibits this response, although this is not in itself an unusual finding. In the present case, it is possible that the major role of rhodocytin is to target GPIa-IIa in tissues other than the platelet that express GPIa-IIa, such as endothelial cells and fibroblasts. Interestingly, the venom of the Malayan pit viper also contains a second GPIa-IIa inhibiting activity of high molecular weight which has yet to be characterised [76].

Metalloproteinases. The class P-III metalloproteinases, jararhagin and catrocollastatin, inhibit platelet adhesion to collagen [25, 77]. Whilst this class of enzymes is best known for its ability to degrade extracellular matrix proteins, it does not use type I or III collagen as substrate [18]. With catrocollastatin, Zhou et al. [25] reported that the mechanism of inhibition of platelet aggregation was through the direct binding of catrocollastatin to collagen. Jararhagin, on the other hand, appears to cleave the integrin GPIa-IIa [78]. GPVI is not affected by jararhagin, as convulxin binding to GPVI and the subsequent downstream signalling events are unaltered in jararhagin-treated platelets [63]. Jararhagin cleaves the GPIIa (β_1) subunit of the integrin receptor, which remains associated by a disulphide bond to the receptor complex [77, 78]. As a consequence of this, the function of the receptor for aggregation or recognition by antibodies targeting the functional I domain of GPIa subunit (α_2), responsible for the recognition of native collagen, is lost.

To bring the enzyme close to the platelet GPIa-IIa, it is crucial that jararhagin binds to platelets. However, the binding of the active enzyme to platelets is weak and can only be demonstrated after enzyme treatment with 1,10-phenanthroline, a general inhibitor of metalloproteinases [77]. It is, however, possible that a transient binding of the active enzyme takes place but that, following cleavage, jararhagin does not remain bound to the receptor. Similarly to integrin recognition by the RGD disintegrins, it is then logical to assume that the disintegrin-like domain of jararhagin also plays a role in the binding.

Table 1. Snake venom toxins that target platelet vWf and collagen and their receptors GPIb-IX-V, GPVI and GPIa-IIa

Venom protein family			Function	References
target	protein	snake species		
C-type lectin-like family				
vWf	aspercetin	*Bothrops asper*	binding to vWF causing agglutination	84
	bitiscetin	*Bitis arietans*	binding to vWF causing agglutination	85
	botrocetin	*Bothrops jararaca*	binding to vWF causing agglutination	8
GPIbα, GPVI	agkicetin	*Agkistrodon acutus*	blocks vWf	86
	alboaggregin B	*Trimeresurus albolabris*	blocks vWf, agglutinates platelets	37, 87, 88
	CHH-A and -B	*Crotalus horridus horridus*	block vWf	37
	crotalin	*Crotalus atrox*	blocks vWf	89
	echicetin	*Echis carinatus*	blocks vWf (and α-thrombin)	35
	flavocetin A and B	*Trimeresurus flavoviridus*	block	
	mamushigin	*Agkistrodon halys blomhoffi*	blocks vWf binding and aggregates platelets under shear	90, 91
	jararaca GPIb-BP	*Bothrops jararaca*	blocks vWf	92
	tokaracetin	*Trimeresurus tokarensis*	blocks vWf	93
GPIbα, GPVI	alboaggregin A	*Trimeresurus albolabris*	activates platelets	37, 38, 40, 41
GPVI	convulxin	*Crotalus durissus terrificus*	activates platelets	9, 60
	trimucytin	*Trimeresurus mucrosquamatus*	activates platelets	70
GPIa-IIa	rhodocetin	*Calloselasma rhodostoma*	inhibits binding of collagen	76
Metalloproteinase-disintegrin family				
Collagen	catrocollastatin	*Crotalus atrox*	binds to collagen, inhibiting aggregation	25
vWf	kauothiagin	*Naja kaouthia*	cleaves vWf leading to loss of binding to collagen and ristocetin-aggregation	94
GPVI	alborhagin	*Trimeresurus albolabris*	activates platelets	72
GPIbα, GPVI	mocarhagin	*Naja mossambica mossambica*	cleaves GPIbα (Glu282/Asp283), abolishes vWf binding	46
GPIa-IIa	jararhagin	*Bothrops jararaca*	cleaves GPIIa, abolishes collagen-induced platelet aggregation	77

vWF = von Willebrand factor.

Indeed, it has been shown that jararhagin C, the proteolytic product of jararhagin, inhibits collagen-induced platelet aggregation [24, 27].

With regard to determining the jararhagin interacting site with GPIa/IIa, synthetic peptides containing the sequence RKKH of the metalloproteinase domain of jararhagin have been shown to bind to the I domain of the GPIa (α_2) subunit of GPIa/IIa [79, 80]. This sequence also inhibits platelet aggregation by collagen [Kamiguti and Fox, unpubl.]. However, synthetic peptides based on the disintegrin-like region of either jararhagin [78] or atrolysin A [28] are also inhibitors, indicating a multidomain function in platelet/collagen interaction by these enzymes. Additional studies are required to fully understand these events, although it is clear that the mechanism through which jararhagin inhibits collagen-induced platelet aggregation involves binding to GPIa and cleavage of the GPIIa subunit.

Snake Venom Toxins Acting on Other Collagen Receptors

A number of other receptors have been proposed for collagen on the platelet surface including CD36 (GPIV) and p65. The role of these proteins in the physiological function of collagen, however, remains unclear. In this context, it is notable that no snake venom toxins have been reported which mediate their actions through these putative collagen receptors.

The Mechanism of Action of Aggretin/Rhodocytin

As discussed above, the C-type lectin aggretin from *C. rhodostoma* was originally proposed to activate platelets through the collagen receptor integrin GPIa-IIa based on the inhibitory effect of antibodies to the integrin [74]. Subsequently, Shin and Morita [81] isolated a C-type lectin from the same snake venom which also activated platelets independently of GPIbα, but whose action was blocked by antibodies to GPIa-IIa [82, 83]. Aggretin and rhodocytin share the same N-terminal amino sequences and are present in a tetrameric structure. However, they do have a number of different problems, making it unclear whether they are the same protein or closely related variants.

Aggretin/rhodocytin have been shown to activate platelets through a tyrosine kinase-dependent pathway that has a number of similarities to the cascade induced by the collagen receptor GPVI including tyrosine phosphorylation of Syk and PLCγ2 [42, 83]. However, it is clear that the toxin does not induce activation through this receptor, since the response is maintained in FcR γ-chain-deficient platelets, which lack the glycoprotein receptor [42, 83]. Furthermore, tyrosine phosphorylation of Syk and PLCγ2 was mediated through a cyclooxygenase-sensitive pathway, in contrast to the response to the collagen receptor GPVI [83]. Navdaev et al. [42] have proposed that aggretin/rhodocytin activates platelets through cross-linking of GPIbα and GPIa-IIa, based on the observation that some but not all antibodies to these surface glycoproteins block platelet activation by the toxin. This conclusion however has been brought into question by the observations that (1) aggretin/rhodocytin does not bind to recombinant GPIa-IIa [76] and (2) the toxin activates GPIa-IIa-deficient murine platelets in which GPIbα has also been removed through treatment with mocarhagin [43]. These observations provide unequivocal evidence that the action of aggretin/rhodocytin is mediated at least partly through a novel surface receptor. The identification of this surface receptor is an area of considerable interest.

Table 1 provides a summary of snake venom toxins that target platelet von Willebrand factor and collagen and their receptors GPIb-IX-V, GPVI and GPIa-IIa.

Concluding Remarks

This review has focused on the action of snake venom toxins, on the actions of two of the major platelet-adhesive ligands: von Willebrand factor and collagen. The contribution that snake venom toxins have made to our understanding of the way these two adhesive ligands interact with platelet receptors has been, and continues to be, immense. In addition, snake venom toxins have made substantial contributions to our understanding of the role of the integrin GPIIb-IIIa in platelet function and in the action of other platelet agonists and their receptors. Given this background, it is remarkable that, for example, it took almost 20 years to identify GPVI as the receptor for convulxin and to exploit the use of the toxin in the eventual purification and cloning of the glycoprotein.

It is interesting to speculate whether snake venom toxins will continue to provide new and novel tools to study receptors for von Willebrand factor and collagen, as well as other platelet ligands. This seems highly likely, bearing in mind that three novel ligands have been described for GPVI in the past 12 months. Moreover, given the large number of venomous snakes in the world and that many venoms have not been rigorously tested, it seems inevitable that new types of activities will be discovered over the coming years, possibly leading to the identification of further, novel activatory or inhibitory platelet receptors, or new tools with which to dissect the function of receptors for adhesive ligands. In this context, rhodocytin/aggretin represents a particularly promising example of a snake venom toxin that appears to mediate at least part of its action through a novel pathway linked to a previously unidentified platelet receptor.

Acknowledgements

S.P.W. thanks the Welcome Trust and British Heart Foundation for support. S.P.W. is a BHF Senior Research Fellow.

References

1 Markland FS: Snake venoms and the hemostatic system. Toxicon 1998;36:1749–1800.
2 Drickamer K: Two distinct classes of carbohydrate-recognition domains in animal lectins. J Biol Chem 1988;263:9557–9560.
3 Spiess M: The asialoglycoprotein receptor: A model for endocytic transport receptors. Biochemistry 1990; 29:10009–10018.
4 Weis WI, Crichlow GV, Murthy HM, Hendrickson WA, Drickamer K: Physical characterization and crystallization of the carbohydrate-recognition domain of a mannose-binding protein from rat. J Biol Chem 1991;266:20678–20686.
5 Weis WI, Drickamer K, Hendrickson WA: Structure of a C-type mannose-binding protein complexed with an oligosaccharide. Nature 1992;360:127–134.
6 Hirabayashi J: Two distinct families of animal lectins: Speculation on their raison d'être; in van Driessche E, Fischer J, Beeckmans S, Bøg-Hansen TC (eds): Lectins. Hellerup, Textop, 1993, vol 10, pp 205–219.
7 Leduc M, Bon C: Cloning of subunits of convulxin, a collagen-like platelet-aggregating protein from *Crotalus durissus terrificus* venom. Biochem J 1998;333:389–393.
8 Usami Y, Fujimura Y, Suzuki M, Ozeki Y, Nishio K, Fukui H, Titani K: Primary structure of two-chain botrocetin, a von Willebrand factor modulator purified from the venom of *Bothrops jararaca*. Proc Natl Acad Sci USA 1993;90:928–932.
9 Jandrot-Perrus M, Lagrue AH, Okuma M, Bon C: Adhesion and activation of human platelets induced by convulxin involve glycoprotein VI and integrin alpha2beta1. J Biol Chem 1997;272:27035–27041.
10 Mizuno H, Fujimoto Z, Koizumi M, Kano H, Atoda H, Morita T: Structure of coagulation factors IX/X-binding protein, a heterodimer of C-type lectin domains. Nat Struct Biol 1997;4:438–441.
11 Mizuno H, Fujimoto Z, Koizumi M, Kano H, Atoda H, Morita T: Crystal structure of coagulation factor IX-binding protein from habu snake venom at 2.6 A: Implication of central loop swapping based on deletion in the linker region. J Mol Biol 1999; 289:103–112.
12 Sen U, Vasudevan S, Subbarao G, McClintock RA, Celikel R, Ruggeri ZM, Varughese KI: Crystal structure of the von Willebrand factor modulator botrocetin. Biochemistry 2001;40:345–352.
13 Huang TF, Holt JC, Lukasiewicz H, Niewiarowski S: Trigramin. A low molecular weight peptide inhibiting fibrinogen interaction with platelet receptors expressed on glycoprotein IIb-IIIa complex. J Biol Chem 1987; 262:16157–16163.
14 Gan ZR, Gould RJ, Jacobs JW, Friedman PA, Polokoff MA: Echistatin. A potent platelet aggregation inhibitor from the venom of the viper, *Echis carinatus*. J Biol Chem 1988;263:19827–19832.
15 Scarborough RM, Rose JW, Naughton MA, Phillips DR, Nannizzi L, Arfsten A, Campbell AM Charo IF: Characterization of the integrin specificities of disintegrins isolated from American pit viper venoms. J Biol Chem 1993;268:1058–1065.
16 Niewiarowski S, McLane MA, Kloczewiak M, Stewart GJ: Disintegrins and other naturally occurring antagonists of platelet fibrinogen receptors. Semin Hematol 1994;31:289–300.
17 Juliano D, Wang Y, Marcinkiewicz C, Rosenthal LA, Stewart GJ, Niewiarowski S: Disintegrin interaction with alpha V beta 3 integrin on human umbilical vein endothelial cells: Expression of ligand-induced binding site on beta 3 subunit. Exp Cell Res 1996;225:132–142.

18 Bjarnason JB, Fox JW: Hemorrhagic metalloproteinases from snake venoms. Pharmacol Ther 1994;62: 325–372.
19 Takeya H, Oda K, Miyata T, Omori-Satoh T, Iwanaga S: The complete amino acid sequence of the high molecular mass hemorrhagic protein HR1B isolated from the venom of *Trimeresurus flavoviridis*. J Biol Chem 1990;265:16068–16073.
20 Paine MJ, Desmond HP, Theakston RD, Crampton JM: Purification, cloning, and molecular characterization of a high molecular weight hemorrhagic metalloprotease, jararhagin, from *Bothrops jararaca* venom. Insights into the disintegrin gene family. J Biol Chem 1992;267: 22869–22876.
21 Hite LA, Shannon JD, Bjarnason JB, Fox JW: Sequence of a cDNA clone encoding the zinc metalloproteinase hemorrhagic toxin e from *Crotalus atrox:* Evidence for signal, zymogen, and disintegrin-like structures. Biochemistry 1992;31:6203–6211.
22 Wolfsberg TG, Straight PD, Gerena RL, Huovila AP, Primakoff P, Myles DG, White JM: ADAM, a widely distributed and developmentally regulated gene family encoding membrane proteins with a disintegrin and metalloprotease domain. Dev Biol 1995;169:378–383.
23 Black RA Rauch CT, Kozlosky CJ, Peschon JJ, Slack JL, Wolfson MF, Castner BJ, Stocking KL, Reddy P, Srinivasan S, Nelson N, Boiani N, Schooley KA, Gerhart M, Davis R, Fitzner JN, Johnson RS, Paxton RJ, March CJ, Cerretti DP: A metalloproteinase disintegrin that releases tumour-necrosis factor-alpha from cells. Nature 1997;385:729–733.
24 Usami Y, Fujimura Y, Miura S, Shima H, Yoshida E, Yoshioka A, Hirano K, Suzuki M, Titani K: A 28 kDa-protein with disintegrin-like structure (jararhagin-C) purified from *Bothrops jararaca* venom inhibits collagen- and ADP-induced platelet aggregation. Biochem Biophys Res Commun 1994;201:331–339.
25 Zhou Q, Dangelmaier C, Smith JB: The hemorrhagin catrocollastatin inhibits collagen-induced platelet aggregation by binding to collagen via its disintegrin-like domain. Biochem Biophys Res Commun 1996; 219:720–726.
26 Smith JB, Dangelmaier C, Selak M: Identification of 50 kDa snake venom proteins which specifically inhibit platelet adhesion to collagen. FEBS Lett 1991;283:307–310.
27 Moura-da-Silva AM, Linica A, Della-Casa MS, Kamiguti AS, Ho PL, Crampton JM, Theakston RD: Jararhagin ECD-containing disintegrin domain: Expression in *Escherichia coli* and inhibition of the platelet-collagen interaction. Arch Biochem Biophys 1999;369:295–301.
28 Jia LG, Wang XM, Shannon JD, Bjarnason JB, Fox JW: Inhibition of platelet aggregation by the recombinant cysteine-rich domain of the hemorrhagic snake venom metalloproteinase, atrolysin A. Arch Biochem Biophys 2000;373:281–286.
29 Berndt MC, Shen Y, Dopheide SM, Gardiner EE, Andrews RK: The vascular biology of the glycoprotein Ib-IX-V complex. Thromb Haemost 2001;86:178–189.
30 Andrews RK, Berndt MC: Snake venom modulators of platelet adhesion receptors and their ligands. Toxicon 2000;38:775–791.
31 Kulkarni S, Dopheide SM, Yap CL, Ravanat C, Freund M, Mangin P, Heel KA, Street A, Harper IS, Lanza F, Jackson SP: A revised model of platelet aggregation. J Clin Invest 2000;105:783–791.
32 Schulte am Esch J 2nd, Cruz MA, Siegel JB, Anrather J, Robson SC: Activation of human platelets by the membrane-expressed A1 domain of von Willebrand factor. Blood 1997; 90:4425–4437.
33 Satoh K, Asazuma N, Yatomi Y, Fujimura Y, Miura S, Titani K, Ozaki Y: Activation of protein-tyrosine kinase pathways in human platelets stimulated with the A1 domain of von Willebrand factor. Platelets 2000;11:171–176.
34 Fujimura Y, Kawasaki T, Titani K: Snake venom proteins modulating the interaction between von Willebrand factor and platelet glycoprotein Ib. Thromb Haemost 1996;76: 633–639.
35 Peng M, Lu W, Beviglia L, Niewiarowski S, Kirby EP: Echicetin: A snake venom protein that inhibits binding of von Willebrand factor and alboaggregins to platelet glycoprotein Ib. Blood 1993;81:2321–2328.
36 Navdaev A, Dormann D, Clemetson JM, Clemetson KJ: Echicetin, a GPIb-binding snake C-type lectin from *Echis carinatus,* also contains a binding site for IgMkappa responsible for platelet agglutination in plasma and inducing signal transduction. Blood 2001;97:2333–2341.
37 Andrews RK, Kroll MH, Ward CM, Rose JW, Scarborough RM, Smith AI, Lopez JA, Berndt MC: Binding of a novel 50-kilodalton alboaggregin from *Trimeresurus albolabris* and related viper venom proteins to the platelet membrane glycoprotein Ib-IX-V complex. Effect on platelet aggregation and glycoprotein Ib-mediated platelet activation. Biochemistry 1996;35:12629–12639.
38 Kowalska MA, Tan L, Holt JC, Peng M, Karczewski J, Calvete JJ, Niewiarowski S: Alboaggregins A and B. Structure and interaction with human platelets. Thromb Haemost 1998;79:609–613.
39 Falati S, Edmead CE, Poole AW: Glycoprotein Ib-V-IX, a receptor for von Willebrand factor, couples physically and functionally to the Fc receptor gamma-chain, Fyn, and Lyn to activate human platelets. Blood 1999;94:1648–1656.
40 Asazuma N, Marshall SJ, Berlanga O, Snell D, Poole AW, Berndt MC, Andrews RK, Watson SP: The snake venom toxin alboaggregin-A activates glycoprotein VI. Blood 2001; 97:3989–3991.
41 Dormann D, Clemetson JM, Navdaev A, Kehrel BE, Clemetson KJ: Alboaggregin A activates platelets by a mechanism involving glycoprotein VI as well as glycoprotein Ib. Blood 2001;97:929–936.

42 Navdaev A, Clemetson JM, Polgar J, Kehrel BE, Glauner M, Magnenat E, Wells TN, Clemetson KJ: Aggretin, a heterodimeric C-type lectin from *Calloselasma rhodostoma* (Malayan pit viper) stimulates platelets by binding to alpha 2beta 1 integrin and glycoprotein Ib, activating Syk and phospholipase Cgamma 2, but does not involve the glycoprotein VI/Fc receptor gamma chain collagen receptor. J Biol Chem 2001;276:20882–20889.
43 Bergmeier W, Bouvard D, Eble JA, Mokhtari-Nejad R, Schulte V, Zirngibl H, Brakebusch C, Fässler R, Nieswandt B: Rhodocytin (aggretin) activates platelets lacking alpha(2)beta(1) integrin, glycoprotein VI, and the ligand-binding domain of glycoprotein Ibalpha. J Biol Chem 2001;276:25121–25126.
44 Sullam PM, Hyun WC, Szollosi J, Dong J, Foss WM, Lopez JA: Physical proximity and functional interplay of the glycoprotein Ib-IX-V complex and the Fc receptor FcgammaRIIA on the platelet plasma membrane. J Biol Chem 1998;273: 5331–5336.
45 Yi W, Suzuki-Inoue K, Satoh K, Asazuma N, Yatomi Y, Berndt MC, Ozaki Y: Role of Fc receptor-chain in platelet glycoprotein Ib-mediated signaling. Blood 2001;97:3836–3845.
46 Ward CM, Andrews RK, Smith AI, Berndt MC: Mocarhagin, a novel cobra venom metalloproteinase, cleaves the platelet von Willebrand factor receptor glycoprotein Ibalpha. Identification of the sulfated tyrosine/anionic sequence Tyr-276-Glu-282 of glycoprotein Ibalpha as a binding site for von Willebrand factor and alpha-thrombin. Biochemistry 1996;35:4929–4238.
47 Dong JF, Li CQ, Lopez JA: Tyrosine sulfation of the glycoprotein Ib-IX complex: Identification of sulfated residues and effect on ligand binding. Biochemistry 1994;33:13946–13953.
48 Marchese P, Murata M, Mazzucato M, Pradella P, De Marco L, Ware J, Ruggeri ZM: Identification of three tyrosine residues of glycoprotein Ib alpha with distinct roles in von Willebrand factor and alpha-thrombin binding. J Biol Chem 1995;270: 9571–9578.
49 de Cristofaro R, de Candia E, Rutella S, Weitz JI: The Asp(272)-Glu(282) region of platelet glycoprotein Ibalpha interacts with the heparin-binding site of alpha-thrombin and protects the enzyme from the heparin-catalyzed inhibition by antithrombin III. J Biol Chem 2000; 275:3887–3895.
50 de Luca M, Dunlop LC, Andrews RK, Flannery JV Jr, Ettling R, Cumming DA, Veldman GM, Berndt MC: A novel cobra venom metalloproteinase, mocarhagin, cleaves a 10-amino acid peptide from the mature N terminus of P-selectin glycoprotein ligand receptor, PSGL-1, and abolishes P-selectin binding. J Biol Chem 1995;270:26734–26737.
51 Romo GM, Dong JF, Schade AJ, Gardiner EE, Kansas GS, Li CQ, McIntire LV, Berndt MC, Lopez JA: The glycoprotein Ib-IX-V complex is a platelet counterreceptor for P-selectin. J Exp Med 1999;190:803–814.
52 Pouyani T, Seed B: PSGL-1 recognition of P-selectin is controlled by a tyrosine sulfation consensus at the PSGL-1 amino terminus. Cell 1995; 83:333–343.
53 Watson SP, Berlanga O, Best D, Frampton J: Update on collagen receptors: Is the two-site, two-state model still valid? Platelets 2000;11: 252–259.
54 Watson SP, Asazuma N, Atkinson B, Berlanga O, Best D, Bobe R, Jarvis G, Marshall S, Snell D, Stafford M, Tulasne D, Wilde J, Wonerow P, Frampton J: The role of ITAM- and ITIM-coupled receptors in platelet activation by collagen. Thromb Haemost 2001;86:276–289.
55 Clemetson K, Clemetson JM: Platelet collagen receptors. Thromb Haemost 2001;86:189–198.
56 Watson SP, Gibbins J: Collagen receptor in platelets: Extending the role of the ITAM. Immunol Today 1998;19:260–265.
57 Prado-Franceschi J: Snake Venoms in the Study of Platelet Function. Thesis, University of Campinas, 1970.
58 Vargaftig BB, Prado-Franceschi J, Chignard M, Lefort J, Marlas G: Activation of guinea-pig platelets induced by convulxin, a substance extracted from the venom of *Crotalus durissus cascavella*. Eur J Pharmacol 1980;68:451–458.
59 Vargaftig BB, Joseph D, Wal F, Marlas G, Chignard M, Chevance LG: Convulxin-induced activation of intact and of thrombin-degranulated rabbit platelets: Specific crossed desensitisation with collagen. Eur J Pharmacol 1982;92:57–68.
60 Polgar J, Clemetson JM, Kehrel BE, Wiedemann M, Magnenat EM, Well TN, Clemetson KJ: Platelet activation and signal transduction by convulxin, a C-type lectin from *Crotalus durissus terrificus* (tropical rattlesnake) venom via the p62/GPVI collagen receptor. J Biol Chem 1997; 272:13576–13583.
61 Francischetti IM, Ghazaleh FA, Reis RA, Carlini CR, Guinmaraes JA: Convulxin induces platelet activation by a tyrosine-kinase dependent pathway and stimulates tyrosine phosphorylation of platelet proteins, including PLCγ2, independently of integrin αIIbβ3. Arch Biochem Biophys 1998;353:239–250.
62 Clemetson JM, Polgar J, Magnenat E, Wells TN, Clemetson KJ: The platelet collagen receptor glycoprotein VI is a member of the immunoglobulin superfamily closely related to FcαR and the natural killer receptors. J Biol Chem 1999;274:29019–29024.
63 Kamiguti AS, Theakston RD, Watson SP, Bon C, Laing GD, Zuzel M: Distinct contributions of glycoprotein VI and alpha(2)beta(1) integrin to the induction of platelet protein tyrosine phosphorylation and aggregation. Arch Biochem Biophys 2000;374:356–362.
64 Hers I, Berlanga O, Tiekstra MJ, Kamiguti AS, Theakston RD, Watson SP: Evidence against a direct role of the integrin alpha2beta1 in collagen-induced tyrosine phosphorylation in human platelets. Eur J Biochem 2000;267:2088–2097.

65 Francischetti IM, Saliou B, Leduc M, Carlini CR, Hatmi M, Randon J, Faili A, Bon C: Convulxin, a potent platelet-aggregating protein from *Crotalus durissus terrificus* venom, specifically binds to platelets. Toxicon 1997;35:1217–1228.
66 Niedergang F, Alcover A, Knight CG, Farndale RW, Barnes MJ, Francischetti IM, Bon C, Leduc M: Convulxin binding to platelet receptor GPVI: Competition with collagen related peptides. Biochem Biophys Res Commun 2000;273:246–250.
67 Asazuma N, Wilde JI, Berlanga O, Leduc M, Leo A, Schweighoffer E, Tybulewicz V, Bon C, Liu SK, McGlade CJ, Schraven B, Watson SP: Interaction of linker for activation of T cells with multiple adapter proteins in platelets activated by the glycoprotein VI-selective ligand, convulxin. J Biol Chem 2000;275: 33427–33434.
68 Cicmil M, Thomas JM, Sage T, Barry FA, Leduc M, Bon C, Gibbins JM: Collagen, convulxin, and thrombin stimulate aggregation-independent tyrosine phosphorylation of CD31 in platelets: Evidence for the involvement of Src family kinases. J Biol Chem 2000;275: 27339–27347.
69 Atkinson B, Stafford M, Pears C, Watson SP: Events underlying platelet aggregation to the GPVI agonist convulxin. Eur J Biochem 2001, in press.
70 Teng CM, Ko FN, Tsai IH, Hung ML, Huang TF: Trimucytin: A collagen-like aggregating inducer isolated from *Trimeresurus mucrosquamatus* snake venom. Thromb Haemost 1993;69:286–292.
71 Du XY, Magnenat E, Wells KJ, Clemetson KJ: Alboluxin, a snake C-type lectin from *Trimeresurus albolabris* venom is a potent platelet agonist acting via both GPIb and GPVI (abstract). Thromb Haemost 2001;86 (Jul suppl): 1160.
72 Andrews RK, Gardiner EE, Asazuma N, Berlanga O, Tulasne D, Nieswandtl B, Smith AI, Berndt MC, Watson SP: A novel viper venom metalloproteinase, alborhagin, is an agonist at the platelet collagen receptor GPVI. J Biol Chem2001;276: 28092–20897.
73 Schulte V, Snell D, Bermeier W, Zirngibl H, Watson SP, Nieswandt B: Evidence for two distinct epitopes within collagen for activation of murine platelets. J Biol Chem 2001;276:364–368.
74 Huang TF, Liu CZ, Yang SH: Aggretin, a novel platelet-aggregation inducer from snake *(Calloselasma rhodostoma)* venom, activates phospholipase C by acting as a glycoprotein Ia/IIa agonist. Biochem J 1995; 309:1021–1027.
75 Wang R, Kini RM, Chung MCM: Rhodocetin, a novel platelet aggregation inhibitor from the venom of *Calloselasma rhodostomai* (Malayan pit viper): Synergistic and noncovalent interaction between its subunits. Biochemistry 1999;38:7584–7593.
76 Eble JA, Beermann B, Hinz JH, Schmidt-Hederich A: Alpha2beta1 integrin is not recognized by rhodocytin but is the specific, high affinity target of rhodocetin, an RGD-independent disintegrin and potent inhibitor of cell adhesion to collagen. J Biol Chem 2001;276:12274–12284.
77 Kamiguti AS, Hay CR, Zuzel M: Inhibition of collagen-induced platelet aggregation as the result of cleavage of alpha 2 beta 1-integrin by the snake venom metalloproteinase jararhagin. Biochem J 1996; 320:635–641.
78 Kamiguti AS, Markland FS, Zhou Q, Laing GD, Theakston RD, Zuzel M: Proteolytic cleavage of the beta1 subunit of platelet alpha2beta1 integrin by the metalloproteinase jararhagin compromises collagen-stimulated phosphorylation of pp72. J Biol Chem 1997;272:32599–32605.
79 Ivaska J, Kapyla J, Pentikainen O, Hoffren AM, Hermonen J, Huttunen P, Johnson MS, Heino J: A peptide inhibiting the collagen binding function of integrin alpha2I domain. J Biol Chem 1999;274:3513–3521.
80 Pentikainen O, Hoffren AM, Ivaska J, Kapyla J, Nyronen T, Heino J, Johnson MS: 'RKKH' peptides from the snake venom metalloproteinase of *Bothrops jararaca* bind near the metal ion-dependent adhesion site of the human integrin alpha(2)I-domain. J Biol Chem 1999; 274:31493–31505.
81 Shin Y, Morita T: Rhodocytin, a functional novel platelet agonist belonging to the heterodimeric C-type lectin family, induces platelet aggregation independently of glycoprotein Ib. Biochem Biophys Res Commun 1998;245:741–745.
82 Inoue K, Ozaki Y, Satoh K, Wu Y, Yatomi Y, Shin Y, Morita T: Signal transduction pathways mediated by glycoprotein la/lla in human platelets: Comparison with those of glycoprotein VI. Biochem Biophys Res Commun 1999;256:114–120.
83 Suzuki-Inoue K, Ozaki Y, Kainoh M, Shin Y, Wu Y, Yatomi Y, Ohmori T, Tanaka T, Satoh K, Morita T: Rhodocytin induces platelet aggregation by interacting with glycoprotein Ia/IIa (GPIa/IIa, integrin alpha 2beta 1). J Biol Chem 2001;276: 1643–1652.
84 Rucavado A, Soto M, Kamiguti AS, Theakston RD, Fox JW, Escalante T, Gutierrez JM: Characterization of aspercetin, a platelet aggregating component from the venom of the snake *Bothrops asper* which induces thrombocytopenia and potentiates metalloproteinase-induced hemorrhage. Thromb Haemost 2001;85: 710–715.
85 Hamako J, Matsui T, Suzuki M, Ito M, Makita K, Fujimura Y, Ozeki Y, Titani K: Purification and characterization of bitiscetin, a novel von Willebrand factor modulator protein from *Bitis arietans* snake venom. Biochem Biophys Res Commun 1996;226:273–279.
86 Chen YL, Tsai IH: Functional and sequence characterization of agkicetin, a new glycoprotein Ib antagonist isolated from *Agkistrodon acutus* venom. offf2p4. Biochem Biophys Res Commun 1995;210:472–477.
87 Peng M, Lu W, Kirby EP: Alboaggregin-B: A new platelet agonist that binds to platelet membrane glycoprotein Ib. Biochemistry 1991;30: 11529–11536.
88 Usami Y, Suzuki M, Yoshida E, Sakurai Y, Hirano K, Kawasaki T, Fujimura Y, Titani K: Primary structure of alboaggregin-B purified from the venom of *Trimeresurus albolabris*. Biochem Biophys Res Commun 1996;219:727–733.

89 Chang MC, Lin HK, Peng HC, Huang TF: Antithrombotic effect of crotalin, a platelet membrane glycoprotein Ib antagonist from venom of *Crotalus atrox*. Blood 1998;91: 1582–1589.
90 Sakurai Y, Fujimura Y, Kokubo T, Imamura K, Kawasaki T, Handa M, Suzuki M, Matsui T, Titani K, Yoshioka A: The cDNA cloning and molecular characterization of a snake venom platelet glycoprotein Ib-binding protein, mamushigin, from *Agkistrodon halys blomhoffii* venom. Thromb Haemost 1998;79: 1199–1207.
91 Taniuchi Y, Kawasaki T, Fujimura Y, Suzuki M, Titani K, Sakai Y, Kaku S, Hisamichi N, Satoh N, Takenaka T, et al: Flavocetin-A and -B, two high molecular mass glycoprotein Ib binding proteins with high affinity purified from *Trimeresurus flavoviridis* venom, inhibit platelet aggregation at high shear stress. Biochim Biophys Acta 1995;1244:331–338.
92 Fujimura Y, Ikeda Y, Miura S, Yoshida E, Shima H, Nishida S, Suzuki M, Titani K, Taniuchi Y, Kawasaki T: Isolation and characterization of jararaca GPIb-BP, a snake venom antagonist specific to platelet glycoprotein Ib. Thromb Haemost 1995; 74:743–750.
93 Kawasaki T, Taniuchi Y, Hisamichi N, Fujimura Y, Suzuki M, Titani K, Sakai Y, Kaku S, Satoh N, Takenaka T, et al: Tokaracetin, a new platelet antagonist that binds to platelet glycoprotein Ib and inhibits von Willebrand factor-dependent shear-induced platelet aggregation. Biochem J 1995;308:947–953.
94 Hamako J, Matsui T, Nishida S, Nomura S, Fujimura Y, Ito M, Ozeki Y, Titani K: Purification and characterisation of kaouthiagin, a von Willebrand factor-binding and cleaving metalloproteinase from *Naja kaouthia* cobra venom. Thromb Haemost 1998;80:499–505.

Substances Affecting Platelet Function

Hæmostasis

Haemostasis 2001;31:173–176

Lebecetin, a C-Lectin Protein from the Venom of *Macrovipera lebetina* That Inhibits Platelet Aggregation and Adhesion of Cancerous Cells

Sameh Sarray[a] Najet Srairi[a] Jose Luis[c] Jacques Marvaldi[c]
Mohamed El Ayeb[a] Naziha Marrakchi[a,b]

[a]Laboratoire des Venins et Toxines, Institut Pasteur de Tunis,
[b]Faculté de Médecine de Tunis, Tunisie; [c]CNRS UPRESA 6032,
Faculté de Pharmacie, Laboratoire de Biochimie Cellulaire, Marseille, France

Key Words

Snake venom · C-lectin protein · Anti-platelet aggregation · Anti-agglutination protein

Abstract

A novel C-lectin protein, lebecetin, was purified and characterized from the venom of *Macrovipera lebetina*. It is a disulfide-linked heterodimer of 15 and 16 kD. The subunits are homologous to each other and to the other snake venom proteins of the C-type (Ca^{2+}-dependent) lectin superfamily. Lebecetin shows a potent inhibitory effect on whole blood and washed platelets induced by different agonists. It inhibits the agglutination of human fixed platelets in the presence of ristocetin. Lebecetin also interferes with the adhesion of IGR39 melanoma and HT29D4 adenocarcinoma cells. These two lines adhere to lebecetin used as matrix. Lebecetin is also able to strongly reduce IGR39 and HT29D4 cell adhesion to fibrinogen and laminin, but not to fibronectin and collagen types I and IV, respectively. Adhesion properties of lebecetin may thus involve integrin receptors.

Introduction

Lectins are defined as nonenzymatic carbohydrate-binding proteins, common in plants including trees, and particulary seeds. They have also been found in bacteria, roe, snails, vertebrates and mushrooms [1]. C-Lectin proteins (CLPs) have also been found in the venom of snakes belonging to the families of Elapidae, Viperidae and Crotalidae [2–4]. These proteins show an M_r of 30,000 D and exist as heterodimers linked by a single inter-

KARGER

Fax + 41 61 306 12 34
E-Mail karger@karger.ch
www.karger.com

0301–0147/01/0316–0173$17.50/0

Accessible online at:
www.karger.com/journals/hae

Dr. Sarray Sameh
Laboratoire des Venins et Toxines, Institut Pasteur de Tunis
13, Place Pasteur, BP 74, 1002 Tunis-Belvédère (Tunisia)
Tel. +216 71 843 755, Fax +216 71 791 833
E-Mail sameh_sarray@voila.fr

chain disulfide bond. They share a structural homology of 30–70%. However, this group of Ca^{2+}-dependent lectin-related proteins exhibits different effects on blood coagulation and platelet aggregation. Several CLPs induce various effects on platelet functions by modulating the interactions between von Willebrand factor (vWF) and platelet GPIb. For example, botrocetin from *Bothrops jararaca* venom, binds to vWF and forms an activated complex that induces platelet agglutination [5]. Alboaggregins (A and B) from *Trimeresurus albolabris* venom [6, 7], echicetin from *Echis carinatus* venom [8] and agkicetin from *Agkistrodon acutus* venom [9] all bind to platelet GPIb and function as receptor blockers for vWF binding. However, alboaggregins induce platelet agglutination whereas echicetin and agkicetin inhibit platelet agglutination.

Lectin-like proteins were present in the venom of *M. lebetina*, mainly as dimeric proteins. A novel CLP, designated lebecetin, was isolated from *Macrovipera lebetina* venom and characterized as a potent platelet aggregation inhibitor with interesting adhesive properties.

Materials and Methods

Purification of Lebecetin

M. lebetina venom was dissolved in 0.2 *M* ammonium acetate (pH 6.8) and applied to a Sephadex G-75 column equilibrated with the same buffer. The fractions containing the main of antiplatelet activity were collected and then loaded on a cation exchange Mono S (HR5/5) column previously equilibrated with 50 m*M* HEPES/HCl, pH 7.5, and eluated with a linear NaCl (0–1 *M*) gradient. RP-HPLC (C8 column) was then used to purify the protein.

Amino Acid Sequence Analysis

Lebecetin was reduced and alkylated and separated into two subunits by RP-HPLC on a C8 column. The N-terminal sequence of each subunit was analyzed by automated Edman degradation (Model 476A, Applied Biosystem).

Platelet Aggregation Assay

Platelet-rich plasma (PRP) was prepared from rabbit blood by mixing whole blood with 0.1 vol of 3.8 of trisodium citrate followed by centrifugation at room temperature [10]. Platelet aggregation studies were carried out with a Chronolog aggregometer using 0.3 ml of PRP at 37 °C.

Platelets were isolated from human or rabbit blood using 5 m*M* EDTA as anticoagulant. Platelets were then separated from blood and washed twice with calcium-free Tyrode's buffer, pH 6.5, containing 0.1% glucose, 0.25% gelatin, 2 m*M* $MgCl_2$ and 0.2 m*M* EGTA as previously described [11]. The final suspension was made in a modified Tyrode's buffer without EGTA, pH 7.4. Platelet aggregation was monitored at 37 °C under stirring (1,100 rpm) in a Chronolog aggregometer. Platelets ($3 \cdot 10^8$ cells/ml) were incubated for 2 min with 2 m*M* $CaCl_2$ in the aggregometer cuvette in the presence of lebecetin before the addition of agonists.

Preparation of Formalin-Fixed Platelets

Fixed human platelets were prepared as previously described [12]. They were incubated with 2% of formalin in Tris-saline buffer overnight at 4 °C. Finally, they were washed twice with Tris-saline buffer. The final platelet pellets were suspended in an equal volume of Tris-saline buffer containing BSA (20 mg/ml).

The agglutination of fixed washed platelets induced by bovine vWF was measured using an aggregometer as previously described [11].

Cell Adhesion Assay

Flat bottom 96-well microtiter plates were coated for 2 h at 37 °C with 50 µl of one of the following purified extracellular matrix (ECM) proteins: fibronectin (10 µg/ml), vitronectin (10 µg/ml), laminin (2 µg/ml), fibrinogen (50 µg/ml), collagen type I (10 µg/ml) and collagen type IV (10 µg/ml). Coated wells were blocked with 1% BSA in PBS for 1 h. The melanoma cell line IGR39 and adenocarcinoma cell line HT29D4 were routinely cultured in Dulbecco's modified Eagle's medium (DME medium) containing 4.5 g/l glucose and 10% fetal calf serum in a humidified atmosphere of 5% CO_2. Cells were harvested in single cell suspensions by treatment with 0.53 m*M* EDTA in PBS, washed twice with DME medium containing 0.2% BSA (adhesion buffer) and resuspended in the same medium in the presence of lebecetin. After incubation for 30 min at room temperature, cells were added to coated wells in a volume of 50 µl (10^6 cells/ml) and allowed to adhere to the substrate for 1 h in a cell culture incubator. Unattached cells were removed by gently washing three times with adhesion buffer. Residual attached cells were fixed by 1% glutaraldehyde, stained by 0.1% crystal violet and lysed with 1% SDS. The absorbance was measured at 600 nm by a microplate reader (model Σ960 from Metertech).

Results

Purification and Characterization of Lebecetin

Lebecetin was purified to homogeneity from the venom of *M. lebetina* in three steps: first by gel filtration chromatography, then by cation exchange chromatography and finally by RP-HPLC. Each fraction was assayed for inhibition of aggregation activity on rabbit whole blood. On SDS-PAGE (15%), lebecetin behaved as a molecule of 29 kD under nonreducing conditions and separated into bands of 15 and 16 kD under reducing conditions. Lebecetin (5 nmol) was reduced, alkylated and separated into two peaks by RP-HPLC on a C8 column. MALDI TOF spectrometry data confirmed results obtained by SDS-PAGE analysis and in particular the heterodimeric nature of lebecetin. The N-terminal amino acid sequence of both chains was determined and aligned with those of other members of C-type lectin-like protein from snake venom. α- and β-Subunits show 42% identity between them and respectively 28–67% [13, 14] and 40–55% [14, 15] identity with those of the other corresponding α and β CLP-related sequences.

Effect of Lebecetin on Platelet Aggregation

Lebecetin showed a potent inhibitory effect on platelet aggregation in rabbit whole blood induced by ADP (1 m*M*) and Ca^{2+}- ionophore (10^{-3} m*M*) with an IC_{50} of about 4 n*M*. This CLP was also able to inhibit rabbit and human washed platelet aggregation induced by thrombin (0.04 IU/ml). The IC_{50} was about 0.1 n*M*. Individually, both α- and β-subunits exhibited a distinct strong inhibitory effect. These results may suggest that individual lebecetin subunits contained interaction sites that enable them to bind to platelets.

Lebecetin inhibited both the aggregation of platelet-rich plasma induced by the antibiotic ristocetin and the agglutination of fixed platelets induced by bovine vWF in a dose-dependent manner. The concentration of lebecetin required for 50% inhibition of agglutination of the fixed platelets was approximately 0.5 n*M*. EDTA and EGTA significantly reduced the agglutination inhibiting activity of lebecetin. These results suggest the importance of Ca^{2+} for the expression of lebecetin activity.

Effect of Lebecetin on Cancerous Adhesion Cells

The inhibitory effect of lebecetin on the adhesion of cancerous cells to immobilized ECM (e.g., fibrinogen, collagen I and IV, laminin and vitronectin) was tested. Lebecetin potently inhibited the adhesion of HT29D4 cells (derived from adenocarcinoma) to laminin matrix as well as the adhesion of IGR39 melanoma cells to fibrinogen matrix. Moreover IGR39 cells and HT29D4 cells adhered tightly to immobilized lebecetin. The adhesion of lebecetin to IGR39 cells was dose-dependent with an ED_{50} of 30 μg/ml.

Discussion

Crude *M. lebetina* venom prevents platelet aggregation induced by collagen, thrombin or PAF acether with similar IC_{50} around 100 μg/ml [16]. Based on this observation, we isolated and described a potent inhibitor of platelet aggregation and agglutination named lebecetin. It is composed of two covalently linked subunits. Lebecetin has a high level of homology with N-terminal amino acid sequences of CLPs. Thus lebecetin belongs to the CLP class of snake venom proteins.

Lebecetin inhibits platelet aggregation induced by different agonists such as ADP, Ca^{2+}-ionophore and thrombin. It is also able to inhibit PRP and washed fixed platelets in the presence of ristocetin. On the other hand, we found that lebecetin predominantly inhibits adhesion of HT29D4 to immobilized laminin and IGR39 to immobilized fibrinogen.

Work is in progress to determine the integrins on the surface of these cells interacting with lebecetin.

Acknowledgements

The authors recognize the friendly efforts of Dr. Cassian Bon from I. P. P. They also thank Prof. K. Dellagi for his support during this study.

References

1 Goldsteĭn IJ, Hayer CE: The lectins: carbohydrate-binding proteins of plants and animals. Adv Carbohydr Chem Biochem 1978;35:127–131.
2 Devi A: The protein and nonprotein constituents of snake venom; in Bucherl W, Buckley E, Deulofeu V (eds): Venomous Animals and Their Venom. New York, Academic Press, 1968, pp 129–131.
3 Gartner TK, Oglive ML: Isolation and characterization of three Ca^{2+} dependent β-galactoside specific lectins from snake venom. Biochem J 1984;24:301–307.
4 Aragon-Ortiz F, Brenes-Brenes JR, Gubensek F: Characterization of a lectin-like protein isolated from *Lachesis muta* snake venom. Rev Biol Trop 1989;37:79–84.
5 Fujimura Y, Titani K, Usami Y, Suzuki M, Oyama R, Matsui T, Fukui H, Sugimoto M, Ruggeri ZM: Isolation and characterization of two structurally and functionally distinct from botrocetin, the platelet coagglutinin from the venom of *Bothrops jararaca.* Biochemistry 1991; 30:1957–1964.
6 Peng M, Lu W, Kirby EP: Alboaggregins-B: A new platelet agonist that binds to platelet membrane glycoprotein Ib. Biochemistry 1991;30: 11529–11536.
7 Peng M, Lu W, Kirby EP: Characterization of three alboaggregins purified from *Trimeresurus albolabris* venom. Thromb Haemost 1992;67: 702–707.
8 Peng M, Lu W, Beviglia L, Niewiarowski S, Kirby EP: A snake venom protein that inhibits binding of von Willebrand Factor and alboaggregins to platelet glycoprotein Ib. Blood 1993;81:2321–2328.
9 Chen YL, Tsai IH: Functional and sequence characterization of agkicetin, a new glycoprotein Ib antagonist isolated from *Agkistrodon acutus* venom. Biochem Biophys Res Commun 1995;210:472–477.
10 De Marco L, Girolami A, Zimmerman TS, Ruggeri ZM: Von Willebrand factor interaction with the glycoprotein IIb/IIa complex. Its role in platelet function as demonstrated in patients with congenital afibrinogenemia. J Clin Invest 1986; 77:1272–1277.
11 Ardlie NG, Packman MA, Mustard JF: Adenosine diphosphate induced platelet aggregation in suspensions of washed rabbit platelets. Br J Haematol 1970;19:7–17.
12 Kirby EP, Mills DCB: The interaction of bovine factor VIII with human platelets. J Clin Invest 1975;56: 491–502.
13 Taniuchi Y, Kawasaki T, Fujimura Y, Suzuki M, Titani K, Sakai K, Kaku S, Hisamichi N, Satoh V, Takenaka T, Handa M, Swai Y: Flavocetin-A and -B, two high molecular mass glycoprotein Ib binding proteins with high affinity purified from *Trimeresurus flavoridis* venom, inhibit platelet aggregation at high shear stress. Biochem Biophys Acta 1995;1244:331–338.
14 Cheng X, Qian Y, Liu Q, Li BX, Zhang M, Li J: Purification, characterization, and cDNA cloning of a new fibrinogenolytic venom protein, agkisacutacin, from *Agkistrodon acutus* venom. Biochem Biophys Res Commun 1999;265:530–535.
15 Kawasaki T, Taniuchi Y, Hisamichi N, Fujimura Y, Suzuki M, Titani K, Kaku S, Satoh N, Takenaka T, Handa M, Swai Y: Tokaracetin, a new platelet antagonist that binds to platelet glycoprotein Ib and inhibits von Willebrand factor-dependent shear-induced platelet aggregation. Biochem J 1995;308:947–953.
16 Barbouche R, Marrakchi N, Mansuelle P, Krifi M, Fenouillet E, Rochat H, El Ayeb M: Novel anti-platelet aggregation polypeptides from *Vipera lebetina* venom: Isolation and characterization. FEBS Lett 1996;392:6–10.

Substances Affecting Platelet Function

Hæmostasis

Haemostasis 2001;31:177–182

New Insights on Disintegrin-Receptor Interactions: Eristostatin and Melanoma Cells

M.A. McLane M.A. Kuchar C. Brando D. Santoli
C.A. Paquette-Straub M.E. Miele

Department of Medical Technology, University of Delaware, Newark, Del., and The Wistar Institute, Philadelphia, Pa., USA

Key Words

Integrin · Disintegrin · Eristostatin · Melanoma · Metastasis · NK cells

Abstract

Viper venom disintegrins have been used frequently to study the cellular receptors which characterize various types of cells, including platelets, endothelial cells and cancer cells. While the majority of such analyses have pointed to involvement of integrin receptors αvβ3, α5β1 or αIIbβ3, this may not always be so. Eristostatin, from *Eristocophis macmahoni*, is a potent inhibitor of ADP-induced platelet aggregation as well as of human and murine melanoma metastases in mouse model systems. This disintegrin requires an RGDW motif, as well as an intact C-terminus, in order to interact with both platelets and four different types of melanoma cells. Eristostatin causes nonmetastatic SBc12 melanoma cells to show higher susceptibility to specific killing by NK-like TALL-104 cells. While it is known that eristostatin binds to αIIbβ3 on platelets, the receptor with which eristostatin binds to the melanoma cells has not yet been identified.

Introduction

Disintegrins are low molecular weight, cysteine-rich proteins isolated from viper venom [1]. Previous investigations have demonstrated that these proteins interact with many types of cells, such as platelets [2], neutrophils [3], human umbilical vein [4] and bovine capillary [5] endothelial cells, and many types of cancer cells, including breast [6], neuroglial [7], bladder [8] and melanoma [9]. Whenever

KARGER

Fax + 41 61 306 12 34
E-Mail karger@karger.ch
www.karger.com

0301–0147/01/0316–0177$17.50/0

Accessible online at:
www.karger.com/journals/hae

Mary Ann McLane, PhD, CLS(NCA)
Department of Medical Technology
McKinly Lab 057, University of Delaware
Newark, DE 19716 (USA), Tel. +1 302 831 8737 (office)
Fax +1 302 831 4180, E-Mail mclane@udel.edu

the investigations have attempted to identify the receptor through which the interaction is achieved, an integrin is usually involved. This is understandable since all disintegrins possess binding motifs (RGD, KGD, MVD, MLD, VGD) similar to those found in native ligands, such as fibronectin, fibrinogen and VCAM-1 [10–12]. Disintegrins, as a family of proteins, simultaneously share significant sequence homology and cysteine arrangement, yet demonstrate substantial diversity in their binding interactions with cellular integrins. We previously demonstrated that certain non-RGD motifs present in these adhesive ligands will also contribute to their binding selectivity for αIIbβ3, αvβ3 and α5β1 integrins [13]. We have used a panel of recombinant mutants of eristostatin (based on the native protein from *Eristocophis macmahoni*) to investigate whether the binding motifs required for interacting with 'normal' cells are also required with transformed, cancerous cells. We also hypothesize a mechanism for the ability of eristostatin to inhibit melanoma cell metastasis.

Material and Methods

Chemicals and Media

Standard laboratory chemicals were obtained from Sigma (St. Louis Mo., USA). Eight-chambered coverslip slides were from Nalge Nunc (Naperville, Ill., USA). Dulbecco's media and PBS were obtained from Mediatech (Herndon, Va., USA). Fetal calf serum was from GibcoBRL (Rockville Md., USA). $H_3{}^{51}CrO_4$ was obtained from Perkin-Elmer (Boston, Mass., USA). Antibodies against α4 integrin subunit were purchased from Immunotech (Marseille, France) and antibodies against CD48, CD27, CD70, CD137, CD137-ligand, CD134, and FAS were purchased by Pharmigen (San Diego, Calif., USA). For flow cytometry analysis, fluorescinated secondary antibodies were purchased from Organon Tecnika Corporation (Durham, N.C., USA).

Cells Used

Human melanoma cell lines 1205-LU, WM164 and SBc12 were kindly provided by Dr. Meenhard Herlyn (The Wistar Institute, Philadelphia, Pa., USA), while C8161 and MV3 cells were obtained from Danny Welch (Penn State University, Hershey, Pa., USA) and Stefan Niewiarowski, (Temple University, Philadelphia, Pa., USA), respectively. Cells were cultured in Dulbecco's Modified Eagle's medium/F12 containing *L*-glutamine and supplemented with 10% fetal calf serum (FCS). Cells were grown at 37 °C, 5% CO_2, detached using 2 mmol/l EDTA and resuspended in Dulbecco's phosphate-buffered saline. Cytotoxic cell line TALL-104 (MHC-nonrestricted, CD8+, CD56+, CD16-, CD161+) were grown in Iscove's Modified Dulbecco's Medium, supplemented and grown as above.

Preparation of Disintegrins

Wild-type eristostatin and five mutations listed in table 1 were expressed in *Escherichia coli* as described previously [13]. ADP-induced platelet aggregation was used to screen each disintegrin, and the IC_{50} values obtained agreed with those previously published [13]. FITC-labeling of eristostatin was performed using a Fluorotag conjugation kit from Sigma, following the manufacturer's directions. The ratio of fluorescence to protein of the labeled disintegrin was 1:6.

Binding Inhibition

Each human melanoma cell line (3E5) was incubated with buffer or unlabeled disintegrin (500–3,000 n*M*). Without washing, the cells were then stained with FITC-labeled eristostatin (for 1205-LU, WM164, C8161) or FITC-labeled anti-α4 (for MV3). All data were acquired on a Zeiss inverted 100M Axioskop equipped with a Zeiss 510 LSM confocal microscope and a krypton-argon laser. Each binding pattern was rated from zero (no inhibition) to +3 (complete inhibition), depending on how much the disintegrin inhibited the binding of the FITC-labeled protein.

Flow Cytometry Analysis

SBc12 melanoma cells were incubated for 3 days in the presence or absence of 500 n*M* eristostatin. After washing, the cells were stained with primary antibodies to cell surface markers, washed and probed with FITC-conjugated secondary antibody. Flow analysis was performed on an Epics XL flow cytometer, and nonspecific fluorescence (using 1 mg FITC-albumin) showed less than 0.5 fluorescence units.

Table 1. Mutations made in RGD loop and C-terminus of eristostatin or echistatin

Forms of eristostatin	Location of mutation within sequence (compared with wildtype eristostatin)
Eristostatin (wildtype)	^{23}CRVARGDWNDDYCTGKSCDCPRNPWNG^{49}
Eristo R27A	A
EristoW30D	D
Eristo N31M	M
Eristo W30D/N31M	DM
EristoV25R/W30D	R D
Eristo (1-43)	~~RNPWNG~~

For each mutation, the amino acid substitution is placed under its corresponding wildtype residue. Those motifs which are removed are indicated by a double crossout.

Cell Cytotoxicity

Sensitivity of human melanoma SBc12 cells, incubated overnight with 500 n*M* eristostatin or media control, to the cytotoxic action of TALL-104 cells was tested in an 18 h ^{51}Cr release assay, performed as previously described [14]. Briefly, SBc12 target cells were resuspended at 1E6/ml in PBS and labeled for 1 h at 37 °C with 0.1 mCi of $H_3{}^{51}CrO_5$. After washing, cells were resuspended at 1E4/50 μl and incubated at 37 °C in 10% CO_2 with serially diluted effector TALL-104 cells. After 18 h, cell supernatant was collected, and released radioactivity was measured with a gamma counter. Percent of specific radioactivity was calculated as: sample radioactivity – spontaneous release/total radioactivity (Triton-lysed cells) – spontaneous release × 100.

Results and Discussion

Eristostatin is unique among the venom-derived disintegrins for a number of reasons. It is a potent inhibitor of ADP-induced platelet aggregation (IC_{50} = 70 n*M*), and binds αIIbβ3 integrin with a K_d of 18 n*M*, whether the platelets are resting or activated by ADP or thrombin [2]. It has been shown to inhibit lymphocyte infiltration of pancreatic islets in nonobese diabetic mice [Clara Brando, unpubl. obs.], and is able to inhibit melanoma metastasis in an in vivo model of experimental metastasis, using either murine [15, 16] or human [9] melanoma cells. It shares an integrin selectivity property with another disintegrin, barbourin [17], since it has been shown to interact with cells primarily through the integrin αIIbβ3 [2], although Danen et al. [9] did demonstrate a possible binding to α4β1 on MV3 melanoma cells. In contrast to most other disintegrins, however, eristostatin does not interact with αvβ3 or α5β1 [13]. It is curious that the murine melanoma cells used in the experimental metastasis assays (BL16F10, BL16F1) showed a strong expression of β3, but the human MV3 cells did not. This would suggest that eristostatin may not use αIIbβ3 consistently, if at all, in its interactions with the melanoma cells whose metastasis it inhibits. As the first step in determining the molecular mechanism used by eristostatin to inhibit human melanoma metastasis, we used our panel of recombinant eristostatin mutants (table 1) to demonstrate which motifs within eristostatin are most critical for the in vitro interaction of eristostatin with four human melanoma cell lines (MV3, C8161, 1205-LU, WM164) by confocal microscopy

Table 2. Summary of all cell lines for inhibition of binding by recombinant disintegrins

Disintegrin	1205-LU	WM164	C8161	MV3
Eristostatin	+2	+3	+3	+3
ErR27A	0	0	0	0
ErW30D	+2	+2	+2	0
ErV25R/W30D	+2	+2	+2	+1
ErW30D/N31M	+3	+3	+3	0
ErN31M	+3	+3	+3	0
Eristostatin(truncated)	0	0	0	0

Each human melanoma cell line (3E5) was incubated with buffer or the indicated disintegrin (500–3000 n*M*). Without washing, the cells were then stained with FITC-labeled eristostatin (for 1205-LU, WM164, C8161) or FITC-labeled anti-α4 (for MV3), and observed by confocal microscopy. Each binding pattern was rated from 0 (no inhibition) to +3 (complete inhibition), depending on how much the disintegrin inhibited the binding of the FITC-labeled protein. The data are representative of a minimum of 2 identical experiments.

(table 2). The four melanoma cell lines were used to determine whether the interaction of eristostatin with human melanoma cells was unique to MV3. In fact, eristostatin bound to all four cell types. This was directly shown with 1205-LU, WM164 and C8161 cells through confocal microscopy using FITC-eristostatin. For MV3 cells, we demonstrated the interaction of eristostatin by showing a dose-dependent inhibition of the binding of anti-α4, confirming the findings of Danen et al. [9] that eristostatin interacts with α4β1 on these cells. As summarized in table 2, binding to 1205-LU, WM164, and C8161 melanoma cells requires an **R**GD sequence and an intact C-terminus. Asparagine is not needed in the RGD loop of eristostatin for binding to these cells. Surprisingly, methionine at position RGDX**X** enhances the ability of eristostatin to bind. For MV3 melanoma cells, an **R**GD sequence, an intact C-terminus, and both asparagine and tryptophan in the RGD loop of eristostatin are required. We hypothesize that eristostatin uses the same molecular mechanism to bind to 1205-LU, WM164, and C8161 melanoma cells. However, the melanoma cell receptor by which eristostatin accomplishes this interaction is as yet unidentified. Based on the pattern of eristostatin mutation results, we propose that the mechanism at work in its interactions with MV3 cells is different from the one used in binding to 1205-LU, WM164, and C8161 cells.

In order to study the consequences of the binding of eristostatin to melanoma cells, we first incubated SBc12 human melanoma cells ± 4000 n*M* eristostatin for 5 days, and observed no difference between the control and treated cells in cell count or gross appearance, confirming the findings of Danen et al. [9] and Morris et al. [16] (data not shown). We then tested whether the presence of eristostatin would make a difference in the ability of cytotoxic cells to kill the SBc12 melanoma cells. As shown in figure 1, TALL-104 cytotoxic cells [18] are capable of killing the SBc12 melanoma cells, in a dose-dependent fashion, without the presence of eristostatin. However,

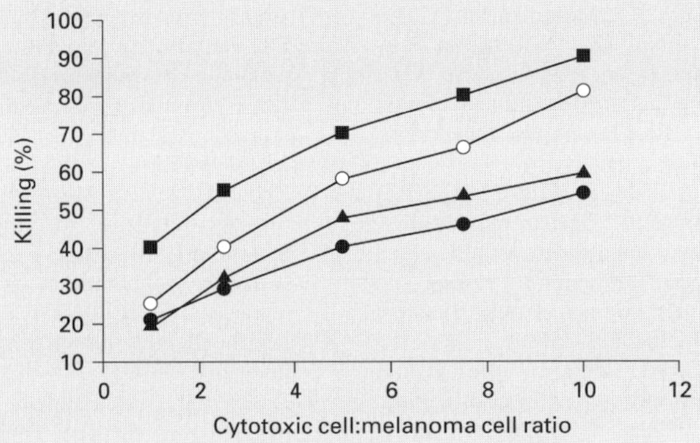

Fig. 1. Human melanoma SBc12 cells were incubated in the presence or absence of eristostatin for 3 days at 37°C. After washing, the cells were counted and 10^6 cells were suspended in media to which was added 100 μCi $H_3{}^{51}CrO_4$ for 1 h. After washing with cold, complete media, these 'target' cells were distributed in 96-well plates, incubated with serial dilutions of effector cells: MHC-nonrestricted TALL-104 cells (CD8+, CD56+, CD16-, CD161+), which had been similarly incubated with or without eristostatin. After overnight incubation, the supernatant was harvested and the released radioactivity was counted with a gamma counter. Calculation of percent killing was expressed as: sample radioactivity – spontaneous release/total radioactivity (Triton-lysed cells) – spontaneous release × 100. Data shown are representative of three identical experiments. ● = Untreated cytotoxic cell/melanoma cell; ▲ = treated cytotoxic cell, untreated melanoma cell; ○ = treated cytotoxic cell, treated melanoma cell; ■ = untreated cytotoxic cell, treated melanoma cell.

when the melanoma cells are incubated with eristostatin, TALL-104 cells show a 2-fold increased cytotoxic effect. One possible explanation is that incubation with eristostatin increases the expression of receptors for cytotoxic substances, secreted by effector cells, on the membrane of the target cell. Thus SBcl2 were screened for the presence of Fas, CD27, CD134, CD137, and CD137-L by flow analysis. No differences were found between eristostatin-treated and control SBcl2 cells.

Another explanation could be that costimulatory molecules for the process of killing are modulated by eristostatin. Thus, cells were analyzed for the presence of CD48, ligand for 2B4, a costimulatory molecule expressed by cytotoxic cells [19]. Expression of CD48 is not affected by treatment with eristostatin. Studies are under way to identify the molecule with which eristostatin binds on these melanoma cells, and the intracellular consequences of this interaction.

References

1 McLane MA, Marcinkiewicz C, Vijay-Kumar S, Wierzbicka-Patynowski I, Niewiarowski S: Viper venom disintegrins and related molecules. Proc Soc Exp Biol Med 1998; 219:109–119.

2 McLane MA, Kowalska MA, Silver L, Shattil SJ, Niewiarowski S: Interaction of disintegrins with the alpha IIb beta 3 receptor on resting and activated human platelets. Biochem J 1994;301:429–436.

3 Coelho AL, de-Freitas MS, Oliveira CA, Moura NV, Zingali RB, Barja FC: Effects of jarastatin, a novel snake venom disintegrin, on neutrophil migration and actin cytoskeleton dynamics. Exp Cell Res 1999; 251:379–387.

4 Juliano D, Wang Y, Marcinkiewicz C, Rosenthal LA: Disintegrin interaction with alpha V beta 3 integrin on human umbilical vein endothelial cells: Expression of ligand-induced binding site on beta 3 subunit. Exp Cell Res 1996;225:132–142.

5 Kang IC, Lee YD, Kim DS: A novel disintegrin salmosin inhibits tumor angiogenesis. Cancer Res 1999;59: 3754–3760.

6 Zhou Q, Hu P, Ritter MR, Swenson SD, Argounova S, Epstein AL, Markland FS: Molecular cloning and functional expression of contortrostatin, a homodimeric disintegrin from southern copperhead snake venom. Arch Biochem Biophys 2000;375:278–288.

7 Abrass CK, Berfield AK, Andress DL: Heparin binding domain of insulin-like growth factor binding protein-5 stimulates mesangial cell migration. Am J Physiol 1997;273: F899–F906.

8 Zhou Q, Nakada MT, Brooks PC, Swenson SD, Ritter MR, Argounova S, Arnold C, Markland FS: Contortrostatin, a homodimeric disintegrin, binds to integrin alphavbeta5. Biochem Biophys Res Commun 2000;267:350–355.

9 Danen EH, Marcinkiewicz C, Cornelissen IM, Van KA, Pachter JA, Ruiter DJ, Niewiarowski S, Van MG: The disintegrin eristostatin interferes with integrin alpha 4 beta 1 function and with experimental metastasis of human melanoma cells. Exp Cell Res 1998;238:188–196.

10 Pytela R, Pierschbacher MD, Argraves S, Suzuki S, Ruoslahti E: Arginine-glycine-aspartic acid adhesion receptors. Methods Enzymol 1987;144:475–489.

11 Plow EF, Pierschbacher MD, Ruoslahti E, Marguerie G, Ginsberg MH: Arginyl-glycyl-aspartic acid sequences and fibrinogen binding to platelets. Blood 1987;70:110–115.

12 Marcinkiewicz C, Calvete JJ, Marcinkiewicz MM, Raida M, Vijay KS, Huang Z, Lobb RR, Niewiarowski S: EC3, a novel heterodimeric disintegrin from *Echis carinatus* venom, inhibits alpha4 and alpha5 integrins in an RGD-independent manner. J Biol Chem 1999;274:12468–12473.

13 Wierzbicka P, Niewiarowski S, Marcinkiewicz C, Calvete JJ, Marcinkiewicz MM, McLane MA: Structural requirements of echistatin for the recognition of alpha(v)beta(3) and alpha(5)beta(1) integrins. J Biol Chem 1999;274:37809–37814.

14 Rosenberg SA, Packard BS, Aebersold PM, Solomon D, Topalian SL, Toy ST, Simon P, Lotze MT, Yang JC, Seipp CA: Use of tumor-infiltrating lymphocytes and interleukin-2 in the immunotherapy of patients with metastatic melanoma. A preliminary report. N Engl J Med 1988;319:1676–1680.

15 Beviglia L, Stewart GJ, Niewiarowski S: Effect of four disintegrins on the adhesive and metastatic properties of B16F10 melanoma cells in a murine model. Oncol Res 1995;7:7–20.

16 Morris VL, Schmidt EE, Koop S, MacDonald IC, Grattan M, Khokha R, McLane MA, Niewiarowski S, Chambers AF, Groom AC: Effects of the disintegrin eristostatin on individual steps of hematogenous metastasis. Exp Cell Res 1995;219: 571–578.

17 Scarborough RM, Rose JW, Hsu MA, Phillips DR, Fried VA, Campbell AM, Nannizzi L, Charo IF, Barbourin A: GPIIb-IIIa-specific integrin antagonist from the venom of *Sistrurus m. barbouri.* J Biol Chem 1991;266:9359–9362.

18 Cesano A, Santoli D: Two unique human leukemic T-cell lines endowed with a stable cytotoxic function and a different spectrum of target reactivity analysis and modulation of their lytic mechanisms. In Vitro Cell Dev Biol 1992;28A: 648–656.

19 Nakajima H, Colonna M: 2B4: An NK cell activating receptor with unique specificity and signal transduction mechanism. Hum Immunol 2000;61:39–43.

Substances Affecting Platelet Function

Hæmostasis

Haemostasis 2001;31:183–191

A Novel Snake Venom Disintegrin That Inhibits Human Ovarian Cancer Dissemination and Angiogenesis in an Orthotopic Nude Mouse Model

Francis S. Markland[a] Kate Shieh[a] Qing Zhou[a]
Vladislav Golubkov[a] Russell P. Sherwin[b] Valda Richters[b]
Richard Sposto[c]

Departments of [a]Biochemistry and Molecular Biology, [b]Pathology and [c]Preventive Medicine, and USC/Norris Comprehensive Cancer Center, University of Southern California, Keck School of Medicine, Los Angeles, Calif., USA

Key Words
Ovarian cancer · Adhesion · Invasion · Integrin · Snake venom · Disintegrin · Angiogenesis · Animal model

Abstract
OVCAR-5 is a human epithelial carcinoma cell line of the ovary, established from the ascitic fluid of a patient with progressive ovarian adenocarcinoma without prior cytotoxic treatment. The unique growth pattern of ovarian carcinoma makes it an ideal model for examining the anticancer activity of contortrostatin (CN), a homodimeric disintegrin from southern copperhead venom. FACS analysis revealed that OVCAR-5 is integrin $\alpha v\beta 3$ negative, but $\alpha v\beta 5$ positive. CN effectively blocks the adhesion of OVCAR-5 cells to several extracellular matrix proteins and inhibits tumor cell invasion through an artificial basement membrane. In a xenograft nude mouse model with intraperitoneal introduction of OVCAR-5 cells, intraperitoneal injection of CN was used for therapy. Tumor dissemination in CN-treated versus control groups was studied by gross examination, and antiangiogenic potential was examined by factor VIII immunohistochemistry and image analysis. CN not only significantly inhibited ovarian cancer dissemination in the nude mouse model, but it also dramatically prevented the recruitment of blood vessels to tumors at secondary sites.

Introduction

Disintegrins are small, disulfide-rich, Arg-Gly-Asp (RGD)-containing peptides from snake venom [1] that bind to cell adhesion

KARGER
Fax + 41 61 306 12 34
E-Mail karger@karger.ch
www.karger.com

0301–0147/01/0316–0183$17.50/0

Accessible online at:
www.karger.com/journals/hae

Dr. Francis S. Markland
Cancer Research Laboratory #106, 1303 N. Mission Road
Los Angeles, CA 90033 (USA)
Tel. +1 323 224 7981, Fax +1 323 224 7679
E-Mail markland@usc.edu

receptors (integrins) on the surface of normal and malignant cells [2–4]. Disintegrins have been characterized from the venom of many snakes and were originally distinguished by their ability to inhibit platelet aggregation [3, 5]. We have purified and characterized the disintegrin, contortrostatin (CN), from southern copperhead snake venom [6]. CN gives a single band of approximately 14–15 kD by nonreducing SDS-PAGE and approximately 6–7 kD by reducing SDS-PAGE. Mass spectrometry indicated a mass of 13,505 for intact CN and 6,956 for the reduced and pyridylethylated protein, indicating that it is a homodimer, a structure unique among all other disintegrins reported to date. We have determined the primary structure of CN [7]. In addition to αIIbβ3 on blood platelets, integrins α5β1, αvβ3 and αvβ5 on tumor cells have been shown to be binding sites for CN [8–10].

Integrins are important cell surface receptors that are involved in cell-cell and cell-matrix interactions [11]. All integrins are α/β heterodimeric glycoproteins [12]; different α-subunits combine with several distinct β-subunits resulting in a range of specificity toward distinct extracellular matrix (ECM) proteins [11]. Several classes of integrins recognize the tripeptide RGD sequence present in ECM proteins [13]. Although originally believed to be inert molecules, integrins are known to link the ECM with cytoskeletal proteins and to be involved in bidirectional signaling that alters cellular functions [14].

In previous investigations, we reported that CN binds to integrins and blocks the adhesion of human breast cancer cells (MDA-MB-435) to the ECM [9]. CN also prevents invasion of MDA-MB-435 cells through an artificial basement membrane. Daily local injection of CN into MDA-MB-435 tumor masses in an orthotopic xenograft nude mouse model significantly inhibits tumor growth and reduces the occurrence of pulmonary metastasis. We have identified αvβ3, an important integrin mediating cell motility and tumor invasion, as one of the binding sites for CN on MDA-MB-435 cells. We hypothesized that CN inhibits the progression of breast cancer, at least in part, by an antiangiogenic action, and we have shown in the xenograft nude mouse model that CN dramatically inhibits breast cancer-induced angiogenesis [9]. In the present study, we report that CN has effective antitumor activity in vitro and in vivo in a human ovarian cancer cell line, OVCAR-5. Studies about OVCAR-5 utilized a xenograft model established by Hamilton et al. [15]. Immunohistochemistry was used to demonstrate that CN significantly inhibited angiogenesis in the ovarian cancer animal model.

Materials and Methods

Materials

Venom from *Agkistrodon contortrix contortrix* was purchased from Miami Serpentarium Laboratories (Punta Gorda, Fla., USA). Contortrostatin was purified by four-step high-performance liquid chromatography [6, 8]. Fibronectin (FN), vitronectin (VN) and Matrigel™ were from Becton Dickinson (Bedford, Mass., USA). Monoclonal antibodies (mAbs) PID6 (anti-α5) and P4C10 (anti-β1) were purchased from Gibco Life Technologies (Gaithersburg, Md., USA), and P1F6 (anti-αvβ5) was from Chemicon International (Temecula, Calif., USA). MAb 7E3 (an anti-αIIbβ3 antibody that has a high degree of cross-reactivity to αvβ3) was kindly provided by Dr. Marian Nakada (Centocor, Malvern, Pa., USA). Goat anti-mouse IgG conjugated with fluorescein isothiocyanate (FITC) was purchased from Jackson ImmunoResearch (West Grove, Pa., USA). The monoclonal rabbit anti-human antibody against factor-VIII-related antigen, goat anti-rabbit secondary antibody, and immunochemistry detection kit (HistoMouse-SP kit) were purchased from Zymed Laboratories, Inc. (Gilroy, Calif., USA).

Cell Culture

OVCAR-5 human ovarian carcinoma cell line was generously provided by Dr. Tom Hamilton (Fox Chase Cancer Center, Philadelphia, Pa., USA). OVCAR-5 cells were derived from a patient with advanced epithelial ovarian cancer with no prior cytotoxic therapy [15]. The cells were grown in RPMI with 10% fetal bovine serum and Premix.

Characterization of Integrin Profile by Flow Cytometry

Standard FACS analysis with specific anti-integrin antibodies was employed to detect integrins displayed by OVCAR-5 cells.

Adhesion of OVCAR-5 Cells to CN and ECM Protein

The level of cell adhesion supported by different CN concentrations (0.1, 0.5, 1, 5 μg/ml) was determined experimentally by immobilizing CN onto Immulon-II 96-well microtiter plates (Dynex Technologies, Inc., Chantilly, Va., USA). Excess protein was washed off and unbound sites were blocked with 1% bovine serum albumin. OVCAR-5 cells (100 μl, 10^5 cells/ml) were seeded in the coated wells (1 h, 37 °C). Unbound cells were washed off with three washings of PBS. Cell adhesion was determined by Cell Titer 96™ AQ_{ueous} Nonradioactive Cell Proliferation Assay kit (Promega, Madison, Wisc., USA).

VN (15 μg/ml) and FN (15 μg/ml) were immobilized onto Immulon-II 96-well microtiter plates. Unbound sites were blocked with 1% bovine serum albumin in PBS. OVCAR-5 cells in serum-free medium (SFM) were treated with different concentrations of CN (0, 1, 10, 100, 1,000 n*M*) for 30 min at room temperature, added to each well of microtiter plates and treated as described above. All tests were repeated at least three times in triplicate.

Invasion Assay

Transwells (Costar, Cambridge, Mass., USA) were coated with Matrigel (1:50) in SFM and dried in the hood overnight. Membranes were rehydrated with 500 μl of SFM 1 h before cells were added. Cells were trypsinized, washed in serum-containing medium, and resuspended in SFM. Different concentrations of CN (0, 10, 100, 1000 n*M*) were used to treat OVCAR-5 cells (30 min at room temperature). MAbs, anti-αvβ3 (7E3) and anti-αvβ5 (P1F6), were used separately to treat cells for 30 min before cell suspensions were added to the top chamber; HT1080-conditioned medium (600 μl) was in the bottom chamber. Untreated OVCAR-5 cells, with HT1080-conditioned medium in the bottom chamber, was used as a positive control, and untreated OVCAR-5 cells, with SFM in the bottom chamber, was used as a negative control. Plates were incubated for 8 h to determine the level of cell migration. After incubation, inserts were removed and stained with 0.2% crystal violet aqueous solution in 20% methanol for 30 min. After washing, noninvaded cells from the upper surface of the membrane were removed with cotton swabs. Invaded stained cells were dissolved in 250 μl of 1% SDS and 200 μl were transferred to a 96-well plate. Optical density at 570 nm was measured in a microplate reader.

Animal Model

Five-week-old female nude mice (BALB/c/nu/nu) were purchased from Simonsen Lab (Gilroy, Calif., US). For the humane treatment of experimental animals, a protocol approved by the Institutional Animal Care and Use Committee, University of Southern California, USA, was followed. Animals were kept in a pathogen-free environment, and fed sterilized food and water. OVCAR-5 cells were washed and resuspended in sterile PBS. Cell suspension (10^6 cells/mouse, 0.2 ml/mouse) was injected into the peritoneal cavity.

The animal study used 22 five-week-old female nude mice. Both the control and treated groups contained 11 mice. Implanted tumors were allowed to grow for 1 week prior to the initiation of treatment. The treated group received twice daily injections of 200 μl of CN (0.1 mg/ml, 40 μg/day). The control group received twice daily injections of physiological saline (200 μl/injection). Treatment was continued for a total of 4 weeks, at which time the nude mice were anesthetized by inhalation of Metofane (Pitman-Moore, Mundelein, Ill., USA) in a closed chamber and euthanized for angiogenic study. Photographs were taken to document tumor load. Tumor tissue from the peritoneal cavity was collected, weighed, and stored in 4% formalin at 4 °C for factor VIII immunohistochemistry.

Microvessel Density Analysis

Tumor masses from the in vivo study were fixed in phosphate-buffered 10% formalin, embedded in paraffin, sectioned at 4 μm, and processed for immunohistochemistry. Slides were prepared for staining by treatment with xylene and a graded alcohol series. Specimens were placed in 3% H_2O_2 to quench endogenous peroxidase activity and treated with citrate buffer. Slides were incubated with 100 μl of rabbit anti-factor VIII mAb overnight (37 °C), treated with secondary goat anti-rabbit antibody, and developed by the standard peroxidase procedure.

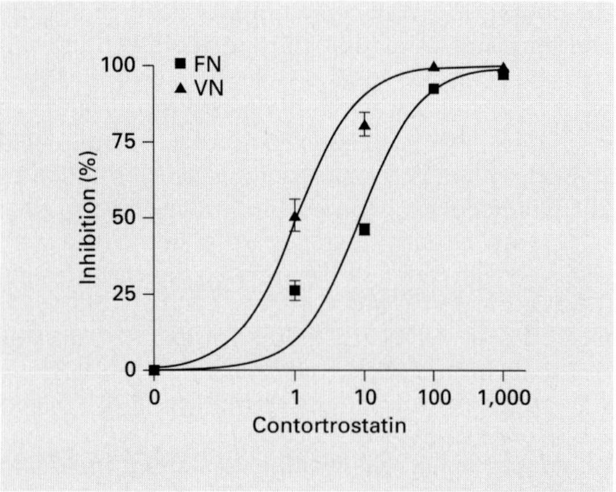

Fig. 1. Inhibition of adhesion of OVCAR-5 cells to immobilized VN and FN. Pretreatment of OVCAR-5 cells with CN for 30 min (0–1,000 n*M* CN) inhibited adhesion of OVCAR-5 cells (100 µl of cells, 10^5 cells/ml) to immobilized ECM proteins following 1 h of incubation. The IC_{50} for CN inhibition of OVCAR-5 adhesion to immobilized VN was 1 n*M*, and to immobilized FN 10 n*M*. CN at similar concentrations did not inhibit adhesion to immobilized laminin or collagen type I.

Image Analysis

'Hot spots' were selected within each sample slide at 100-fold magnification with a Zeiss light microscope and area measurements (pixels) were made using a Leica Q570 image analyzer. 'Hot spots' are defined as areas of high vessel density [9, 16]. To reduce bias in this study, analysis of the slides was done in a blind fashion.

Results

OVCAR-5 Integrin Profile

Primary mAbs against integrins αvβ3 (7E3), αvβ5 (P1F6), α5 (P1D6), and β1 (P4C10) were incubated with OVCAR-5 cells; bound antibodies were detected by secondary FITC-labeled antibody. Flow cytometry results showed that OVCAR-5 cells do not express integrin αvβ3. This is an interesting result because previous studies have pointed to the importance of this integrin for tumor cell migration [17, 18]. However, OVCAR-5 cells do express αvβ5, α5 and β1.

Contortrostatin Inhibits OVCAR-5 Adhesion to Immobilized ECM Proteins

OVCAR-5 cells were pretreated with CN at different concentrations for 30 min before incubation with immobilized VN or FN. Although OVCAR-5 cells adhered to VN and FN, CN inhibited adhesion to both ECM proteins dose dependently: FN IC_{50} = 10 n*M*, VN IC_{50} = 1 n*M* (fig. 1). Although OVCAR-5 adhered to laminin and collagen I, CN was unable to inhibit this adhesion.

Contortrostatin Inhibits OVCAR-5 Invasion

In order to find the optimal invasion time for OVCAR-5 cells across Matrigel™-coated Transwells, several incubation times were tested and a time of 8 h was chosen for analysis. Under these conditions, CN (1,000 n*M*) pretreatment of the OVCAR-5 cells was found to inhibit cell migration by about 60% (fig. 2). 7E3 (mAb to αvβ3) did not inhibit cell migration, but P1F6 (mAb to αvβ5) at 100 n*M* inhibited migration to approximately the same extent as 1,000 n*M* CN.

Animal Model Study

In order to prevent the occurrence of necrosis that appears in the later stages of ovarian tumor growth, a 4-week treatment schedule was planned, beginning 1 week after the initial tumor injection. In the treatment group (11 animals), CN was administered at the level of 20 µg twice per day. Control animals received intraperitoneal saline injections similarly administered. At the conclusion of the study after 4 weeks of treatment, 16 mice (8 in each group) were sacrificed. For various reasons, 3 mice in each group were excluded

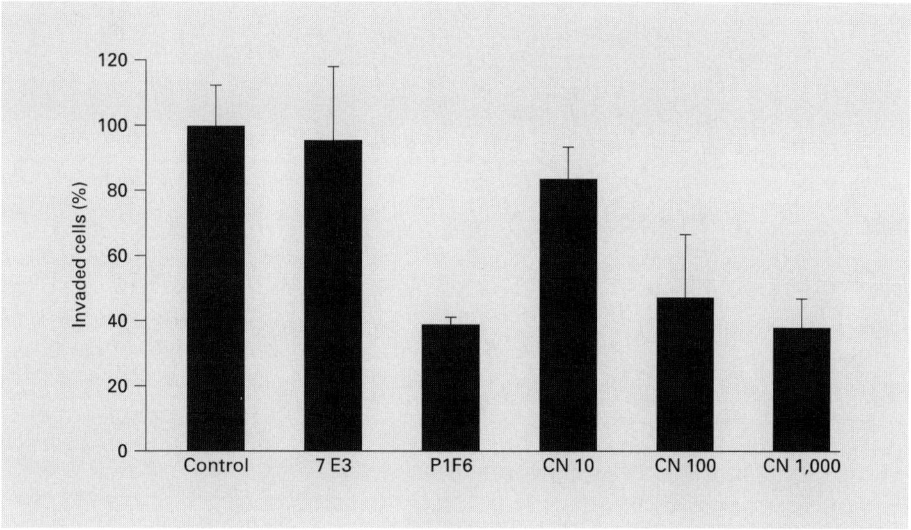

Fig. 2. Inhibition of OVCAR-5 cell invasion through an artificial basement membrane (Matrigel) using a Boyden chamber. Thirty-minute pretreatment of OVCAR-5 cells with varying concentrations of CN (0, 10, 100 and 1,000 n*M*) inhibited migration across a Matrigel barrier, measured after 8 h. HT1080-conditioned medium served as a chemoattractant in the bottom chamber. Untreated OVCAR-5 cells with HT1080-conditioned medium in the bottom chamber was used as a positive control, and untreated OVCAR-5 cells with SFM in the lower chamber was used as a negative control. Pretreatment of OVCAR-5 cells for 30 min with anti-αvβ3 mAb (7E3) did not inhibit OVCAR-5 invasion significantly, whereas an anti-αvβ5 mAb (P1F6) at 100 n*M* inhibited cell invasion by about 60%, approximately the same as that achieved by 1,000 n*M* CN.

from the study. A representative mouse was chosen from each group for photography (fig. 3). It is obvious that CN significantly decreased tumor burden in the CN-treated group. Tissue from 4 representative mice from the remaining mice in each group (control and CN-treated) were formalin fixed for histological examination. Histological sections were stained with antibody against factor-VIII-related antigen as a specific marker of vascular endothelial cells. One of the most remarkable differences between CN-treated and control groups was the complete absence of tumor in various organs of some of the CN-treated animals. Thus, these organs in the animals did not have 'hot spots' of tumor in which microvessel growth could be detected. The paucity of tumor in some organ specimens resulted in a zero microvessel count. Moreover, in the CN-treated specimens in which tumor did grow, the microvessel density was dramatically lower than that found in the untreated group (table 1).

Discussion

Ovarian cancer is usually asymptomatic and not diagnosed until the cancer has spread outside of the ovary. About 70% of patients with epithelial ovarian cancer present as stage III or IV (advanced disease). Because of this, ovarian cancer has been called a silent killer; it is the leading cause of death from gyneco-

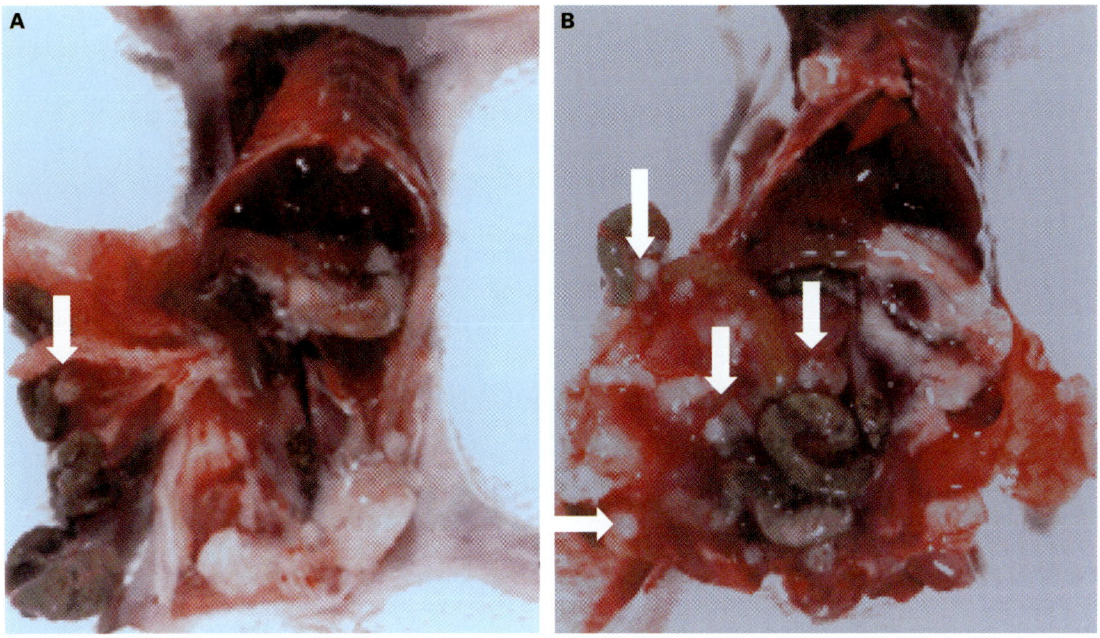

Fig. 3. In vivo model of OVCAR-5 in nude mice. Treated mice (**A**) showed significantly fewer nodules than control mice (**B**). OVCAR-5 (1×10^6 cells/mouse) implanted intraperitoneally. Treatment consisted of 20 μg of CN per mouse twice daily for 4 weeks starting 1 week after tumor injection. Arrows point to tumor nodules.

Table 1. Total factor VIII staining area, by treatment and tumor location, determined by image analysis and area measurement (pixels)

Treatment	Intestine	Ovary	Spleen	Stomach	Total
CN			728.9		728.9
CN			3,128.3		3,128.3
CN			1,771.7	1,042.8	2,814.5
CN		1,779.3			1,779.3
Control (saline)	49,439.5		15,580.8		65,016.3
Control (saline)	2,260.2	25,841.5			28,101.7
Control (saline)				10,369.5	10,369.5
Control (saline)	14,540.6	587.2	25,094.9		40,222.6

A blank indicates that no tumor developed at this site.

logical malignancy resulting in 14,800 deaths per year in the United States [19]. It is apparent that present methods of treatment for advanced-stage ovarian cancer are not effective, and novel therapeutic approaches to control growth and spread are needed for this most devastating cancer in women.

The role of angiogenesis in solid tumor growth and dissemination has long been known [20, 21], and recently the understanding that tumor growth is angiogenesis-dependent [20] has become a dogma. Folkman [22] proposed a two-compartment tumor model composed of vascular endothelial cell and tumor cell compartments. Angiogenesis appears to be an early event in epithelial ovarian cancer [23] and plays an important role in ovarian cancer growth and dissemination [24]. Angiogenesis has been correlated with prognosis in ovarian cancer patients, a higher vessel count shows a trend to worse overall survival in advanced ovarian cancer [23, 25–27]. Antiangiogenic therapy represents a potentially effective therapy for long-term control of ovarian cancer [28].

The importance of VN receptors in angiogenesis is well known. It has been shown that an mAb to integrin αvβ3, as well as a cyclic RGD-containing peptide, perturbed angiogenesis and produced regression of a human cancer growing on the chick embryo chorioallantoic membrane [29]. In previous studies, we have shown that CN effectively blocked adhesion of human umbilical vein endothelial cells to immobilized VN and significantly inhibited invasion of human umbilical vein endothelial cells through a Matrigel barrier [30]. Competitive binding assays and adhesion assays with different integrin antibodies suggested that integrin αvβ3 is a binding site for CN on vascular endothelial cells. We conclude that CN binds to αvβ3, and interferes with the anchorage-dependent survival of the vascular endothelial cells and the mobility of the cells. The consequent suppression of angiogenesis is an important component of the antineoplastic activity of CN [30].

It has also been suggested that integrins play an important role in intraperitoneal dissemination of ovarian cancer [31], as demonstrated by the finding that multiple intraperitoneal injections of an RGD peptide inhibited experimental peritoneal seeding of human ovarian cancer cells in vivo [32]. In vitro studies suggested that β1 and β3 integrins are important in the interaction of ovarian cancer cells with metastatic sites on the mesothelium [33]. We have shown that OVCAR-5 is αvβ3 negative; however, CN is able to block invasion of these cells due to its ability to inhibit the function of αvβ5 [10]. The role of this integrin (αvβ5) in tumor invasion has recently been reported [34]. OVCAR-5 is αvβ5 positive, and it is apparently through this integrin that CN exerts its inhibitory activity on OVCAR-5 adhesion and invasion.

Human ovarian carcinoma cells were found to express integrins at a higher level than normal ovarian cells [35]. Thus, integrins can be targeted in the treatment of ovarian carcinoma. Our in vitro data showing inhibition of ovarian cancer adhesion and invasion have supported the potential for CN as a possible treatment of this disease. However, an animal model was needed to more closely mimic the interactions between cancer cells and the environment. We chose to use the xenograft model for studying ovarian carcinoma that was established by Hamilton et al. [15]; this model allows both tumor cells and treatment to be administered by intraperitoneal injection. The spread of ovarian cancer was observed grossly in our animal model and the level of angiogenesis was quantitated immunohistochemically (factor VIII staining) to determine whether CN had an inhibitory effect on angiogenesis.

In the in vivo studies, image analysis of factor VIII-related antigen detected significantly greater microvessel areas in tissues from control than from CN-treated animals. The findings indicate that the level of angiogenesis was dramatically decreased in CN-treated animals. Moreover, the tumor burden in some of the CN-treated mice was so greatly decreased as to provide no tumor sites for angiogenic study. Thus, the area measurements (pixels) in several of these fields was quantitated to be zero. Gasparini et al. [36] established the term 'hot spot' to describe peripheral areas of the tumor that show greatest microvessel concentration. The use of image analysis for quantitation of microvessel density was found to show less variation than simple manual counts of microvessel density [37]. Although this quantitative method has been used frequently for prognostic purposes [27, 38], ours is one of the first studies to use this technology for quantitating experimental data.

CN is a good candidate for therapeutic use in ovarian cancer because it can inhibit several steps of cancer progression. Unlike other cancers, which progress by the circulatory and lymphatic systems, ovarian cancer is limited to the peritoneum. Intraperitoneal administration of CN facilitates directing its antitumor activity specifically to tumors localized in the peritoneal cavity. Thus, our investigations have demonstrated that it is feasible to design a delivery system for CN that targets potential sites of tumor growth.

Acknowledgements

The research supporting this publication was made possible by funds received from the State of California, Cancer Research Fund, under Interagency Agreement #97-12013 (University of California contract #98-00924V) with the Department of Health Services, Cancer Research Program. The authors would like to acknowledge the contribution of Dr. Steve Swenson, Department of Biochemistry and Molecular Biology, University of Southern California, Keck School of Medicine, in the preparation of the manuscript.

References

1 Huang TF, Ouyang C: Action mechanism of the potent platelet aggregation inhibitor from *Trimeresurus gramineus* snake venom. Thromb Res 1984;33:125–138.
2 Huang TF, Holt JC, Lukasiewicz H, Niewiarowski S: Trigramin. A low molecular weight peptide inhibiting fibrinogen interaction with platelet receptors expressed on glycoprotein IIb-IIIa complex. J Biol Chem 1987; 262:16157–16163.
3 Dennis MS, Henzel WJ, Pitti RM, Lipari MT, Napier MA, Deisher TA, Bunting S, Lazarus RA: Platelet glycoprotein IIb-IIIa protein antagonists from snake venoms: Evidence for a family of platelet-aggregation inhibitors. Proc Natl Acad Sci USA 1990;87:2471–2475.
4 Scarborough RM, Rose JW, Naughton MA, Phillips DR, Nannizzi L, Arfsten A, Campbell AM, Charo IF: Characterization of the integrin specificities isolated from American pit viper venoms. J Biol Chem 1993; 268:1058–1065.
5 McLane MA, Marcinkiewicz C, Vijay-Kumar S, Wierzbicka-Patynowski I, Niewiarowski S: Viper venom disintegrins and related molecules. Proc Soc Exp Biol Med 1998; 219:109–119.
6 Trikha M, Rote WE, Manley PJ, Lucchesi BR, Markland FS: Purification and characterization of platelet aggregation inhibitors from snake venoms. Thromb Res 1994; 73:39–52.
7 Zhou Q, Hu P, Ritter MR, Swenson SD, Argounova S, Epstein AL, Markland FS: Molecular cloning and functional expression of contortrostatin, a homodimeric disintegrin from southern copperhead snake venom. Arch Biochem Biophys 2000; 375:278–288.
8 Trikha M, De Clerck YA, Markland FS: Contortrostatin, a snake venom disintegrin, inhibits beta 1 integrin-mediated human metastatic melanoma cell adhesion and blocks experimental metastasis. Cancer Res 1994;54:4993–4998.

9 Zhou Q, Sherwin RP, Parrish C, Richters V, Groshen SG, Tsao-Wei D, Markland FS: Contortrostatin, a dimeric disintegrin from *Agkistrodon contortrix contortrix*, inhibits breast cancer progression. Breast Cancer Res Treat 2000;61:249–260.
10 Zhou Q, Nakada MT, Brooks PC, Swenson SD, Ritter MR, Argounova S, Arnold C, Markland FS: Contortrostatin, a homodimeric disintegrin, binds to integrin alphavbeta5. Biochem Biophys Res Commun 2000;267:350–355.
11 Hynes RO: Integrins: Versatility, modulation, and signaling in cell adhesion. Cell 1992;69:11–25.
12 Cheresh DA: Structural and biologic properties of integrin-mediated cell adhesion. Clin Lab Med 1992;12: 217–236.
13 Ruoslahti E: Integrins. J Clin Invest 1991;87:1–5.
14 Giancotti FG, Ruoslahti E: Integrin signaling. Science 1999;285:1028–1032.
15 Hamilton TC, Young RC, Ozols RF: Experimental model systems of ovarian cancer: Applications to the design and evaluation of new treatment approaches. Semin Oncol 1984;11:285–298.
16 Gasparini G, Pozza F, Harris AL: Evaluating the potential usefulness of new prognostic and predictive indicators in node-negative breast cancer patients. J Natl Cancer Inst 1993;85:1206–1219.
17 Seftor RE, Seftor EA, Gehlsen KR, Stetler-Stevenson WG, Brown PD, Ruoslahti E, Hendrix MJ: Role of the alpha v beta 3 integrin in human melanoma cell invasion. Proc Natl Acad Sci USA 1992;89:1557–1561.
18 Filardo EJ, Brooks PC, Deming SL, Damsky C, Cheresh DA: Requirement of the NPXY motif in the integrin beta 3 subunit cytoplasmic tail for melanoma cell migration in vitro and in vivo. J Cell Biol 1995;130: 441–450.
19 Parker SL, Tong T, Bolden S, Wingo PA: Cancer statistics, 1996. CA Cancer J Clin 1996;46:5–27.
20 Folkman J: Tumor angiogenesis: Therapeutic implications. N Engl J Med 1971;285:1182–1186.
21 Folkman J: Editorial: Angiogenesis and breast cancer. J Clin Oncol 1994;12:441–443.
22 Folkman J: Tumor angiogenesis and tissue factor. Nat Med 1996;2:167–168.
23 Nakanishi Y, Kodama J, Yoshinouchi M, Tokumo K, Kamimura S, Okuda H, Kudo T: The expression of vascular endothelial growth factor and transforming growth factor-beta associates with angiogenesis in epithelial ovarian cancer. Int J Gynecol Path 1997;16:256–262.
24 Barcz E, Sommer E, Sokolnicka I, Gawrychowski K, Roszkowska-Purska K, Janik P, Skopinska-Rozewska E: The influence of theobromine on angiogenic activity and proangiogenic cytokines production of human ovarian cancer cells. Oncol Rep 1998;5:517–520.
25 Abu-Jawdeh GM, Faix JD, Niloff J, Tognazzi K, Manseau E, Dvorak HF, Brown LF: Strong expression of vascular permeability factor (vascular endothelial growth factor) and its receptors in ovarian borderline and malignant neoplasms. Lab Invest 1996;74:1105–1115.
26 Abulafia O, Triest WE, Sherer DM: Angiogenesis in primary and metastatic epithelial ovarian carcinoma. Am J Obstet Gynecol 1997;177: 541–547.
27 Schoell WM, Pieber D, Reich O, Lahousen M, Janicek M, Guecer F, Winter R: Tumor angiogenesis as a prognostic factor in ovarian carcinoma: Quantification of endothelial immunoreactivity by image analysis. Cancer 1997;80:2257–2262.
28 Pluda JM: Tumor-associated angiogenesis: Mechanisms, clinical implications, and therapeutic strategies. Semin Oncol 1997;24:203–218.
29 Brooks PC, Montgomery AM, Rosenfeld M, Reisfeld RA, Hu T, Klier G, Cheresh DA: Integrin alpha v beta 3 antagonists promote tumor regression by inducing apoptosis of angiogenic blood vessels. Cell 1994; 79:1157–1164.
30 Zhou Q, Nakada MT, Arnold C, Markland FS: Contortrostatin, a dimeric disintegrin from *Agkistrodon contortrix contortrix*, inhibits angiogenesis. Angiogenesis 1999;3:259–269.
31 Buczek-Thomas JA, Chen N, Hasan T: Integrin-mediated adhesion and signalling in ovarian cancer cells. Cellular Signal 1998;10:55–63.
32 Yamamoto K, Murae M, Yasuda M: RGD-containing peptides inhibit experimental peritoneal seeding of human ovarian cancer cells. Nippon Sanka Fujinka Gakkai Zasshi 1991; 43:1687–1692.
33 Gardner MJ, Jones LM, Catterall JB, Turner GA: Expression of cell adhesion molecules on ovarian tumour cell lines and mesothelial cells, in relation to ovarian cancer metastasis. Cancer Lett 1995;91:229–234.
34 Brooks PC, Klemke RL, Schon S, Lewis JM, Schwartz MA, Cheresh DA: Insulin-like growth factor receptor cooperates with integrin alpha v beta 5 to promote tumor cell dissemination in vivo. J Clin Invest 1997;99:1390–1398.
35 Bartolazzi A, Kaczmarek J, Nicolo G, Risso AM, Tarone G, Rossino P, Defilippi P, Castellani P: Localization of the alpha 3 beta 1 integrin in some common epithelial tumors of the ovary and in normal equivalents. Anticancer Res 1993;13:1–11.
36 Gasparini G, Brooks PC, Biganzoli E, Vermeulen PB, Bonoldi E, Dirix LY, Ranieri G, Miceli R, Cheresh DA: Vascular integrin alpha(v)beta3: A new prognostic indicator in breast cancer. Clin Cancer Res 1998;4:2625–2634.
37 Wild R, Ramakrishnan S, Sedgewick J, Griffioen AW: Quantitative assessment of angiogenesis and tumor vessel architecture by computer-assisted digital image analysis: Effects of VEGF-toxin conjugate on tumor microvessel density. Microvasc Res 2000;59:368–376.
38 Charpin C, Devictor B, Bergeret D, Andrac L, Boulat J, Horschowski N, Lavaut MN, Piana L: CD31 quantitative immunocytochemical assays in breast carcinomas. Correlation with current prognostic factors. Am J Clin Pathol 1995;103:443–448.

Substances Affecting Platelet Function

Hæmostasis

Haemostasis 2001;31:192–206

Viper Venom Components Affecting Angiogenesis

T.F. Huang C.H. Yeh W.B. Wu

Pharmacological Institute, College of Medicine, National Taiwan University, Taipei, Taiwan, ROC

Key Words
Angiogenesis · Endothelial cell · Cytokine · Integrin · Disintegrin · Glycoprotein Ib · Metalloproteinase

Abstract
Angiogenesis is a complex process consisting of the proliferation, migration and differentiation of endothelial cells, and it is essential for the progression of malignant solid tumors. In this report, we examine the effects of disintegrins (e.g. rhodostomin and accutin) and glycoprotein-Ib-binding proteins (e.g. agkistin) on each step in angiogenesis using in vitro and in vivo models. Rhodostomin (but not agkistin) inhibited the viability of human umbilical vein endothelial cells (HUVECs) and capillary tube formation of HUVECs. Rhodostomin also inhibited HUVEC migration and invasion evoked by basic fibroblast growth factor (bFGF). In in vivo studies, rhodostomin inhibited bFGF-, but not vascular-endothelial-growth-factor (VEGF)-associated angiogenesis in chick chorioallantoic membrane model, blocked both bFGF and B16F10 melanoma cell-induced neovascularization, and suppressed the growth of subcutaneously inoculated B16F10 solid tumor, leading to a prolonged survival of the C57BL/6 mice treated with rhodostomin. The antiangiogenic effects of rhodostomin on bFGF-treated HUVECs may be mainly related to the blockade of the interaction of endothelial $\alpha_v\beta_3$ and extracellular matrix.

Introduction

Angiogenesis, which consists of a complex process of the development of new capillaries from preexisting vessels, plays a critical role in a variety of normal physiological events, such as female menstrual cycle, bone remodeling and wound healing. In contrast, many pathological conditions are also characterized by undesirable neovascularization, including

KARGER
Fax + 41 61 306 12 34
E-Mail karger@karger.ch
www.karger.com

0301–0147/01/0316–0192$17.50/0

Accessible online at:
www.karger.com/journals/hae

Dr. T.F. Huang, Pharmacological Institute, College of Medicine
National Taiwan University
No. 1, Jen-Ai Road, Sec. 1, Taipei (Taiwan, ROC)
Tel. +886 2 23 123 456 ext. 8332, Fax +886 2 23 417 930
E-Mail turfu@ccms.ntu.edu.tw

diabetic retinopathy, tumor growth, metastasis and various inflammation diseases [1]. Tumor angiogenesis involves several processes, including endothelial proliferation, migration, invasion and tube formation that are regulated by cell adhesion receptors and specific angiogenic growth factors produced by tumor cells and the surrounding stroma [2, 3]. Integrins are a family of heterodimeric transmembrane receptors that mediate cell-cell and cell-extracellular matrix (ECM) interaction [4]. The function of integrins during angiogenesis has been studied extensively with $\alpha_v\beta_3$, which is one of several Arg-Gly-Asp (RGD)-dependent integrins [5]. In vivo, expression of this integrin is strongly upregulated in angiogenic endothelial cells, where it plays a critical role in angiogenesis induced by basic fibroblast growth factor (bFGF), tumor necrosis factor-α (TNF-α) or fragments of human tumors in corneal or chick chorioallantoic membrane (CAM) model [6]. Friedlander et al. [7] demonstrated that angiogenesis depends on cytokines and that vascular cell integrin $\alpha_v\beta_3$ is required when angiogenesis was stimulated by bFGF or TNF-α, whereas angiogenesis, induced by vascular endothelial growth factor (VEGF), transforming growth factor-β or phorbol ester, depends on $\alpha_v\beta_5$.

Glycoprotein (GP) Ib is an adhesive receptor expressed both on platelets and endothelial cells. Binding of plasma von Willebrand factor (vWF) to platelet GPIb plays a critical role in the earliest phase of primary hemostasis, which consists of the anchoring of platelets to injured vessel walls [8]. In contrast to platelets, the functions of endothelial GPIb for vWF is not well established. Several studies have demonstrated the presence of GPIb complex expressed on the endothelial cell surface and the upregulation of endothelial GPIb by cytokines, i.e., TNF and interferon [9–12]. It has been shown that endothelial GPIb participates in HUVEC binding to sickle cell erythrocyte [13] and also promotes platelet bridging to endothelial cells [14]. Recently, Lian et al. [15] assumed that endothelial GPIb may contribute to mediate the process of endothelial cell migration during wound repair in vivo.

Snake venoms contain many unique components that affect cell-matrix interaction. Disintegrins represent a family of low molecular weight, cysteine-rich polypeptides that bind specifically to integrins $\alpha_{IIb}\beta_3$, $\alpha_5\beta_1$ and $\alpha_v\beta_3$ expressed on platelets and other cells, including vascular endothelial cells and some tumor cells, leading to inhibition of platelet aggregation, inhibition of cell adhesion, migration and angiogenesis [16]. On the other hand, some GPIb-binding venom proteins modulate vWF-GPIb interaction, leading to platelet activation, inhibition of platelet aggregation [17] or mediating HUVEC adhesion [18]. Agkistin (also named agkicetin [19]), belonging to C-type lectin polypeptides derived from the viper venom of *Agkistrodon acutus,* specifically inhibited human platelet aggregation and agglutination triggered by ristocetin in the presence of vWF in vitro by acting as platelet GPIb antagonist. Furthermore, agkistin is a potent antithrombotic agent because it pronouncedly blocked platelet plug formation in an experimental model in vivo [20].

In this study, we will describe the effects of disintegrin (e.g. rhodostomin, accutin) and GPIb antagonists (e.g. agkistin) on each distinct step of angiogenesis, including proliferation, migration, invasion and differentiation in HUVECs in vitro and in the chick CAM model in vivo. On the other hand, the antitumor activity of disintegrin was also evaluated by examining its effect on B16F10 melanoma tumor-induced angiogenesis and on tumor mass, and the survival rate in tumor-bearing C57BL/6 mice.

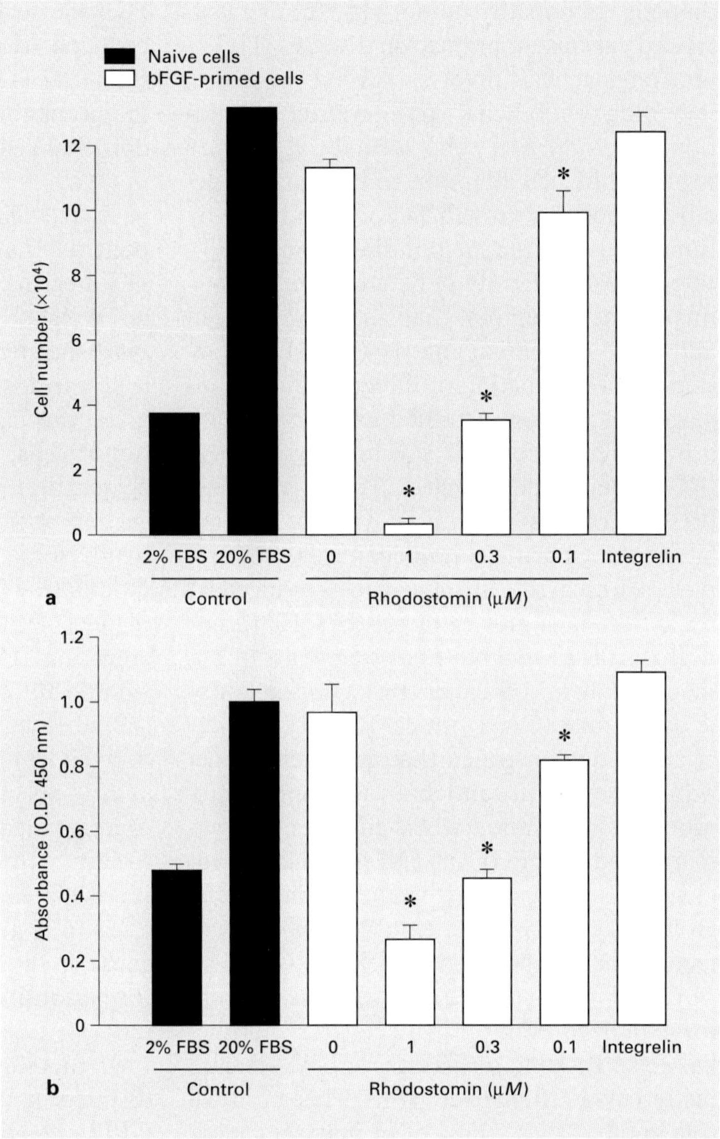

Fig. 1. Effect of rhodostomin on bFGF-induced HUVEC proliferation and on B16F10 melanoma cell viability. **a–c** HUVECs (4×10^4/ml) were seeded on 24-well plates. After attachment, cells were treated with the indicated concentration of test compounds for 48 h in the presence or absence (naive) of bFGF (30 ng/ml). Cell proliferation was measured by cell number counting (**a**), metabolic activity using WST-1 reagent (**b**), and BrdU incorporation (**c**). For control, HUVECs were grown in M199 containing 2% FBS or 20% FBS for 48 h and then assayed for cell proliferation. **d** The inhibition of B16F10 melanoma cell viability was analyzed in the absence (control) or presence of various concentrations of rhodostomin. After the incubation for 24, 48 or 72 h, cells were stained with MTT and lysed. The absorbance was read by an ELISA reader. Data are presented as means ± SEM (n = 3). * $p < 0.01$ vs. bFGF-primed cells in the absence of rhodostomin (**a–c**).

Materials and Methods

Rhodostomin and agkistin were purified from the snake venom of *Agkistrodon rhodostoma* and *A. acutus*. HUVECs were isolated from human umbilical cord veins and identified by positive immunofluorescence staining of human vWF antigen. The B16F10 murine melanoma cells were maintained in DMEM containing 10% FBS. Some in vitro studies were used in this work, including proliferation and viability assays, migration (chemotaxis and haptotaxis) and invasion assays using a Transwell apparatus, and Matrigel tube formation assays. For in vivo studies, the chick CAM and Matrigel plug angiogenesis assays were performed. Flow cytometry studies were used for the determination of endothelial integrins and GPIb expression in either naive or cytokine-primed HUVECs. Finally, the effect of rhodostomin on B16F10 tumor model in C57BL/6 mice were performed to examine the antitumor activity of disintegrin.

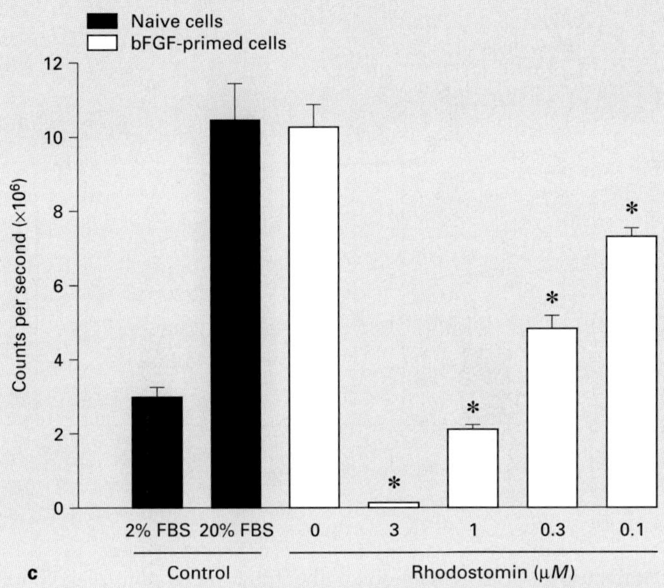

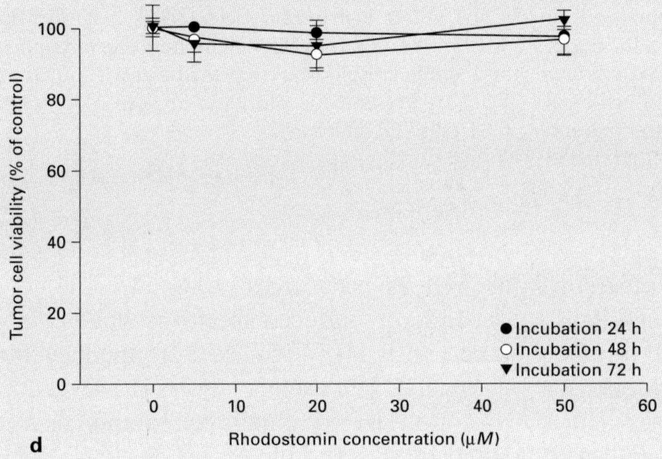

Results

Effects of Disintegrin and GPIb Antagonist on HUVEC Proliferation

The assays for cell proliferation were performed by counting cells with a hemocytometer and by determining the cell metabolic activity with tetrazolium salt (i.e., WST-1). Furthermore, DNA synthesis was measured by BrdU incorporation and used as a proliferation marker. As shown in figure 1, endothelial cells incubated with bFGF (30 ng/ml) in 2% FBS for 48 h, as well as those grown in medium-containing 20% FBS, showed a 2.5- to 3.5-fold increase in three proliferation assays compared with control cells (grown in

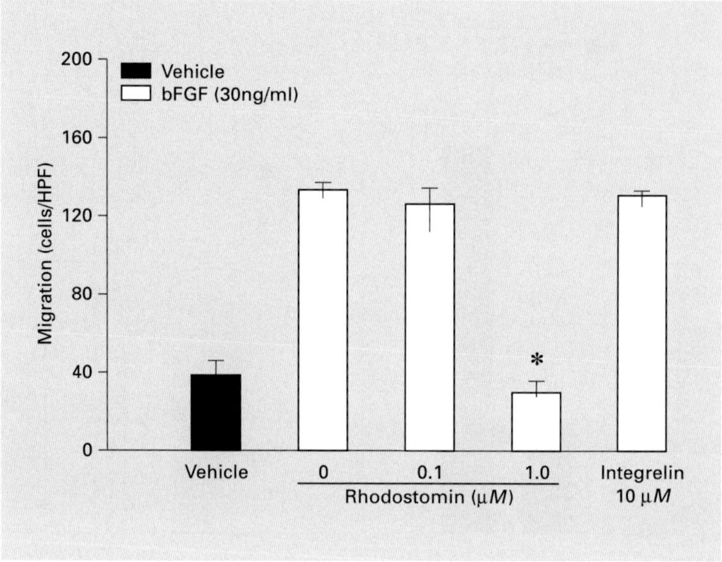

Fig. 2. Rhodostomin inhibited HUVEC migration toward bFGF. HUVECs (5 × 10^4/ml) were treated with the indicated concentration of test compounds or PBS (vehicle) for 30 min and then placed in the upper chamber of Transwell containing a gelatin-coated filter membrane. Chemotaxis was induced by adding bFGF (30 ng/ml) to the lower chamber for 16 h. After fixation and removing nonmigrated cells, cells that migrated to the underside of the filter membrane were photographed and quantified by phase-contrast light microscope under high power field (HPF, magnification: ×100). All experiments were conducted in triplicate and similar results were repeated at least four times. Data are presented as means ± SEM (n = 4–5). * $p < 0.01$ vs. bFGF alone.

2% FBS). Rhodostomin (but not integrelin, 10 μ*M*) inhibited bFGF-induced HUVEC proliferation as examined by cell number counting (fig. 1a), cell metabolic activity (fig. 1b) and BrdU-labeling index (fig. 1c), in a dose-dependent manner with IC_{50} values of 0.24, 0.36 and 0.27 μ*M*, respectively. Moreover, at a higher concentration (i.e., 1 μ*M*), rhodostomin significantly inhibited HUVEC proliferation, compared with that of resting cells (control in 2% FBS, $p < 0.01$ for fig. 1a, b; $p < 0.05$ for fig. 1c). However, the viability of B16F10 melanoma tumor cells was not significantly decreased upon incubation with rhodostomin (5–50 μ*M*) for 24–72 h, as examined by an MTT assay (fig. 1d). Furthermore, rhodostomin (0.1–0.6 μ*M*, data not shown), as well as accutin (0.1–0.8 μ*M* [21]) dose dependently induced apoptosis when coincubated with HUVECs, as examined by flow cytometric cell cycle analysis. However, rhodostomin did not impair cell vitality, as measured by trypan blue exclusion, if HUVECs were incubated with rhodostomin in suspension (>90% exclusion, data not shown). These results indicate that the antiproliferative activity of rhodostomin is specifically restricted to anchorage-dependent cells (e.g. HUVECs).

Effects of Disintegrin and GPIb Antagonist on HUVEC Migration and Invasion

Angiogenesis is highly dependent on endothelial cell motility and invasion. Therefore, we examined the effects of rhodostomin on

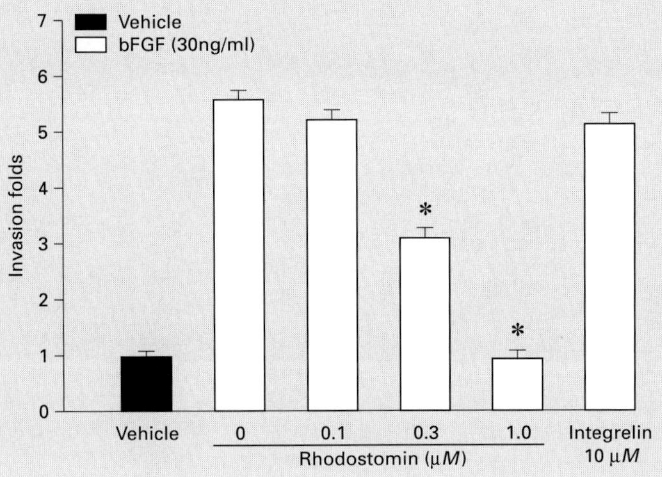

Fig. 3. Rhodostomin inhibited bFGF-induced HUVEC invasion through Matrigel. Invasion assay was performed by using Transwell containing a Matrigel-coated filter membrane with or without addition of bFGF (30 ng/ml) in the lower chamber. HUVECs (5×10^4/ml) were placed in the upper chamber in the presence or absence (vehicle) of test compounds. After a 16-hour incubation, cells that invaded to the underside of the filter membrane were quantified. All experiments were conducted in duplicate and repeated at least four times. Data are presented as means ± SEM (n = 4). * $p < 0.01$ vs. bFGF alone.

HUVEC migration toward bFGF. Chemotaxic migration was performed by Transwell containing insert filter coated with 0.25% gelatin. Adhesion of HUVECs toward gelatin was not affected by rhodostomin, even at a high dose of 2 μ*M* (data not shown). As shown in figure 2, few control cells (without bFGF) were found in lower chamber, whereas a 3.4-fold increment of the migrated cells could be detected if bFGF was present as a chemotaxic agent. After a 16-hour incubation, rhodostomin (1 μ*M*) showed a significant inhibition of HUVEC migration (77.6%), but 0.1 μ*M* rhodostomin and 10 μ*M* integrelin exerted little effect (5.1and 2.1% inhibition, respectively). For the invasion assay, HUVECs were stimulated by bFGF and penetrated through a uniform layer of Matrigel, which consists of collagen type IV, heparan sulfate proteoglycan, entactin and laminin. bFGF (30 ng/ml) induced a 5.5-fold increase in cell invasion through Matrigel, as compared with control (in the absence of bFGF). Moreover, rhodostomin inhibited HUVEC invasion evoked by bFGF in a dose-dependent manner with an IC_{50} of 0.37 μ*M* (fig. 3).

To establish the role of GPIb in vWF-induced HUVEC migration, haptotaxis assays were performed that allowed HUVECs to migrate to immobilized vWF (10 μg/ml), which was coated on the reverse side of the filters. Either rhodostomin (0.25 μ*M*) or agkistin (1 μ*M*) partially blocked the haptotactic migration response of HUVECs toward immobilized vWF (43 and 37%, respectively). Moreover, pretreatment of both rhodostomin and agkistin significantly inhibited this migration behavior (87%), but 6F1 raised against integrin $\alpha_2\beta_1$ (20 μg/ml) produced less than 10% inhibition. These results suggest that both endothelial integrin $\alpha_v\beta_3$ and GPIb receptors, but not $\alpha_2\beta_1$, contribute to HUVEC migration toward immobilized vWF, and that agkistin is an effective inhibitor on this HUVEC haptotactic migration.

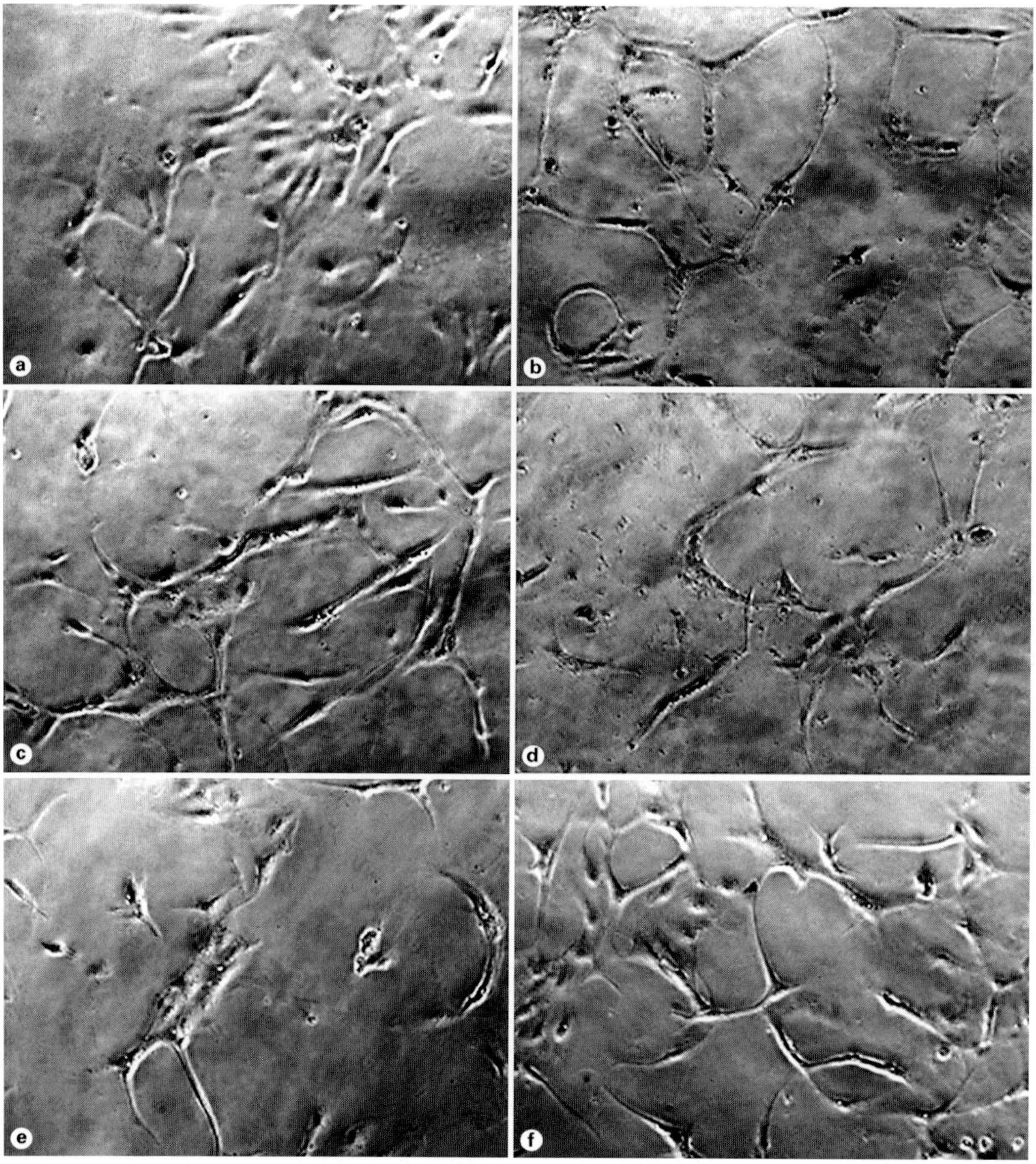

Fig. 4. Rhodostomin inhibited bFGF-induced HUVEC tube formation in diluted Matrigel. HUVECs (2×10^5/well) were treated with or without the indicated concentration of test compounds and were then plated on diluted Matrigel (1 mg/ml) in the absence or presence of bFGF for 24 h. After washing and fixation, cells were observed under the microscope at ×40 magnification and photographed. This is a representative experiment of four similar ones. Data are presented as means ± SEM (n = 4). * $p < 0.01$ vs. bFGF alone. **a** Vehicle. **b** bFGF 30 ng/ml. **c** Rhodostomin 0.1 μM. **d** Rhodostomin 0.3 μM. **e** Rhodostomin 1 μM. **f** Integrelin 10 μM.

Effects of Disintegrin and GPIb Antagonist on Angiogenesis in vitro and in vivo

In vitro angiogenesis assays were carried out by using bFGF-treated HUVECs in a three-dimensional gel consisting of diluted Matrigel (1 mg/ml). As shown in figure 4, if HUVECs were plated on diluted Matrigel without the addition of exogenous growth factor, they showed only a few spontaneous tube formations and most of them were still in a highly proliferating state with a cobblestone shape (fig. 4a). However, when HUVECs were plated on the diluted Matrigel with addition of bFGF (30 ng/ml), cells displayed high motility, cell-cell communication, aligned and formed an anastomosing capillary-like network within 24 h (fig. 4b). Rhodostomin significantly decreased the total tube length in each of four randomly chosen fields in a dose-dependent manner (fig. 4c–e), in contrast to the minimal effect of integrelin (10 μM, fig. 4f). Furthermore, in vivo chick CAM assays were performed under the induction of angiogenesis by both bFGF and VEGF (200 ng/embryo) for 48 h with or without the addition of rhodostomin (fig. 5). As a control (PBS), formation of new blood vessels under disks on chick embryo CAMs could be observed (fig. 5a). Either bFGF or VEGF induced a pronounced angiogenic response in this model (fig. 5b and e, respectively). Rhodostomin, at 2 μM, displayed a significant inhibition on bFGF-induced angiogenesis (fig. 5c), as compared to that of bFGF alone, but it slightly affected VEGF-associated angiogenesis (fig. 5f). Similarly, the control peptide integrelin (2 μM) exerted little effect on bFGF-induced angiogenic response in the same model (fig. 5d).

Assays of tube formation with agkistin were carried out by using cytokine-primed (TNF-α plus IFN-γ, for 48 h) or naive (2% FBS) HUVECs in a thick gel consisting of diluted Matrigel (1 mg/ml). If HUVECs were grown on diluted Matrigel without the addition of exogenous vWF, they showed little tube formation. On the other hand, when cytokine-primed HUVECs were plated on diluted Matrigel with the addition of vWF (10 μg/ml), they aligned with one another and formed an anatomotic capillary-like network within 24 h. Agkistin (1 μM) significantly inhibited this tube formation of cytokine-primed HUVECs on diluted Matrigel in the presence of exogenous vWF (30 ng/ml), but not in the absence of vWF. Moreover, neither agkistin (5 μM) nor AP1 (40 μg/ml), an mAb raised against GPIb, affected the spontaneous tube formation of HUVECs induced by a higher concentration of Matrigel (4 mg/ml), even in the presence of exogenous vWF.

Elucidation of Targeted Receptors of Rhodostomin on Growth Factor-Activated Endothelial Cells

To identify which integrins expressed on HUVEC interact with rhodostomin, we first examined whether rhodostomin displaces the binding of anti-integrin mAb toward HUVEC by flow cytometry. As shown in figure 6, the expression of integrin $\alpha_v\beta_3$ or $\alpha_v\beta_5$ on HUVECs in naive cells was probed by 7E3 or P1F6, respectively, and presented as the mean fluorescence intensity. Exposure to either bFGF or VEGF did not significantly affect the expression of $\alpha_v\beta_3$, but a 2-fold increment of $\alpha_v\beta_5$ expression was detected in VEGF-primed cells. Moreover, a conformational change of $\alpha_v\beta_3$ following bFGF or VEGF exposure was observed, as probed by an anti-LIBS1 mAb raised against ligand-induced binding sites of β_3 integrins [22]. Rhodostomin inhibited 7E3 binding toward both naive and bFGF-primed HUVECs in a dose-dependent manner with IC_{50} values of 2.95 and 1.88 μM, respectively, but had little effects on P1F6-binding reaction under these conditions.

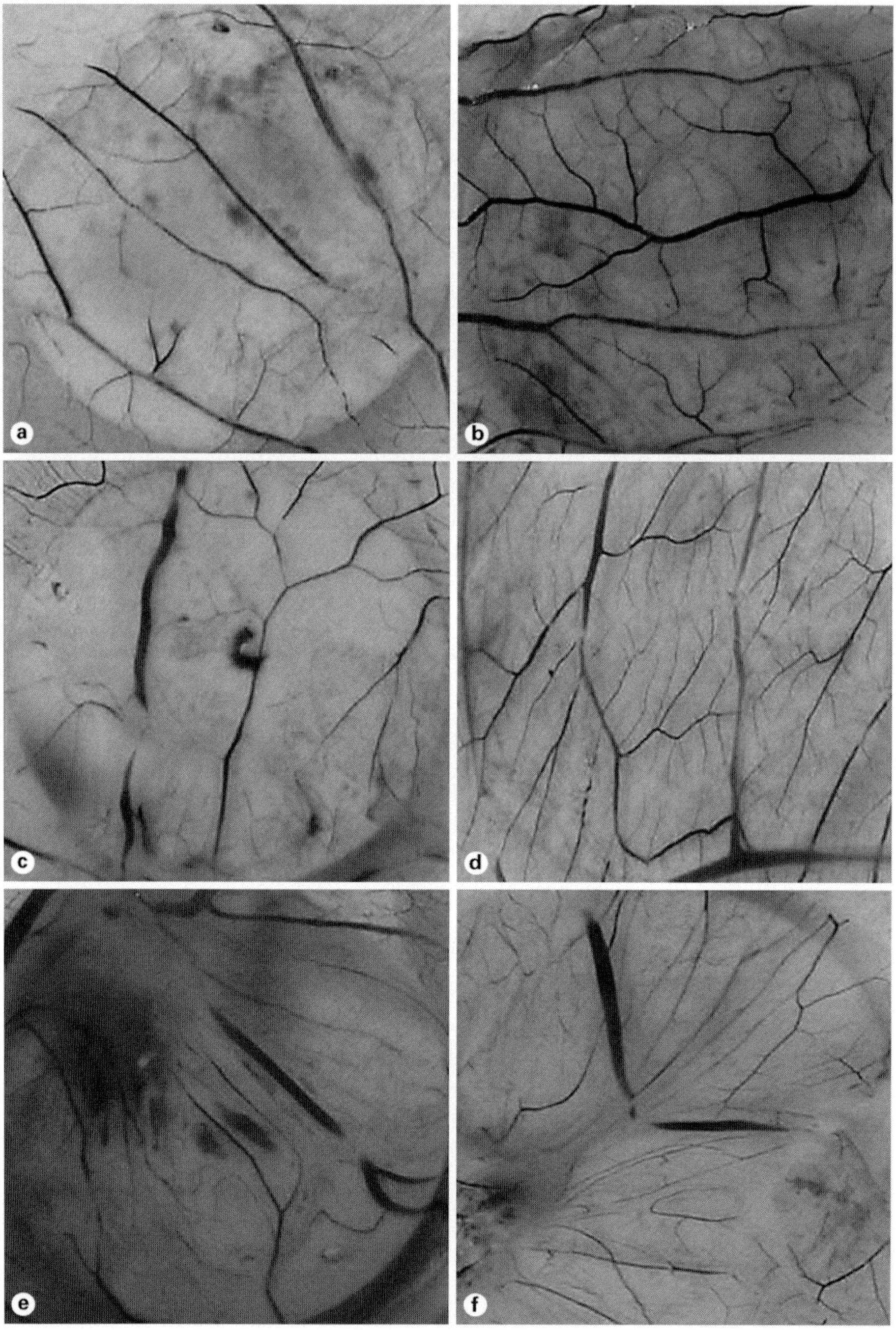

5

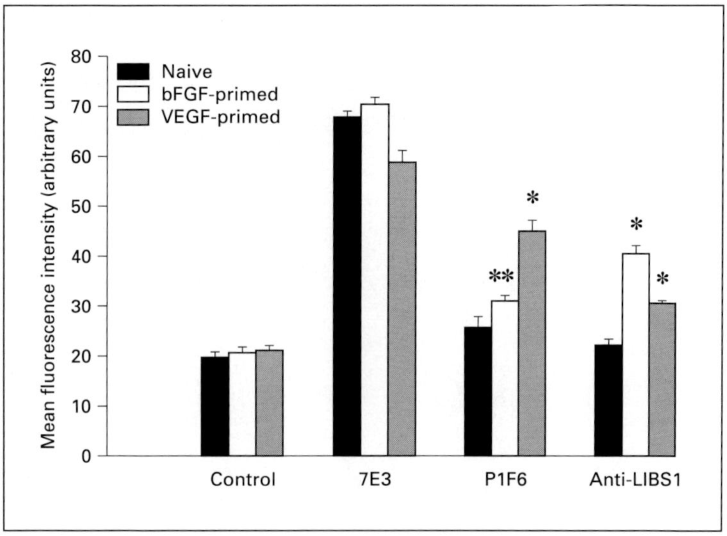

Fig. 6. Effect of growth factors on HUVEC integrin expression and their binding affinity. HUVECs exposed to the standard medium (naive), bFGF or VEGF (both in 30 ng/ml, for 48 h) were incubated with primary probe 7E3, P1F6 (both in 20 μg/ml) or anti-LIBS1 mAb (0.17 μM) and analyzed by flow cytometry. Nonspecific binding was performed by incubating HUVECs with nonimmune IgG (control, 1:50 dilution). Results are represented as mean fluorescence intensity ± SEM (n = 4). * $p < 0.01$, ** $p < 0.05$ vs. naive.

Effect of Rhodostomin on bFGF- and Tumor-Associated Angiogenesis in Matrigel Plug Assays

Matrigel serves as a vehicle for the slow release of angiogenic factor since it exists as a liquid at 4°C, but forms a gel in vivo. The formed gel, after subcutaneous injection of Matrigel alone, was readily distinguished from surrounding tissue and produced little local reaction or angiogenic response [23]. In the presence of bFGF (500 ng/mouse, fig. 7a) or B16F10 murine melanoma cells (3 × 10^6/mouse, fig. 7c), Matrigel plug induced intense vascularization in vivo 14 days after implantation, as examined by immunohistological staining for vWF, an endothelial-specific antigen. Rhodostomin (50 μM) significantly inhibited both bFGF (fig. 7b) and B16F10 cell-induced angiogenesis (fig. 7d) in the murine experimental model.

Fig. 5. Rhodostomin inhibited bFGF-induced angiogenesis in chick CAM assays. Filter disks soaked in buffer (**a**, vehicle), bFGF or VEGF alone (**b**, **e**, respectively, both in 200 ng/embryo) or growth factor plus the indicated compounds in a total volume of 20 μl were applied on 10-day-old CAMs. After a 48-hour incubation, each CAM under disk was resected, fixed and photographed. This is a representative experiment of three similar ones. **a** PBS. **b** bFGF only. **c** bFGF + rhodostomin 2 μM. **d** bFGF + integrelin 2 μM. **e** VEGF only. **f** VEGF + rhodostomin 2 μM.

Effect of Rhodostomin on Subcutaneous B16F10 Tumor Growth and Survival Rate of Tumor-Bearing Mice

To examine the antitumor activity of disintegrin, we examined the effect of rhodostomin on the B16F10 tumor model in C57BL/6 mice. As shown in figure 8, the B16F10 tumor-bearing mice exhibited aggressive tumor growth kinetics, as observed in the control group (PBS treatment). After coadministration of B16F10 cells with rhodostomin (5 and 50 μM), both tumor growth rate and tumor volumes were significantly reduced, compared with those coinjected with B16F10 cells with PBS or integrelin (50 μM). It was necessary to sacrifice the animals receiving coinjection because the tumors were large (i.e.,

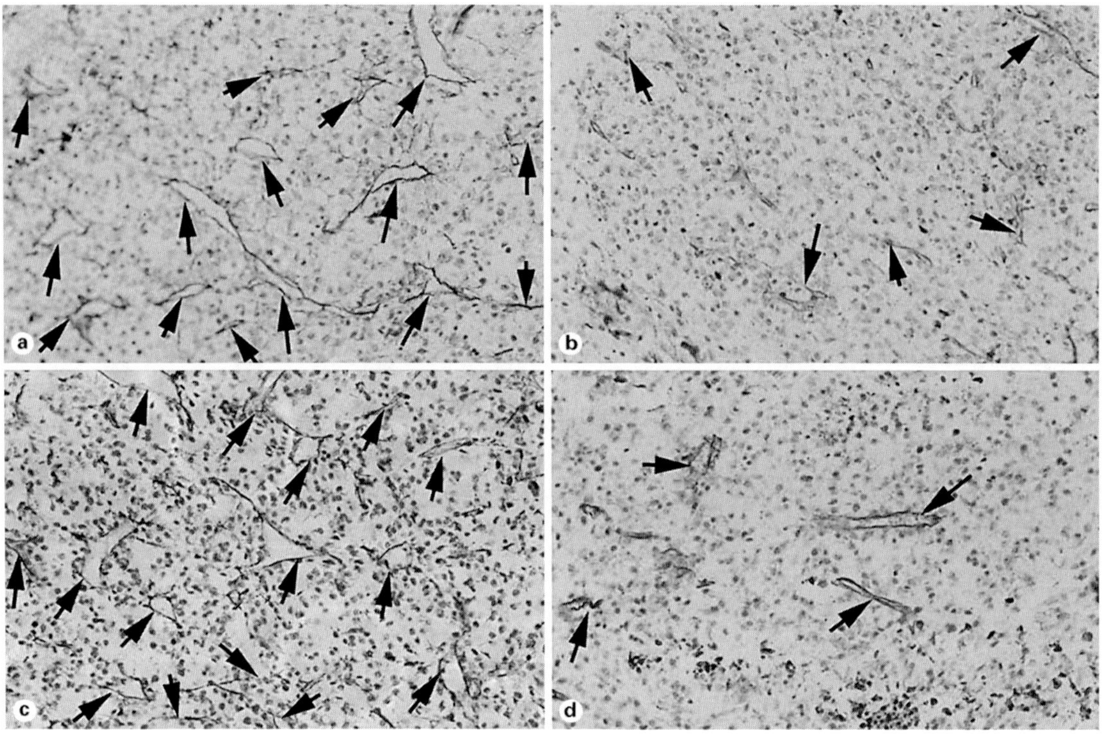

Fig. 7. Effect of rhodostomin on bFGF or B16F10 tumor melanoma cell-induced angiogenesis in Matrigel plug murine model. An aliquot of Matrigel (0.5 ml) supplemented with bFGF (500 ng, **a**, **b**) or B16F10 melanoma cells (3×10^6, **c**, **d**) in the absence (**a**, **c**) or presence of rhodostomin (50 μ*M*, **b**, **d**) was subcutaneously injected into the dorsal region of B57CL/6 mice. Analyses of the vasculature of respective Matrigel from the mice 14 days after the implantation were performed by immunohistochemical staining for vWFantigen that were indicated by arrowheads in the figure. Magnification: ×180.

4,806.56 ± 614.58 and 4,530.7 ± 719.56 mm^3, respectively). In contrast, cells coinjected with rhodostomin (5 and 50 μ*M*) grew significantly smaller tumors in mice (i.e., 2,908.42 ± 445.39 and 485.29 ± 134.83 mm^3, respectively), compared with those arising from the control group ($p < 0.05$ and $p < 0.01$, respectively). Moreover, the survival rates of B16F10 tumor-bearing animals receiving rhodostomin (5 and 50 μ*M*) also increased, compared with those mice receiving vehicle (PBS) or integrelin (50 μ*M*, data not shown). Thus, rhodostomin appears to be effective in the treatment of mice suffering from solid tumor progression of B16F10 melanoma, not only by reducing the tumor growth rate but also by reducing the consequent mortality.

Discussion

The role of integrin $\alpha_v\beta_3$ in mediating crucial angiogenic processes has led to the conclusion that this integrin facilitates the survival of stimulated endothelial cells [24]. We

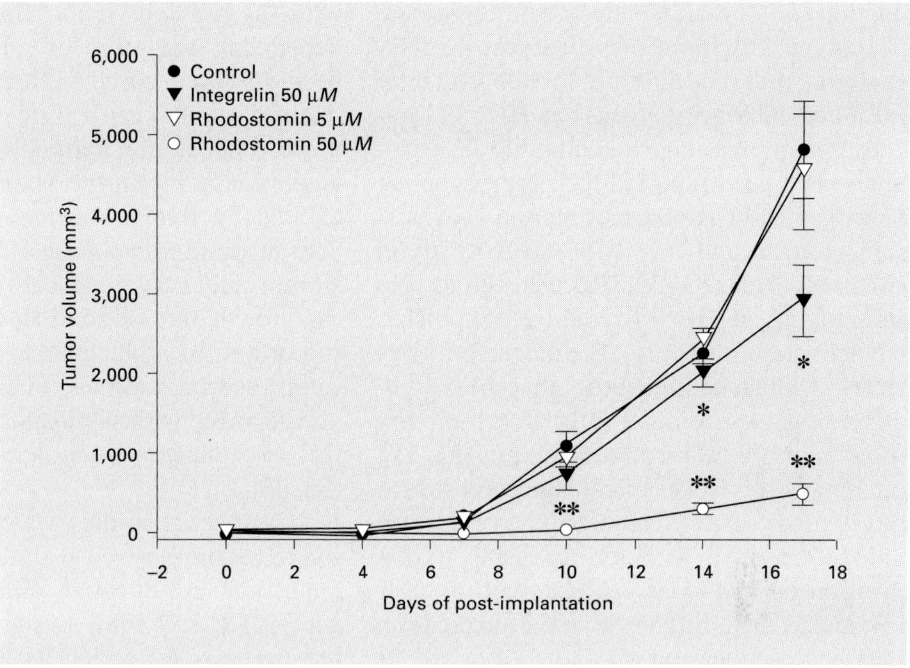

Fig. 8. Effect of rhodostomin on subcutaneous B16F10 tumor growth. B16F10 melanoma cells (3×10^6/mouse) were subcutaneously injected in the dorsal region of B57CL/6 mice, in the absence (control: ●) or presence of rhodostomin: 5 (▽) or 50 μ*M* (○) or integrelin (50 μ*M*, ▼) in a final volume of 0.3 ml. Tumor volumes were monitored by periodic caliper measurements and presented as means ± SEM, n = 9–11. * $p < 0.05$; ** $p < 0.01$, compared with control group on the same day.

showed here that rhodostomin inhibited bFGF-induced human endothelial cell proliferation at submicromolar concentrations, as examined by using various methods, including determination of total cell number, metabolic activity and BrdU labeling index. However, rhodostomin did not decrease B16F10 melanoma cell viability, even up to 50 μ*M*, as determined by an MTT assay. Several lines of evidence indicate that endothelial cell apoptosis plays a critical regulatory role in angiogenesis [25]. Moreover, primary cultured endothelial cells are anchorage dependent. The selective antiproliferating effect of rhodostomin on primary cultured endothelial cells is related to the induction of apoptosis, as evidenced by flow cytometric cell cycle analysis, but it did not cause any cytotoxic effect, as revealed by trypan blue exclusion. This is consistent with the previous observations that RGD peptides and disintegrin (i.e., accutin) induce an apoptotic effect toward HUVECs [26, 21], but not toward melanoma tumor cells [27, 28].

A close interaction exists among cell motility, integrin receptors, and soluble angiogenic factors during each step of the angiogenic processes. Upon angiogenesis, integrin $\alpha_v\beta_3$ and the receptor for bFGF cooperate to promote signaling events, thereby facilitating the induction and/or maintenance of the angiogenic phenotype [29]. In the present study, we demonstrated that rhodostomin dose dependently exhibited inhibitory effects on HUVEC migration and invasion evoked by bFGF. Cytokines have been reported to trigger signaling pathways leading to the modulation of expression and/or ligand affinity of integrin $\alpha_v\beta_3$ on endothelial cells [5, 30]. Recent studies also showed that integrin $\alpha_v\beta_3$ could exist in different activation states [31]. Unlike the previous studies, which demonstrate that bFGF increases the expression of integrin $\alpha_v\beta_3$ in cultured microvascular endothelial cells [32, 33], our data in this work indicate that exposure to bFGF promotes the ligand affinity of $\alpha_v\beta_3$ for anti-LIBS1 mAb without increasing its expression on HUVECs, as examined by flow cytometry. The difference in $\alpha_v\beta_3$ expression may be dependent on the tissue of origin, as well as the maturity of endothelial cells.

Several models have been used to study the role of cell adhesion molecules in angiogenesis, including the Matrigel tube formation model, chick CAM assay and the murine Matrigel plug model [34]. Rhodostomin is an effective inhibitor of bFGF-associated angiogenesis in vitro and in vivo, as examined by multiple models described above. However, rhodostomin did not interfere with VEGF-associated formation of blood vessels in CAM model. Regarding the binding site of rhodostomin to HUVEC, rhodostomin dose dependently inhibited 7E3 (i.e., an anti-$\alpha_{IIb}\beta_3$ and $\alpha_v\beta_3$ mAb), but not P1F6 (i.e., an anti-$\alpha_v\beta_5$ mAb) binding to both naive and bFGF-primed HUVECs. Therefore, the action of rhodostomin in suppressing bFGF-elicited angiogenesis is due to the specific blockade of $\alpha_v\beta_3$ ligation rather than $\alpha_v\beta_5$ ligation of the activated endothelial cells. In addition to the modulation of $\alpha_v\beta_3$ and $\alpha_v\beta_5$ integrins by bFGF and VEGF, respectively, as examined by flow cytometry, these results are consistent with the previous report that two cytokine-dependent pathways of angiogenesis have been proposed, based on their dependence on distinct vascular cell α_v integrins [7]. Furthermore, rhodostomin markedly inhibited tumor angiogenesis in murine Matrigel plug models elicited by B16F10 melanoma tumor cells. The antitumor angiogenesis effects of rhodostomin may play a critical role in inhibiting the growth rate of solid B16F10 melanoma tumor and in prolonging survival kinetics of tumor-bearing animals. Similar results were also reported with salmosin and contortrostatin, two snake venom-derived disintegrins [35, 36].

In our serial studies, we showed that both snake venom-derived disintegrins and GPIb antagonist are effective inhibitors of angiogenesis [21, 37, 38]. Disintegrins inhibited angiogenesis by acting as an integrin $\alpha_v\beta_3$ antagonist and by inducing apoptosis [21], while GPIb antagonist disrupted interaction of vWF-endothelial GPIb and inhibited angiogenesis [37], although both of them suppress angiogenesis by blocking endothelial adhesive receptors ligation with ECMs. However, there are several differences between these two families of antiangiogenic snake proteins. First, disintegrins exerted a potent inhibition on growth factor-induced HUVEC proliferation [38], whereas GPIb antagonist did not. Second, disintegrins significantly inhibited Matrigel-induced tube formation of HUVECs in the absence of exogenous substances [21], but GPIb antagonist showed little effect. Third, disintegrins inhibited angiogenesis in vivo without affecting preexisting blood vessel in a chick CAM model [21]; however, GPIb antagonist significantly disrupted pre-existing blood vessels, as evidenced by the disappearance of large vessels in the same model [37]. Thus, the actions of antiangiogenic activity of GPIb antagonist are quite different from those of disintegrins. These observations also suggest that GPIb and integrin $\alpha_v\beta_3$ may play different roles in angiogenesis. A combination of rhodostomin and agkistin exhibited an additive or synergistic effect by suppressing HUVEC migration, raising a hypothesis that the blockade of endothelial GPIb and $\alpha_v\beta_3$ may have additive or synergistic effects on angiogenesis. However, this needs to be confirmed.

The results of this study demonstrate that exposure of HUVECs to TNF-α plus IFN-γ upregulated both the expression of functional GPIb protein and of mRNA in these cells, thereby increasing a GPIb-dependent endothelial cell adhesion to immobilized agkistin. The suppression mechanisms of agkistin in cytokine-induced multiple angiogenic process mainly result from its inhibitory effect on endothelial cell migration and tube formation, but not on proliferation. The elucidation of the antiangiogenic mechanism of agkistin based on structure-activity relationship studies may be helpful in designing a new therapeutic strategy in angiogenesis-related diseases.

Acknowledgments

We appreciate very much the generous supply of mAbs, 7E3 from Dr. B.S. Coller (Mount Sinai Hospital, New York, N.Y., USA), anti-LIBS1 and AP1 from Dr. M.H. Ginsberg and Dr. R. Montgomery, respectively (Research Institute of Scripps, La Jolla, Calif., USA). This work was supported by NSC-90-2320-B002-084.

References

1 Folkman J: Seminars in Medicine of the Beth Israel Hospital, Boston. Clinical applications of research on angiogenesis. N Engl J Med 1995; 333:1757–1763.
2 Folkman J, Shing Y: Angiogenesis. J Biol Chem 1992;267:10931–10934.
3 Sastry SK, Horwitz AF: Adhesion-growth factor interactions during differentiation: An integrated biological response. Dev Biol 1996;180: 455–467.
4 Hynes RO: Integrins: Versatility, modulation, and signaling in cell adhesion. Cell 1992;69:11–25.
5 Byzova TV, Rabbani R, D'Souza SE, Plow EF: Role of integrin $\alpha_v\beta_3$ in vascular biology. Thromb Haemost 1998;80:726–734.
6 Brooks PC, Clark RAF, Cheresh DA: Requirement of vascular integrin $\alpha_v\beta_3$ for angiogenesis. Science 1994;264:569–571.
7 Friedlander M, Brooks PC, Shaffer RW, Kincaid CM, Varner JA, Cheresh DA: Definition of two angiogenic pathways by distinct α_v integrins. Science 1995;270:1500–1502.
8 Sadler JE: von Willebrand factor. J Biol Chem 1991;266:22777–22780.
9 Wu G, Essex DW, Meloni FJ, et al: Human endothelial cells in culture and in vivo express on their surface all four components of the glycoprotein Ib/IX/V complex. Blood 1997; 90:2660–2669.
10 Beacham DA, Tran LP, Shapiro SS: Cytokine treatment of endothelial cells increases glycoprotein Ibα-dependent adhesion to von Willebrand factor. Blood 1997;89:4071–4077.
11 Konkle BA, Shapiro SS, Asch AS, Nachman RL: Cytokine-enhanced expression of glycoprotein Ibα in human endothelium. J Biol Chem 1990;265:19833–19838.
12 Rajagopalan V, Essex DW, Shapiro SS, Konkle BA: Tumor necrosis factor-α modulation of glycoprotein Ibα expression in human endothelial and erythroleukemia cells. Blood 1992;80:153–161.
13 Wick TM, Moake JL, Udden MM, McIntire LV: Unusually large von Willebrand factor multimers preferentially promote young sickle and nonsickle erythrocyte adhesion to endothelial cells. Am J Hematol 1993;42:284–292.
14 Bombeli T, Schwatz BR, Harian JM: Adhesion of activated platelets to endothelial cells: Evidence for a GPIIbIIIa-dependent bridging mechanism and novel roles for endothelial intercellular adhesion molecule I (ICAM-1), $\alpha_v\beta_3$ integrin and GPIbα. J Exp Med 1998;187:329–339.
15 Lian J, Guoping C, Shapiro SS, Tran LP, Beacham DA: Glycoprotein Ib_α can mediate endothelial cell migration on von Willebrand factor-containing substrata. Exp Cell Res 1999;252:114–122.
16 Gould RJ, Polokoff MA, Friedman PA, Huang TF, Holt JC, Cook JJ, Niewiarowski S: Disintegrins: A family of integrin inhibitory proteins from viper venoms. Proc Soc Exp Biol Med 1990;195:168–171.
17 Fujimura Y, Kawasaki T, Titani K: Snake venom proteins modulating the interaction between von Willebrand factor and platelet glycoprotein Ib. Thromb Haemost 1996;76: 633–639.

18 Tan L, Kowalska MA, Romo GM, Lopez JA, Darzynkiewicz Z, Niewiarowski S: Identification and characterization of endothelial glycoprotein Ib using viper venom proteins modulating cell adhesion. Blood 1999;93:2605–2616.
19 Chen YL, Tsai IH: Functional and Sequence characterization of agkicetin, a new glycoprotein Ib antagonist isolated from *Agkistrodon acutus* venom. Biochem Biophys Res Commun 1995;210:472–477.
20 Yeh CH, Chang MC, Peng HC, Huang TF: Pharmacological characterization and antithrombotic effect of agkistin, a platelet glycoprotein Ib antagonist. Br J Pharmacol 2001; 132:843–850.
21 Yeh CH, Peng HC, Huang TF: Accutin, a new disintegrin, inhibits angiogenesis in vitro and in vivo by acting as integrin $\alpha_v\beta_3$ antagonist and inducing apoptosis. Blood 1998; 92:3268–3276.
22 Frelinger AL 3rd, Cohen I, Plow EF, Smith MA, Robert J, Lam SCT, Ginsberg MH: Selective inhibition of integrin function by antibodies specific for ligand-occupied receptor conformers. J Biol Chem 1990;265: 6346–6352.
23 Passaniti A, Taylor RM, Pili R, Gou Y, Long PV, Haney JA, Pauly RR, Grant DS, Martin GR: A simple, quantitative method for assessing angiogenesis and antiangiogenic agents using reconstituted basement membrane, heparin, and fibroblast growth factor. Lab Invest 1992;67: 519–529.
24 Eliceiri BP, Cheresh DA: The role of α_v integrins during angiogenesis: Insights into potential mechanisms of action and clinical development. J Clin Invest 1999;103:1227–1230.
25 Dimmeler S, Zeiher AM: Endothelial cell apoptosis in angiogenesis and vessel regression. Circ Res 2000;87:434–439.
26 Brooks PC, Montgomery AMP, Rosenfeld M, Reisfeld RA, Hu T, Klier G, Cheresh DA: Integrin $\alpha_v\beta_3$ antagonists promote tumor regression by inducing apoptosis of angiogenic blood vessels. Cell 1994;79:1157–1164.
27 Mitjans F, Meyer T, Fittschen C, Goodman S, Jonczyk A, Marshall JF, Reyes G, Piulats J: In vivo therapy of malignant melanoma by means of antagonists of α_v integrins. Int J Cancer 2000;87:716–723.
28 Allman R, Cowburn P, Mason M: In vitro and in vivo effects of a cyclic peptide with affinity for the $\alpha_v\beta_3$ integrin in human melanoma cells. Eur J Cancer 2000;36:410–422.
29 Plopper GE, McNames HP, Dike LE, Bojanowski K, Ingber DE: Convergence of integrin and growth factor receptor signaling pathways within the focal adhesion complex. Mol Biol Cell 1995;6:1349–1365.
30 Yeh CH, Peng HC, Huang TF: Cytokines modulate integrin $\alpha_v\beta_3$-mediated human endothelial cell adhesion and calcium signaling. Exp Cell Res 1999;251:57–66.
31 Byzova TV, Plow EF: Activation of $\alpha_v\beta_3$ on vascular cells controls recognition of prothrombin. J Cell Biol 1998;143:2081–2092.
32 Enenstein J, Waleh NS, Kramer RH: Basic FGF and TGF-beta differentially modulate integrin expression of human microvascular endothelial cells. Exp Cell Res 1992;203: 499–503.
33 Klein S, Giancotti FG, Presta M, Albelda SM, Buck CA, Rifkin DB: Basic fibroblast growth factor modulates integrin expression in microvascular endothelial cells. Mol Biol Cell 1993;4:973–982.
34 Stromblad S, Cheresh DA: Cell adhesion and angiogenesis. Trends Cell Biol 1996;6:462–468.
35 Kang IC, Lee YD, Kim D: A novel disintegrin salmosin inhibits tumor angiogenesis. Cancer Res 1999;59: 3754–3760.
36 Zhou Q, Nakada MT, Arnold C, Shieh KY, Markland FS: Contortrostatin, a dimeric disintegrin from *Agkistrodon contortrix contortrix*, inhibits angiogenesis. Angiogenesis 1999;3:259–269.
37 Yeh CH, Wang WC, Hsieh TT, Huang TF: Agkistin, a snake venom-derived glycoprotein Ib antagonist, disrupts von Willebrand factor-endothelial cell interaction and inhibits angiogenesis. J Biol Chem 2000;275:18615–18618.
38 Yeh CH, Peng HC, Yang RS, Huang TF: Rhodostomin, a snake venom disintegrin, inhibits angiogenesis elicited by basic fibroblast growth factor and suppresses tumor growth by a selective $\alpha_v\beta_3$ blockade of endothelial cells. Mol Pharmacol 2001; 59:1333–1342.

Substances Affecting Platelet Function

Hæmostasis

Haemostasis 2001;31:207–210

Lebetin Peptides: Potent Platelet Aggregation Inhibitors

N. Marrakchi[a] K. Mabrouk[b] I. Regaya[a] S. Sarray[a] M. Fathallah[b] H. Rochat[b] M. El Ayeb[a]

[a]Laboratoire des Venins et Toxines, Institut Pasteur de Tunis, Tunisie; [b]CNRS UMR 6560, Marseille, France

Key Words
Snake venom · Lebetin peptides · Antiplatelet aggregation · Thrombocytopenia

Abstract
Lebetins from *Macrovipera lebetina* snake venom constitute a new class of inhibitors of platelet aggregation. There are two groups of peptides: lebetin 1 (L1; 11- to 13-mer) and lebetin 2 (L2; 37- to 38-mer). The short lebetins are identical to the N-terminal segments of the longer ones. They inhibit platelet aggregation induced by various agonists (e.g. thrombin, PAF-acether or collagen). The shortest lebetin (11-mer) shows potent inhibition of rabbit (IC_{50} = 7 n*M*) and human (IC_{50} = 5 n*M*) platelets. They prevent collagen-induced thrombocytopenia in rats. N- and C-terminal-truncated synthetic L1γ (sL1γ; 11-mer) is less active in inhibiting platelet aggregation than the native peptide. Results from Ala scan studies of the sL1γ peptide indicated that replacement of the residues (P3, G7, P8, P9 or N10) resulted in a remarkable drop in the activity, whereas replacement of residues K2, P4 or K6 by Ala resulted in enhancement of the antiplatelet activity by at least 10-fold. To examine the activity of multimeric L1γ, several multimeric peptides were synthesized using the multiple-antigen peptide system assembled on a branched lysine core and their antiplatelet activity was evaluated in vitro. The largest multimeric peptides showed a 1,000-fold increase in antiplatelet activity.

Introduction

Platelet aggregation plays a fundamental role in hemostasis as it arrests bleeding through the initial hemostatic plug that forms at the site of vascular injury [1]. Platelet aggregation involves a complex network of cell surface adhesion proteins, one of which is GPIIb/

KARGER
Fax + 41 61 306 12 34
E-Mail karger@karger.ch
www.karger.com

0301–0147/01/0316–0207$17.50/0

Accessible online at:
www.karger.com/journals/hae

Dr. N. Marrakchi
Laboratoire des Venins et Toxines, Institut Pasteur de Tunis
13, Place Pasteur, BP 74, 1002 Tunis-Belvédère (Tunisia)
Tel. +216 71 783 022 or +216 71 843 755, Fax +216 71 791 833
E-Mail naziha.marrakchi@pasteur.rns.tn

Table 1. Amino acid sequences and molecular masses of lebetins

	Sequences	Measured mass
L1α	GDNKPPKKGPPNG	1,305.74 ± 0.38
L1β	DNKPPKKGPPNG	1,248.69 ± 0.31
L1γ	NKPPKKGPPNG	1,191.45 ± 0.43
L2α	GDNKPPKKGPPNGGCFGHKIDRIGSHSGLGCNKVDDNKG	3,943.74 ± 0.46
L2β	DNKPPKKGPPNGGCFGHKIDRIGSHSGLGCNKVDDNKG	3,886.92 ± 0.38

IIIa. GPIIb/IIIa binds fibrinogen; their interaction is inhibited by proteins isolated from snake venom containing an RGD sequence [2]. Disintegrins, short RGD containing peptides such as echistatin, inhibit several integrins (e.g. $\alpha_{IIb}\beta_3$, $\alpha_v\beta_3$, $\alpha_5\beta_1$), whereas eristotatin is selective for $\alpha_{IIb}\beta_3$ [3, 4]. The heterodimeric disintegrins are more specific for leukocyte integrins. Two of them, EC3 and EMF10, inhibit $\alpha_4\beta_1/\alpha_4\beta_7$ and $\alpha_5\beta_1$, respectively [5, 6].

We isolated a new family of platelet aggregation inhibitors, lebetin 1 (L1) and lebetin 2 (L2), from *Macrovipera lebetina* venom. Despite the absence of RGD-related sequences or other sequences with known antiaggregating activity, lebetins displayed strong in vitro antiplatelet activity and were able to prevent collagen-induced thrombocytopenia in rats.

Purification and Characterization of L1 and L2

L1 and L2 were purified to homogeneity from *M. lebetina* venom in two steps of gel filtration followed by reverse-phase HPLC on a C8 column [7]. N-terminal sequencing of native L2 revealed two homologous sequences representing 80% (α) and 20% (β) of the total yield. The N-terminal Gly residue was deleted in the latter (table 1). Their molecular weights determined by ES-MS were 3,943.74 and 3,886.96, respectively. A minor L1γ was also found in *M. lebetina* venom (table 1). Complete sequences of L2α and L2β were deduced by sequencing the alkylated peptides and the C-terminal peptide obtained by digestion with Arg-C protease. The complete L2α and L2β isoforms are shown in table 1. L1α, L1β and L1γ contained 13, 12 and 11 amino acid residues, respectively, whereas L2α and L2β contained 38 and 37 amino acid residues, respectively. The L1 isoforms are identical to the N-terminal sequences of the L2 isoforms. The lebetin peptides lack the RGD motif and do not share any significant structural similarity with known disintegrins.

In vitro Platelet Aggregation Inhibition by Lebetins

Rabbit platelet aggregation induced by thrombin, PAF-acether or collagen was inhibited by L1 with IC_{50} values of 125, 48 and 27 n*M*, respectively, and L2 with IC_{50} values of 8, 48 and 5 n*M*, respectively. L1 and L2 also inhibited thrombin-induced aggregation of washed human platelets with IC_{50} values of 590 and 100 n*M*, respectively.

In vivo Activity of Lebetins

Intravenous administration of L1 and L2 inhibited collagen-induced thrombocytopenia with ED_{50} values of 2 and 4.2 nmol/kg, respectively. L1 (20 nmol/kg) and L2 (13 nmol/kg) reduced the thrombocytopenia by 80–90%.

Structure-Activity Relationships

To determine the relative activity of each peptide, L1α, L1β, L1γ, L2α, L2β were synthesized and their antiplatelet activities determined. The synthetic short and long lebetin peptides inhibited thrombin-induced rabbit and human platelet aggregation in the nanomolar range (table 2). Various L1γ-derived synthetic peptides, truncated at either their N- or C-terminus by one or more amino acid residues, were synthesized. The antiplatelet activity of truncated peptides was drastically decreased showing that the integrity of synthetic L1γ (sL1γ) is essential for its potent effect. The results from Ala scan of sL1α demonstrated that substitution of P3, G7, P8, P9 or N10 drastically decreased the antiplatelet activity of the peptides. However, a significant antiplatelet activity is retained when N1, K5 are individually replaced by Ala (IC_{50} = 50 n*M* instead of 7 n*M* for sL1α). The antiplatelet activity increased by at least 10-fold when K2, P4 or K6 were replaced by Ala.

Another way to dramatically increase the activity is to prepare multimeric peptides. Several sizes of multimeric sL1γ peptides were chemically synthesized using the multiple-antigen peptide system assembled on a branched lysine core and fully characterized. The multimeric peptides were found to be 1,000-fold more active than the sL1γ peptide in inhibiting rabbit thrombin-induced platelet aggregation, with IC_{50} values ranging from 2 to 25 p*M*. The larger the peptide, the more inhibitory effect there was on platelet aggregation.

Table 2. Effect of synthetic lebetin on rabbit and human platelet aggregation induced by thrombin (0.04 IU/ml)

Synthetic peptides	Rabbit platelets IC_{50}, n*M*	Human platelets IC_{50}, n*M*
sL1α	23	140
sL1β	10	32
sL1γ	7	5
sL2α	0.2	2.5
sL2β	0.9	2.8

IC_{50} = Concentration of inhibitor required to inhibit 50% aggregation.

Conclusion and Perspectives

Native and synthetic lebetins possess potent antiplatelet activity and prevent thrombocytopenia induced by collagen in rats. Structure-activity relationship studies of sL1γ showed the importance of its integrity and the possibilities of enhancing the potency by alanine replacement or by synthesis of multimeric forms. However, the antiplatelet mechanism of action of lebetin remains to be elucidated and its in vivo antithrombotic effects evaluated.

Acknowledgements

The authors acknowledge the friendly collaboration of Dr. Cassian Bon from I.P.P. They also thank Prof. K. Dellagi for his support during this study.

References

1 Kroll MH, Schafer AI: Biochemical mechanism of platelet activation. Blood 1989;74:1181–1195.

2 Savage B, Marzec UM, Chao BH, Harker LA, Maragnore JM, Ruggeri ZM: Binding of the snake venom derived proteins applagin and echistatin to the arginine-glycine-aspartic acid recognition sites of platelet glycoprotein IIb/IIIa complex receptor function. J Biol Chem 1990;265: 11766–11772.

3 Niewiarowski S, Mc Lane MA, Kloczewiak M, Stewart GJ: Disintegrins and other naturally occurring antagonists of platelet fibrinogen receptors. Semin Hematol 1994;314: 289–300.

4 Marcinkiewicz C, Rosenthal LA, Mosser DM, Kunicki TJ, Niewiarowski S: Immunological characterization of eristostatin and echistatin binding sites on $\alpha_{IIb}\beta_3$ and $\alpha_V\beta_3$ integrins. Biochem J 1996;317:817–825.

5 Marcinkiewicz C, Calvete JJ, Marcinkiewicz MM, Raida M, Vijay-Kumar S, Huang Z, Lobb RR, Marcinkiewicz C: EC3, a novel heterodimeric disintegrin from *Echis carinatus* venom, inhibits α_4 and α_5 integrins in an RGD-independent manner. J Biol Chem 1999;274:12468–12473.

6 Marcinkiewicz C, Calvete JJ, Vijay-Kumar S, Marcinkiewicz MM, Raida M, Schick P, Lobb RR, Niewiarowski S: Structural and functional characterization of EMF10, a heterodimeric disintegrin from *Eristocophis macmahoni* venom that selectively inhibits $\alpha_5\beta_1$ integrin. Biochemistry 1999;38:13302–13309.

7 Barbouche R, Marrakchi N, Mansuelle P, Krifi M, Fenouillet E, Rochat, H, El Ayeb M: Novel antiplatelet aggregation polypeptides from *Vipera lebetina* venom: Isolation and characterization. FEBS Lett 1996;392:6–10.

Procoagulant Proteins and Diagnostic Agents

Hæmostasis

Haemostasis 2001;31:211–217

Diagnostic Uses of Snake Venom

Neville A. Marsh

School of Life Sciences, Queensland University of Technology, Brisbane, Qld., Australia

Key Words

Snake venom toxins · Snake venom thrombin-like enzymes · Disintegrins · Haemostasis · Platelet function tests · Coagulation assays

Abstract

Snake venom toxins are invaluable for the assay of coagulation factors and for the study of haemostasis generally. Thrombin-like enzymes (SVTLE) are used for fibrinogen and fibrinogen breakdown product assays as well as detecting dysfibrinogenaemias. Since SVTLE are not inhibited by heparin, they can be used for assaying antithrombin III in samples containing heparin. Snake venom prothrombin activators are utilised in prothrombin assays, whilst Russell's viper venom (RVV) can be used to assay clotting factors V, VII, X and lupus anticoagulants (LA). Activators from the taipan, Australian brown snake and saw-scaled viper have also been used to assay LA. Protein C (PC) and activated PC (APC) resistance can be measured by means of RVV, Protac™ (from Southern copperhead snake venom) and STA-Staclot (from *Crotalus viridis helleri*) whilst von Willebrand factor can be studied with Botrocetin™ *(Bothrops jararaca)*. Finally, snake venom C-type lectins and metalloproteinase disintegrins are being used to study platelet glycoprotein receptors and show great potential for use in the routine coagulation laboratory.

Introduction

Snake venoms contain a rich variety of materials having profound effects on the haemostatic mechanism, and these have been extensively reviewed [1–4]. These materials can be classified into five groups: coagulant (thrombin-like and coagulation-activating toxins), anticoagulants (PC activators), toxins inhibiting platelet function, activators of fibrinolysis and haemorrhagins. Some of these

KARGER

Fax + 41 61 306 12 34
E-Mail karger@karger.ch
www.karger.com

0301–0147/01/0316–0211$17.50/0

Accessible online at:
www.karger.com/journals/hae

Prof. N.A. Marsh
School of Life Sciences, Queensland University of Technology
2 George Street, GPO Box 2434, Brisbane, Qld. 4001 (Australia)
Tel. +61 7 3864 2495, Fax +61 7 3864 1534
E-Mail n.marsh@qut.edu.au

snake venom materials have become invaluable in laboratory diagnosis of haemostatic disorders and these practical applications are reviewed in this paper.

Thrombin-Like Enzymes and Investigation of Fibrinogen

Over 90 thrombin-like enzymes from 35 snake species have been recorded [5] being serine proteases with active site residues recognisable in positions H57-D102-S195. The group does not share all properties common to thrombin since enzymes typically cleave fibrinopeptide A. The most widely studied enzymes are from *Bothrops atrox* (batroxobin, reptilase), *Callosellasma rhodostoma* (ancrod) and *Agkistrodon c. contortrix* (ACTE).

Since snake venom thrombin-like enzymes (SVTLEs) are not inhibited by heparin, they can be used to test plasma samples containing this anticoagulant or to remove fibrinogen from samples containing heparin. The reptilase time is a simple alternative to the thrombin time for rapid fibrinogen assay in samples containing heparin [6] and is particularly useful in the assay of antithrombin III where plasma can be prepared free of fibrinogen [7, 8]. The presence of fibrin degradation products (FDPs), hypofibrinogenaemia and defects in fibrin polymerisation will prolong the reptilase time, thus if this prolongation is less than that of the thrombin time, then the presence of FDPs is indicated [9]. SVTLEs are useful for investigating fibrin function including magnetic birefringence studies of fibrin assembly, [10] and other aspects of the clotting mechanism [11, 12]. Batroxobin and ACTE can also be used in preparing desaa- and desbb-fibrinogen, the former for use as a fibrin stimulant in the functional assay of tissue plasminogen activator [13].

Prothrombin: Assay and Functional Aspects

Commercially available prothrombin activators include ecarin from saw-scaled viper *(Echis carinatus)* venom, textarin from the Australian brown snake *(Pseudonaja textilis)* and the enzyme from the taipan *(Oxyuranus s. scutellatus)* [14]. Crude taipan venom contains a high proportion of enzyme and can be used without further purification in a one-stage assay of prothrombin [15]. However, the use of these enzymes is of limited value in warfarin-treated patients since they cleave functionally abnormal types of prothrombin which are present in addition to normal prothrombin [16]. These forms include the descarboxy variety or PIVKA (proteins induced by vitamin K antagonists). Nonetheless, this property has been harnessed in the assay of PIVKA in liver disease [17], in dysprothrombinaemias [18, 19] and in the diagnosis of disseminated intravascular coagulation [20]. Snake venom prothrombin-activating enzymes have also been used to prepare meizothrombin [21, 22], non-enzymic forms of thrombin and meizothrombin [14] and in studies of recombinant prothrombin cleavage [23], and prothrombin activation [24].

Factor V

Activation of factor V (FV) can be achieved by a serine protease from the venom of Russell's viper *(Daboia russelli)* (RVV-V) which cleaves the single-chain glycoprotein at Arg^{1545}. The enzyme can be used for the routine assay of FV by virtue of its selective activation of FV [25] and although the use of RVV-V for FV assay is somewhat limited, the reagent remains a useful tool in the study of FV [26] (see further comments below on assay of FV Leiden, PC and APC resistance).

Factors VII and X

Russell's viper venom also contains a potent activator of factor X (FX) (RVV-X), and the enzyme has become widely used for FX itself [27], for distinguishing between factor VII (FVII) and FX deficiency [28] and lupus anticoagulant assay [29]. The clotting time of plasma using RVV-X is known as the Stypven time™ [30] and a normal Stypven time used in conjunction with a prothrombin time (PT) suggests FVII deficiency whereas a prolonged Stypven time indicates FX deficiency [28]. Measurement of FX can be performed with RVV-X in functional clotting assays [27], with a chromogenic substrate [31] and by a solid-phase enzyme-linked coagulation assay [32]. In the clotting assay, FX is converted to FXa by RVV-X in the presence of calcium ions, FVa and phospholipid. This complex activates prothrombin initiating clot formation, thus clotting time is proportional to FX concentration. The chromogenic assay utilises a specific substrate (Pefachrome FXa™) which is cleaved by the RVV-activated FXa. Not only has RVV-X an established place in FX assay, but is also useful for identifying FX recognition sites [33], determining the effects of deglycosylation on FX activation [34] and for assay of platelet factor 3 [35, 36].

Lupus Anticoagulants

Lupus anticoagulants (LA) are a heterogeneous population of immunoglobulins which interfere with phospholipid-dependent clotting tests, i.e. the activated partial thromboplastin time (APTT), PT and kaolin clotting time (KCT). Because of the molecular heterogeneity associated with LA, no single standalone test can provide a satisfactory screen and in the past, the APTT and KCT have been used for this purpose. Screening for LA has been reviewed elsewhere [37, 38] and snake venom activators are now used including RVV-X, and activators from venom of the taipan, Australian brown snake and saw-scaled viper. The dilute Russell's viper venom time (dRVVT) is quick, sensitive and inexpensive [29]. It has been found to be less sensitive than the KCT [39], but new formulations of the dRVVT give increased sensitivity [40, 41]. A test has also been developed based on the taipan time combined with a platelet neutralisation procedure [42] and a further snake venom-based confirmatory test for LA has been developed utilising Textarin™ and Ecarin™ [43]. This has proved to be a sensitive and relatively specific test and the T/E ratio has proved a useful confirmatory test for the diagnosis of LA patients with unexplained prolonged APTT [44].

PC and APC Resistance

PC is a vitamin-K-dependent protein which plays a key role in haemostasis, most notably in its active form, APC, as an anticoagulant proteinase inactivating FVa and VIIIa. Functional assay of PC is hampered by incomplete activation using thrombin and by interference of the same thrombin in chromogenic assays. These problems are overcome by the use of a potent, fast-acting PC activator (Protac) from Southern copperhead snake venom, *A. c. contortrix* [45, 46]. Protac is used in a chromogenic method [47, 48], in functional clotting assays [49] and in a new global clotting assay (ProC Global) [50]. Results from these assays compare favourably with those from immunological methods [51] and the Protac-based method seems preferable to thrombin-based assays since the latter incur a degree of misclassification of both normal and congenitally PC-deficient patients [52]. Likewise a global test based on the ratio of APTT in the absence and presence of Protac successfully identified all patients with defects in the PC anticoagulant pathway [53] and may enable recognition of dysfunctional PC by differential results from chromogenic and coagulometric assays [54]. Use of Protac has greatly simplified the assay of PC, as well as the assay of protein S, the co-factor required for the inactivation of FVa and FVIIIa by APC, the latter being assayed by both amidolytic [55] and clotting methods [56]. In addition, this snake venom enzyme has minimized the associated costs of these assays, permitting the scaling down of sample sizes for use on microtitre plates [47].

APC resistance is probably the major cause of thrombophilia and its molecular basis has been determined (FV Leiden) [58]. Functional assays using snake venom activators can accurately detect the genotypic abnormality including the dRVVT (PC Impedance Test, Gradipore), the textarin time (Pentapharm) and STA-Staclot (Diagnostica Stago), using a specific activator from *Crotalus viridis helleri* venom [59]. The dRVVT offers improved sensitivity and may make DNA analysis unnecessary [60]. In addition, the textarin time may used [61], but with both this method and the ACV *(A. contortrix* venom) test, conclusive DNA analysis is strongly recommended [62].

Von Willebrand Factor Studies

Botrocetin™ is a platelet-aggregating protein found widely in *Bothrops jararaca* venom. Botrocetin depends on the presence of von Willebrand factor (VWF) for its effect on platelets, and this property has been utilised in VWF assay [63, 64]. Together with ristocetin, the two agonists can be used to differentiate molecular variants of VWF: botrocetin partially aggregates platelets from patients suffering from Bernard-Soulier disease where glycoprotein receptor Ib (GPIb) is absent. However, ristocetin will not induce platelet aggregation in plasma from these patients since it depends on GPIb. The same distinction can be made in type IIa von Willebrand disease (VWD) where high molecular weight VWF multimers are absent [65]. A solid-phase inhibition assay for VWF-binding to platelet GPIb has now been proposed using botrocetin [66]. This ristocetin-botrocetin combination has proved invaluable in the characterization of human platelet VWF [67] and in the detection of a missense mutation in Type B VWD [68]. A second VWF-dependent snake venom platelet aggregant, alboaggregin-B isolated from *Trimeresurus albolabris* venom has been used to quantitate VWF receptors on the GPIb molecule [69].

Platelet Glycoprotein Studies

Snake venoms contain many materials affecting platelet function [70, 71] including the metalloproteinase-disintegrins which inhibit platelet aggregation via interference with surface glycoprotein receptors and the C-type lectins [72]. These compounds offer a unique opportunity for the study of platelet-platelet and platelet-endothelium interactions, although as yet, no routine test of haemostasis utilises a disintegrin.

Conclusions

Snake venoms are widely used in the coagulation laboratory for routine assay of coagulation factors and as research tools. Some applications have been adopted as the preferred option to conventional coagulation assays and may even be more appropriate than some DNA-based tests. Other compounds like the disintegrins, with powerful in vitro and in vivo actions as anti-platelet aggregants, are useful tools in the study of platelet physiology and potentially, as reagents in the routine coagulation laboratory.

References

1 Marsh NA: Snake venoms affecting the haemostatic mechanism – a consideration of their mechanisms, practical applications and biological significance. Blood Coagul Fibrinol 1994;5:399–410.

2 Markland FS: Snake venoms. Drugs 1997;54(suppl 3):1–10.

3 Braud S, Bon C, Wisner A: Snake venom proteins acting on hemostasis. Biochimie 2000;82:851–859.

4 Matsui T, Fujimura Y, Titani K: Snake venom proteases affecting hemostasis and thrombosis. Biochim Biophys Acta 2000;1477:146–156.

5 Pirkle H: Thrombin-like enzymes from snake venoms: An updated inventory. Thromb Haemost 1998;79: 675–683.

6 Funk C, Gmur J, Herold R, Straub PW: Reptilase-R, a new reagent in blood coagulation. Br J Haematol 1971;21:43–52.

7 Howie PW, Prentice CRM, McNicol GP: A method of antithrombin estimation using plasma defibrinated with ancrod. Br J Haematol 1973;25:101–110.

8 Canoso RT, Hutton RA, Deykin D: The haemostatic defect of chronic liver disease. Gastroenterology 1979;76:540–547.

9 Latallo Z, Teisseyre E: Evaluation of reptilase-R and thrombin clotting time in the presence of fibrinogen degradation products and heparin. Scand J Haematol Suppl 1971;13: 261–266.

10 Torbet J: Fibrin assembly after fibrinopeptide A release in model systems and human plasma studied with magnetic birefringence. Biochem J 1987:244:633–637.

11 Kumar R, Beguin S, Hemker HC: The influence of fibrinogen and fibrin on thrombin generation – evidence for feedback activation of the clotting system by clot bound thrombin. Thromb Haemost 1994; 72:713–721.

12 Carr ME, Carr SL, Greilich PE: Heparin ablates force development during platelet mediated clot retraction. Thromb Haemost 1996;75: 674–678.

13 Ranby M, Norrman B, Wallen P: A sensitive assay for tissue plasminogen activator. Thromb Res 1982;27: 743–749.

14 Rosing J, Tans G: Structural and functional properties of snake venom prothrombin activators. Toxicon 1992;30:1515–1527.

15 Denson KWE, Borrett R, Biggs R: The specific assay of prothrombin using the taipan snake venom. Br J Haematol 1971;21:219–226.

16 Kornalik F, Schieck A, Habermann E: Isolation, biochemical and pharmacological characterization of a prothrombin activating principle from *Echis carinatus* venom. Naunyn Schmiedebergs Arch Pharmacol 1969;264:259–260.

17 Kornalik F, Vorlova Z: Ecarin test in diagnosis of dicoumarol therapy, liver diseases and DIC. Folia Haematol (Leipzig) 1988;115:483.

18 Weinger RS, Rudy C, Moake JL, Olson JD, Cimo PL: Prothrombin Houston: A dysprothrombin identifiable by crossed immunoelectrofocussing and abnormal *Echis carinatus* venom activation. Blood 1980;55:811–815.

19 Collados MT, Fernandez J, Paramo JA, Montes R, Borbolla JR, Montano LF, Rocha E: Purification and characterization of a variant of human prothrombin – prothrombin Segovia. Thromb Res 1997;85:465–477.

20 Sakuragawa N, Takahashi K, Hoshiyama M, Jimbo Ch, Matsuoka M, Ohnishi Y: Significance of a prothrombin assay method using *Echis carinatus* venom for diagnostic information in disseminated intravascular coagulation syndrome. Thromb Res 1975;7:643–653.

21 Novoa E, Seegers WH: Mechanisms of α-thrombin and β-thrombin-E formation: Use of ecarin for isolation of meizothrombin 1. Thromb Res 1980;18:657–668.

22 Rosing J, Tans G: Meizothrombin, a major product of factor Xa-catalysed prothrombin activation. Thromb Haemost 1988;60:355–360.

23 Stevens WK, Cote HCF, MacGillivray RTA, Nesheim ME: Calcium ion modulation of meizothrombin autolysis at Arg55-Asp56 and catalytic activity. J Biol Chem 1996;271: 8062–8067.

24 Yamada D, Sekiya F, Morita T: Isolation and characterization of carinactivase, a novel prothrombin activator in *Echis carinatus* venom with a unique catalytic mechanism. J Biol Chem 1996;271:5200–5207.

25 Kisiel W, Canfield WM: Snake venom proteases that activate blood coagulation factor V; in Lorand L (ed): Methods in Enzymology. New York, Academic Press, 1981, vol 80, pp 275–285.

26 Keller FG, Ortel TL, Quinn-Allen MA, Kane WH: Thrombin-catalysed activation of recombinant human factor V. Biochemistry 1995; 34:4118–4124.

27 Bachmann F, Duckert F, Koller F: The Stuart-Prower assay and its clinical significance. Thromb Diath Haemorrh 1958;2:24.

28 Quick AJ: Thromboplastin generation: Effect of the Bell-Alton reagent and Russell's viper venom on prothrombin consumption: Am J Clin Pathol 1971;55:555–560.
29 Thiagarajan P, Pengo V, Shapiro SS: The use of dilute Russell viper venom time for the diagnosis of lupus coagulants. Blood 1986;68:869–874.
30 Denson KWE: The specific assay of Prower Stuart factor and factor VII. Acta Haematol 1961;25:105.
31 Svendsen L, Brogli M, Lindeberg G, Stocker K: Differentiation of thrombin and factor Xa-related amidolytic activity in plasma by means of a synthetic thrombin inhibitor. Thromb Res 1984;34:457–462.
32 Durkee KH, Cheng TM, Doellgast GJ: Enzyme-linked coagulation assay. V. Amplified blotting assays using snake venom conjugates. Anal Biochem 1990;184:375–380.
33 Chattopadhyay A, Fair DS: Molecular recognition in the activation of human blood coagulation factor X. J Biol Chem 1989;264:11035–11043.
34 Inoue K, Morita T: Identification of O-linked oligosaccharide chains in the activation peptides of blood coagulation factor X. Eur J Biochem 1993;218:153–163.
35 Hardisty RM, Hutton RA: Platelet aggregation and the availability of platelet factor 3. Br J Haematol 1966;12:664.
36 Jy WC, Horstman LL, Wang F, Duncan RC. Ahn YS: Platelet factor 3 in plasma fractions – its relation to microparticle size and thromboses. Thromb Res 1995;80:471–482.
37 Exner T: Some recent developments with lupus anticoagulants. Blood Coagul Fibrinolysis 1994;5:281–289.
38 Brandt JL, Barna LK, Triplett DA: Laboratory identification of lupus anticoagulants – results of the Second International Workshop for Identification of Lupus Anticoagulants. Thromb Haemost 1995;74: 1597–1603.
39 Martin BA, Branch DW, Rodgers GM: Sensitivity of the activated partial thromboplastin time, the dilute Russell's viper venom time, and the kaolin clotting time for the detection of the lupus anticoagulant: A direct comparison using plasma dilutions. Blood Coagul Fibrinolysis 1996;7: 31–38.
40 Joyner KA, Ortel TL: A sensitive DRVVT reagent system for the detection of lupus anticoagulants. Clin Appl Thromb/Hemost 1995;1:73–75.
41 Moore GW, Savidge GF, Smith MP: Improved detection of lupus anticoagulants by the dilute Russell's viper venom time. Blood Coagul Fibrinolysis 2000;11:767–774.
42 Rooney AM, McNally T, Mackie IJ, Machin SJ: The taipan snake venom time: A new test for lupus anticoagulant. J Clin Pathol 1994;47:497–501.
43 Triplett DA, Stocker, Unger GA, Barna LK: The textarin/ecarin ratio: A confirmatory test for lupus anticoagulants. Thromb Haemost 1993; 70:925–931.
44 Forastiero RR, Cerrato GS, Carreras LO: Evaluation of recently described tests for the detection of the lupus anticoagulant. Thromb Haemost 1994;72:728–733.
45 Klein JD, Walker FJ: Purification of a protein C activator from the venom of the Southern copperhead snake *(Agkistrodon contortrix contortrix)*: Biochemistry, 1986;25: 4175–4179.
46 Stocker K, Fischer H, Meier J, Brogli M, Svendsen L: Characterization of the protein C activator Protac® from the venom of the southern copperhead *(Agkistrodon contortrix)* snake. Toxicon 1987;25:239–252.
47 Nathan IV, Ping H, Pradhan MM: Protein C functional assay using snake venom activator. Thromb Res 1987;47:85–91.
48 Stocker K, Fischer H, Meier J: Practical application of the protein C activator Protac® from *Agkistrodon contortrix* venom. Folia Haematol (Leipzig) 1988;115:260–264.
49 Martinoli JL, Stocker K: Fast functional protein C assay using Protac®, a novel protein C activator. Thromb Res 1986;43:253–264.
50 Toulon P, Halbmeyer WM, Hafner G, Schmitt Y, Randgard B, Odpadlik M, van den Eynden C, Wagner C: Screening for abnormalities of the protein C anticoagulant pathway using the ProC Global assay. Results of a European multicentre evaluation. Blood Coagul Fibrinolysis 2000;11:447–454.
51 Takahashi H, Hanano M, Tatewaki W, Shibata A: Fast functional assay of protein C in whole plasma using a snake venom activator: Evaluation in patients with congenital and acquired protein C deficiencies: Clin Chim Acta 1988;175:217–225.
52 Franchi F, Tripodi A, Valsecchi C, Mannucci PM: Functional assays of protein C: Comparison of two snake venom assays with two thrombin assays. Thromb Haemost 1988;60: 145–147.
53 Robert A, Eschwege V, Hameg H, Drouet L, Aillaud MF: Anticoagulant response to *Agkistrodon contortrix* venom (ACV test): A new global test to screen for defects in the anticoagulant protein C pathway. Thromb Haemost 1996;75:562–566.
54 Girolami A, Simioni P, Girolami B, Marchiori A, Millar DS, Bignell P, Kakkar VV, Cooper DN: A novel dysfunctional protein C (Protein C Padua 2) associated with a thrombotic tendency: Substitution of Cys for Arg-1 results in a strongly reduced affinity for binding of Ca^{++}: Br J Haematol 1993;85:521–527.
55 Francis RB, Seyfert U: Rapid amidolytic assay of protein C in whole plasma using an activator from the venom of *Agkistrodon contortrix*. Am J Clin Pathol 1987;87:619–625.
56 Suzuki K, Nishioka J: Plasma protein S activity measured using Protac, a snake venom derived activator of protein C. Thromb Res 1988;49: 241–251.
57 Lee TK, Bangalore N, Velander W, Drohan WN, Lubon H: Activation of recombinant human protein C. Thromb Res 1996;82:225–234.
58 Bertina RM, Koelman BPC, Koster T: Mutation in blood coagulation factor V associated with resistance to activated protein C. Nature 1994; 369:64–67.

59 Quehenberger P, Handler S, Mannhalter C, Pabinger-Fasching I, Speiser W: Evaluation of a highly specific functional test for the detection of factor V Leiden. Int J Clin Lab Res 2000;30:113–117.
60 Solano C, Hogg AL, Saal RJ, Scott DC, Cobcroft RG: The use of two different APC resistance assay systems provides optimal sensitivity and specificity for diagnosing genetic APC resistance. Blood Coagul Fibrinolysis 1997;8:268–273.
61 Hoagland LE, Triplett DA, Peng F, Barna L: APC-resistance as measured by a textarin time assay – comparison to the APTT-based method. Thromb Res 1996;83:363–373.
62 Gemmati D, Serino ML, Mari R, Verzola I, Moratelli S, Ballerini G: Different anticoagulant response to activated protein C (APC test) and to *Agkistrodon contortrix* venom (ACV test) in a family with FV-R506Q substitution. Clin App Thrombosis/Hemostasis 1997;3:168–173.
63 Read MS, Potter JY, Brinkhous KM: Venom co-agglutinin for detection on von Willebrand factor activity in animal plasmas. J Lab Clin Med 1983;101:74–82.
64 Brinkhous KM, Smith SV, Read MS: Botrocetin and von Willebrand factor; in Pirkle H, Markland FS (eds): Hemostasis and Animal Venoms. New York-Basel, Marcel Dekker, 1988, pp 377–398.
65 Bloom AL: Progress in the clinical management of haemophilia. Thromb Haemost 1991;66:166–177.
66 Miura S, Nishida S, Makita K, Sakurai Y, Shimoyama T, Sugimoto M, Yoshioka A, Ishii K, Kito M, Kobayashi T, Fujimura Y: Inhibition assay for the binding of biotinylated von Willebrand factor to platelet-bound microtiter wells in the presence of ristocetin or botrocetin. Anal Biochem 1996;236:215–220.
67 Williams SB, McKeown LP, Krutzsch H, Hansmann K, Gralnick HR: Purification and characterization of human platelet von Willebrand factor. Br J Haematol 1994;88:582–591.
68 Rabinowitz I, Tuley EA, Mancuso DJ, Randi AM, Firkin BG, Howard M, Sadler JE: Von Willebrand disease type B: A missense mutation selectively abolishes ristocetin-induced von Willebrand factor binding to platelet glycoprotein Ib. Proc Natl Acad Sci USA 1992;89:9846–9849.
69 Yoshida E, Fujimura Y, Ikeda Y, Takeda I, Yamamoto Y, Nishikawa K, Miyataka K, Oonuki M, Kawasaki T, Katayama M: Impaired high-shear-stress-induced platelet aggregation in patients with chronic renal failure undergoing dialysis. Br J Haematol 1995;89:861–867.
70 Ouyang C, Teng CM, Huang TF: Characterization of snake venom components acting on blood coagulation and platelet function. Toxicon 1992;30:945–966.
71 Lu X, Rahman S, Deadman JJ, Chino N, Claeson G, Kakkar VV, Williams JA: Natural integrin antagonists from snake venoms: Ligand-specific inhibitors. Blood Coagul Fibrinolysis 1993;4:361–362.
72 Andrews RK, Berndt MC: Snake venom modulators of platelet adhesion receptors and their ligands. Toxicon 2000;38:775–791.

Procoagulant Proteins and Diagnostic Agents

Hæmostasis

Haemostasis 2001;31:218–224

Procoagulant Proteins from Snake Venoms

R. Manjunatha Kini[a,c] Veena S. Rao[a] Jeremiah S. Joseph[b]

[a]Department of Biological Sciences, Faculty of Science, [b]Department of Biochemistry, Faculty of Medicine, National University of Singapore, Singapore;
[c]Department of Biochemistry and Molecular Biophysics, Medical College of Virginia, Virginia Commonwealth University, Richmond, Va., USA

Key Words
Blood coagulation · Clotting cascade · Venom proteinase

Abstract
Several procoagulant proteins from snake venoms have been isolated and characterized. They are either serine proteinases or metalloproteinases, which activate specific zymogens of coagulation factors and initiate the coagulation cascade. These procoagulant proteins are useful in treating various thrombotic and hemostatic conditions and contribute to our understanding of molecular details in the activation of specific coagulation factors. Recent studies have shown that the prothrombin activators with serine proteinase activity are structurally and functionally similar to mammalian coagulation factors. Their structural studies should provide us insight into prothrombinase complex formation. Here, we present an overview of snake venom procoagulant factors.

Introduction

Snake venoms are rich sources of pharmacologically active proteins and peptides. These toxins interfere in several vital physiologic functions; some act as agonists and behave similarly to a natural ligand or activator of a specific step, whereas others act as antagonists and interfere in the function of a natural ligand or activator. In general, the toxins exhibit high specificity and have evolved to attack the specific, and most likely, key step in the physiologic pathway. Thus, they affect various physiologic functions with high specificity and high potency. Therefore, snake ven-

KARGER

Fax + 41 61 306 12 34
E-Mail karger@karger.ch
www.karger.com

0301–0147/01/0316–0218$17.50/0

Accessible online at:
www.karger.com/journals/hae

R. Manjunatha Kini
Department of Biological Sciences, Faculty of Science
Block S2, Room 04-11, 14 Science Drive
National University of Singapore, Singapore 117543
Tel. +65 874 5235, Fax +65 779 2486, E-Mail dbskinim@nus.edu.sg

om toxins provide several highly specific 'tools', which are useful in deciphering molecular details in various physiological processes. Some of these toxins are also useful in developing prototypes of therapeutic agents. Further, the study of toxins helps in developing protective measures to counter mortality and morbidity associated with snake envenomation.

The circulation of blood is essential for life and the integrity of the process is crucial for the survival of the organism. The integrity of this closed system is strictly regulated by the interplay of platelet aggregation, blood coagulation and vasoconstriction. These processes are intertwined and synergistic, and occur almost simultaneously to prevent blood loss from the injured vessel. Several snake venoms and their toxins have evolved to interfere in the mechanism of thrombosis and hemostasis. Some of these toxins affect platelet aggregation [1–4], whereas others affect blood coagulation [5–7]. In this review, we will describe structural and functional properties of procoagulant toxins.

Procoagulant Proteins from Snake Venom

Procoagulant venoms contain components that hasten clot formation. All the procoagulants characterized to date are proteinases; they activate a zymogen of specific coagulation factors in the coagulation cascade. Some venom proteinases also activate the protein cofactor, factor V. For example, Russell's viper *(Daboia russelli)* venom (RVV-V) and thrombocytin *(Bothrops atrox)* activate factor V, the cofactor in the prothrombinase complex [reviewed in ref. 8 and 9].

Factor VII Activation

So far, snake venom proteins that specifically activate only factor VII are not known. However, oscutarin, a group C prothrombin activator (see below) from *Oxyuranus scutellatus* venom activates factor VII [10]. Gel electrophoresis analysis of the cleavage products indicated that activation occurred at a site similar to the natural cleavage site. The activation of factor VII, similar to that of prothrombin, was greatly potentiated by Ca^{2+} ions and phospholipids, and was not dependent upon the presence of its factor Va-like subunit [10]. The structural details of oscutarin and related proteins are given below.

Factor X Activation

Venoms from Viperidae, Crotalidae and Elapidae contain a variety of proteinases capable of activating factor X [for reviews, see ref. 11 and 12]. They are either metalloproteinases or serine proteinases. The factor X activator from RVV-X, first purified by Williams and Esnouf [13], is a metalloproteinase (93 K_d glycoprotein) with a heavy chain (catalytic subunit) and two distinct light chains [14]. The heavy chain (427 residues) consists of a metalloproteinase, a disintegrin and a cysteine-rich domain [15]. The light chains are homologous to C-type lectins, particularly to factor IX/X-binding protein from *Trimeresurus flavoviridis* venom [16]. They act as regulatory subunits and recognize the Ca^{2+}-bound conformation of the Gla domain of factor X [11]. This structure of RVV-X is very similar to that of the prothrombin activator, carinactivase-1, from *Echis carinatus leukogaster* (see below). RVV-X specifically cleaves the Arg_{52}-Ile_{53} bond in the heavy chain of factor X, and it requires Ca^{2+} ions at millimolar concentrations for optimal activity. RVV-X is unable to activate factor X bound to negatively charged phospholipid bilayers; hence, it can be used to determine the binding of factor X to phospholipid vesicles [17]. Other metalloproteinase factor X activators from *B. atrox* and *Cerastes cerastes* venoms have structures similar to RVV-X [11]. Hence, they probably have similar catalytic mechanisms.

A few serine proteinases that activate factor X have also been isolated from the venoms of *Ophiophagus hannah*, *Bungarus fasciatus*, *C. cerastes* and *C. vipera* [11]. We have also isolated a serine proteinase factor X activator from the venom of the Malayan krait, *Bungarus candidus* [Joseph and Kini, unpubl. results]. However, currently no structural details are available on any of these proteins.

Prothrombin Activators

A large number of snake species contain prothrombin activators in their venoms [for an inventory, see ref. 18 and for reviews, see ref. 19–21]. Based on their structural properties, functional characteristics and cofactor requirements, they have been categorized into four groups [19, 22–23]. For a recent review on prothrombin activators, [see ref. 24].

Group A Prothrombin Activators

These metalloproteinases efficiently activate prothrombin without the requirement of any cofactors, such as Ca^{2+} ions, phospholipids or factor Va [19]. They are widely distributed in many kinds of viper venoms. The best-known example is ecarin, from the venom of the saw-scale viper *E. carinatus* [25]. The sequence of this protein was deduced from the sequences of cDNA clones [26]. The mature protein is a metalloproteinase with 426 amino acids and shares 64% identity with the heavy chain of RVV-X. It consists of three domains: a metalloproteinase, disintegrin and a cysteine-rich domain. It is a highly efficient enzyme with a low K_m for prothrombin and a high K_{cat} [21]. It cleaves the Arg_{320}-Ile_{321} bond in prothrombin and produces meizothrombin, which is converted to α-thrombin by autolysis. Ecarin also activates descarboxyprothrombin that accumulates in plasma during warfarin therapy. Other prothrombin activators in this class [18, 19], for example those isolated from the *Bothrops* species, also have similar properties [21].

Group B Prothrombin Activators

These are Ca^{2+}-dependent prothrombin activators, such as carinactivase-1 [23] and multiactivase [27] from *E. carinatus leukogaster* and *E. multisquamatus* venom, respectively. They consist of two subunits held noncovalently: a 62 K_d metalloproteinase and a 25 K_d C-type lectin-like disulfide-liked dimer. In contrast to the group A activators, carinactivase-1 requires millimolar concentrations of Ca^{2+} for activity; it has virtually no activity in the absence of Ca^{2+}. Also, unlike ecarin, it does not activate prothrombin derivatives in which Ca^{2+}-binding has been perturbed: prethrombin-1 and descarboxyprothrombin. This property is used to develop a chromogenic assay for normal prothrombin in the plasma of warfarin-treated individuals [28]. The metalloproteinase catalytic subunit taken in isolation does not require Ca^{2+} for activity and its reconstitution with the C-type lectin-related subunit restores Ca^{2+} dependence. They recognize the Ca^{2+}-bound conformation of the Gla domain in prothrombin via the 25 K_d regulatory subunit.

Group C Activators

These are serine proteinases found apparently exclusively in the venoms of Australian elapids [18, 19], and they require only Ca^{2+} ions and negatively charged phospholipids, but not factor Va, for maximal activity. They have been purified and characterized from *O. scutellatus* [29–31] and *Pseudonaja textilis* venoms [32] [Rao and Kini, unpubl. obs.]. The native activators have a molecular mass of approximately 300 K_d and consist of a factor Xa-like catalytic subunit (60 K_d) and a factor Va-like cofactor subunit (approximately 200 K_d). The two chains of the factor Xa-like catalytic subunit are linked by disulfide bonds, whereas the two chains of the factor Va-like cofactor are held together by noncovalent interactions. The catalytic subunit, similar to factor Xa, has weak catalytic activity in isolation, which is greatly stimulated by the presence of the factor Va-like subunit. Bovine factor Va can also substitute for the cofactor subunit of the enzyme [31, 32] [Rao and Kini, unpubl. obs.]. These activators cleave at both Arg_{271}-Thr_{272} and Arg_{320}-Ile_{321} bonds of prothrombin [31], converting it to mature thrombin, in contrast to group A and B activators, which only convert prothrombin to meizothrombin. Recent structural studies indicate that the group C prothrombin activators structurally and functionally resemble the mammalian factor Xa-Va complex.

Group D Prothrombin Activators

These are also serine proteinases found exclusively in the venoms of Australian elapids and their activities are strongly stimulated by Ca^{2+} ions, factor Va and negatively charged phospholipid vesicles [33–35]. They are serine proteinases with molecular mass ranging from 45,000 to 47,000 (based on mass spectrometry and primary structure [36], or from 52,000 to 58,000, SDS-PAGE, [37–41]). They have two chains held together by disulfide bonds. The serine proteinase active site is located in the heavy chain [36, 37].

They exhibit potent procoagulant effects, comparable to mammalian factor Xa, through activation of prothrombin [36, 37]. Similar to factor Xa, they cleave prothrombin at two sites, Arg_{274}-Thr_{275} and Arg_{323}-Ile_{324} [36, 37]. Their poor activator activity is stimulated by more than a million-fold by negatively charged phospholipids, factor Va and Ca^{2+} ions [36, 37]. They also hydrolyze factor Xa-specific chromogenic substrates [36, 37]. Recently, we determined the complete amino acid sequences of trocarin [34] and hopsarin D [Rao et al., unpubl. data]. They share high identity (53–60%) and homology (62–70%) and hence similar domain architecture with factor Xa [36]. Their light chains are homologous to the light chains of vitamin K-dependent coagulation factors, especially that of factor Xa (53–59% identity, 60–67% homology with factor Xa) [36]. Their light chains consist of an N-terminal Gla domain (residues 1–39), followed by two epidermal growth factor (EGF)-like domains, EGF-I (residues 50–81) and EGF-II (residues 89–124). Thus, group D venom prothrombin activators are true structural and functional homologues of blood coagulation factors.

Thrombin-Like Enzymes

These fibrinogen-clotting enzymes are widely distributed within several pit viper genera *(Agkistrodon, Bothrops, Lachesis* and *Trimeresurus)*, as well as some true vipers *(Bitis* and *Cerastes)* and the colubrid, *Dispholidus typus* (for an inventory, see [42] and for recent review, see [43, 44]). The thrombin-like enzymes (TLEs) are single-chain serine proteinases (for example [45]), except for the enzyme from *C. cerastes*, which is reported to consist of two identical disulfide-linked chains [46]. Some of them are glycoproteins. They share a high degree of sequence identity among themselves (approximately 67%). However, they show less than 40% homology with human thrombin. The presence of Asp 189 in all the TLEs explains their cleavage of arginyl bonds in fibrinogen, as this is the specificity pocket residue common to all serine proteinases that hydrolyze peptide bonds after basic residues [44]. They preferentially release either fibrinopeptide A or B, rarely both with equal efficiency, unlike thrombin [43, 47]. Secondary cleavages are also often seen. They also show a fair degree of species specificity in the efficiency of fibrinogen conversion. Classical serine proteinase inhibitors inhibit TLEs, but most of them are not inhibited by thrombin inhibitors like antithrombin III and hirudin [5, 43, 47]. They act on blood plasma usually forming friable and translucent clots, presumably due to a lack of cross-linking of fibrin by factor XIIIa. TLEs often act on coagulation factor XIII as well, but appear to degrade rather than activate it [5]. Unlike thrombin, TLEs do not activate other coagulation factors [47]. Thus, although TLEs 'resemble' thrombin to an extent, they are structurally and functionally dissimilar to the coagulation factor [5, 43, 24]. These unique properties enable their clinical use as defibrinogenating agents; for example, ancrod (Arvin®; *Calloselasma rhodostoma*) and batroxobin (Defibrase®; *B. atrox moojeni*; [48]), [reviewed in ref. 43 and 49]. Since the fibrin formed is not cross-linked, it is readily degraded by the fibrinolytic system. TLEs are clinically well tolerated with no or only minimal side effects [43].

Importance of Procoagulant Venom Proteins

Procoagulant venom proteins activate the coagulation cascade at specific steps. This property has enabled their use in diagnosis of coagulation factor deficiencies by the determination of coagulation times of blood plasma treated with these proteins. As mentioned above, TLEs are also used for clinical defibrinogenation of blood plasma. The unique sequence specificity of cleavage of group D prothrombin activators can be exploited in cleavage of recombinant fusion proteins in molecular biology. Potentially, these proteins can also be used to selectively detect fully carboxylated prothrombin and lupus anticoagulant [49, 50] and possibly factor Va deficiency in human plasma. Finally, characterization of structure-function relationships in group C and D prothrombin activator will enable us to better understand interactions within the prothrombinase complex and the mechanisms of prothrombin activation.

Acknowledgements

This work was supported by National University of Singapore academic research grants R-154-000-093-112 and R-368-000-004-112.

References

1 Kini RM, Chow G: Exogenous inhibitors of platelet aggregation from animal sources. Thromb Haemost 2001;85:179–181.
2 Chow G, Kini RM: Exogenous factors from animal sources that induce platelet aggregation. Thromb Haemost 2001;85:177–178.
3 Kini RM, Evans HJ: Effect of phospholipase A_2 enzymes on platelet aggregation; in Kini RM (ed): Venom Phospholipase A_2 Enzymes: Structure, Function and Mechanism. Chichester, Wiley & Sons, 1997, pp 369–387.
4 Kini RM, Evans HJ: Effects of snake venom proteins on blood platelets. Toxicon 1990;28:1387–1422.
5 Hutton RA, Warrell DA: Action of snake venom components on the haemostatic system. Blood Rev 1993;7:176–189.
6 Marsh NA: Snake venoms affecting the haemostatic mechanism – a consideration of their mechanisms, practical applications and biological significance. Blood Coagul Fibrinolysis 1994;5:399–410.

7 Markland FS Jr: Snake venoms. Drugs 1997;54(suppl 3):1–10.
8 Tokunaga F, Iwanaga S: Proteases activating factor V; in Bailey GS (ed): Enzymes from Snake Venom. Fort Collins, Alaken, 1998, pp 209–225.
9 Rosing J, Govers-Riemslag JWP, Yukelson L, Tans G: Factor V activation and inactivation by venom proteases. Haemostasis 2001;31: 241–246.
10 Nakagaki T, Lin P, Kisiel W: Activation of human factor VII by the prothrombin activator from the venom of *Oxyuranus scutellatus* (Taipan snake). Thromb Res 1992;65: 105–116.
11 Morita T: Proteases which activate factor X; in Bailey GS (ed): Enzymes from Snake Venom. Fort Collins, Alaken, 1998, pp 179–208.
12 Tans G, Rosing J: Snake venom activators of factor X: An overview. Haemostasis 2001;31:225–233.
13 Williams WJ, Esnouf MP: The fractionation of Russell's viper *(Vipera russelli)* venom with special reference to the coagulant protein. Biochem J 1962;84:52–62.
14 Gowda DC, Davidson EA: Isolation and characterization of novel mucin-like glycoproteins from cobra venom. J Biol Chem 1994;269: 20031–20039.
15 Takeya H, Nishida S, Miyata T, Kawada S, Saisaka Y, Morita T, Iwanaga S: Coagulation factor X activating enzyme from Russell's viper venom (RVV-X). A novel metalloproteinase with disintegrin (platelet aggregation inhibitor)-like and C-type lectin-like domains. J Biol Chem 1992;267:14109–14117.
16 Atoda H, Hyuga M, Morita T: The primary structure of coagulation factor IX/factor X-binding protein isolated from the venom of *Trimeresurus flavoviridis*. Homology with asialoglycoprotein receptors, proteoglycan core protein, tetranectin, and lymphocyte Fcε receptor for immunoglobulin E. J Biol Chem 1991; 266:14903–14911.
17 van Dieijen GV, Tans G, van Rijn JV, Zwaal RF, Rosing J: Simple and rapid method to determine the binding of blood clotting factor X to phospholipid vesicles. Biochemistry 1981;20:7096–7101.
18 Rosing J, Tans G: Inventory of exogenous prothrombin activators. Thromb Haemost 1991;65:627–630.
19 Rosing J, Tans G: Structural and functional properties of snake venom prothrombin activators. Toxicon 1992;30:1515–1527.
20 Tans G, Rosing J: Prothrombin activation by snake venom proteases. J Toxicol Toxin Rev 1993;12:155–173.
21 Petrovan R, Tans G, Rosing J: Proteases activating prothrombin; in Bailey GS (ed): Enzymes from Snake Venom. Fort Collins, Alaken, 1998, pp 227–252.
22 Kini RM, Morita T, Rosing J: Classification and nomenclature of prothrombin activators isolated from snake venoms. Thromb Haemost, in press.
23 Yamada D, Sekiya F, Morita T: Isolation and characterization of carinactivase, a novel prothrombin activator in *Echis carinatus* venom with a unique catalytic mechanism. J Biol Chem 1996;271:5200–5207.
24 Kini RM, Joseph JS and Rao VS: Prothrombin activators from snake venoms; in Mènez A (ed): Perspectives in Molecular Toxinology. Chichester, Wiley, in press.
25 Kornalik F, Blomback B: Prothrombin activation induced by Ecarin – a prothrombin converting enzyme from *Echis carinatus* venom. Thromb Res 1975;6:57–63.
26 Nishida S, Fujita T, Kohno N, Atoda H, Morita T, Takeya H, Kido I, Paine MJ, Kawabata S, Iwanaga S: cDNA cloning and deduced amino acid sequence of prothrombin activator (ecarin) from Kenyan *Echis carinatus* venom. Biochemistry 1995;34:1771–1778.
27 Yamada D, Morita T: Purification and characterization of a Ca^{2+}-dependent prothrombin activator, multactivase, from the venom of *Echis multisquamatus*. J Biochem (Tokyo) 1997;122:991–997.
28 Yamada D, Morita T: CA-1 method, a novel assay for quantification of normal prothrombin using a Ca^{2+}-dependent prothrombin activator, carinactivase-1. Thromb Res 1999;94:221–226.
29 Owen WG, Jackson CM: Activation of prothrombin with *Oxyuranus scutellatus scutellatus* (Taipan snake) venom. Thromb Res 1973;3:705–714.
30 Walker FJ, Owen WG, Esmon CT: Characterization of the prothrombin activator from the venom of *Oxyuranus scutellatus scutellatus* (Taipan venom). Biochemistry 1980;19: 1020–1023.
31 Speijer H, Govers-Riemslag JWP, Zwaal RFA, Rosing J: Prothrombin activation by an activator from the venom of *Oxyuranus scutellatus* (Taipan snake). J Biol Chem 1986; 261:13258–13267.
32 Masci PP, Whitaker AN, de Jersey J: Purification and characterization of the prothrombin activator of the venom of *Pseudonaja textilis*; in Gopalakrishnakone P, Tan CK (eds): Progress in Venom and Toxin Research. Proceedings of the First Asia-Pacific Congress on Animal, Plant and Microbial Toxins. Singapore, 1987, pp 209–219.
33 Jobin F, Esnouf MP: Coagulant activity of tiger snake (*Notechis scutatus scutatus*) venom. Nature 1966; 211:873–875.
34 Chester A, Crawford GP: In vitro coagulant properties of venoms from Australian snakes. Toxicon 1982;20:501–504.
35 Marshall LR, Herrmann RP: Coagulant and anticoagulant actions of Australian snake venoms. Thromb Haemost 1983;50:707–711.
36 Joseph JS, Chung MC, Jeyaseelan K, Kini RM: Amino acid sequence of trocarin, a prothrombin activator from *Tropidechis carinatus* venom: Its structural similarity to coagulation factor Xa. Blood 1999;94:621–631.
37 Tans G, Govers-Riemslag JWP, van Rihn JL, Rosing J: Purification and properties of a prothrombin activator from the venom of *Notechis scutatus scutatus*. J Biol Chem 1985; 260:9366–9372.
38 Morrison JJ, Masci PP, Bennett EA, Gauci M, Pearn J. Whitaker AN, Korschenko LP: Studies of the venom and clinical features of the Australian rough-scaled snake *(Tropidechis carinatus)*; in Gopalakrishnakone P, Tan CK (eds): Progress in Venom and Toxin Research. Proceedings of the First Asia-Pacific Congress on Animal, Plant and Microbial Toxins, Singapore, 1987, pp 220–233.

39 Williams V, White J: Purification and properties of a procoagulant from peninsula tiger snake *(Notechis ater niger)* venom. Toxicon 1989; 27:773–779.
40 Stocker K, Hauer H, Muller C, Triplett DA: Isolation and characterization of Textarin, a prothrombin activator from eastern brown snake *(Pseudonaja textilis)* venom. Toxicon 1994;32:1227–1236.
41 Marsh NA, Fyffe TL, Bennett EA: Isolation and partial characterization of a prothrombin-activating enzyme from the venom of the Australian rough-scaled snake *(Tropidechis carinatus)*. Toxicon 1997;35:563–571.
42 Pirkle H, Stocker K: Thrombin-like enzymes from snake venoms: an inventory. Thromb Haemost 1991;65: 444–450.
43 Bell WR Jr: Defibrinogenating enzymes. Drugs 1997;54 (suppl 3):18–30.
44 Pirkle H, Theodor I: Thrombin-like enzymes; in Bailey GS (ed): Enzymes from Snake Venom. Fort Collins, Alaken, 1998, pp 39–69.
45 Au LC, Lin SB, Chou JS, Teh GW, Chang KJ, Shih CM: Molecular cloning and sequence analysis of the cDNA for ancrod, a thrombin-like enzyme from the venom of *Calloselasma rhodostoma*. Biochem J 1993; 294:387–390.
46 Laraba-Djebari B, Martin-Eauclaire MF, Marchot P: A fibrinogen-clotting serine proteinase from *Cerastes cerastes* (horned viper) venom with arginine-esterase and amidase activities. Purification, characterization and kinetic parameter determination. Toxicon 1992;30:1399–1410.
47 Aronson DL: Comparison of the actions of thrombin and the thrombin-like venom enzymes ancrod and batroxobin. Thromb Haemost 1976; 36:9–13.
48 Stocker K, Barlow GH: The coagulant enzyme from *Bothrops atrox* venom (batroxobin). Methods Enzymol 1976;45:214–223.
49 Stocker K: Research, diagnostic and medicinal uses of snake venom enzymes; in Bailey GS (ed): Enzymes from Snake Venom. Fort Collins, Alaken, 1998, pp 705–736.
50 Triplett DA, Stocker KF, Unger GA, Barna LK: The Textarin/Ecarin ratio: A confirmatory test for lupus anticoagulants. Thromb Haemost 1993;70:925–931.

Procoagulant Proteins and Diagnostic Agents

Hæmostasis

Haemostasis 2001;31:225–233

Snake Venom Activators of Factor X: An Overview

Guido Tans Jan Rosing

Cardiovascular Research Institute Maastricht (CARIM), Maastricht University, Maastricht, The Netherlands

Key Words

Snake venom · Blood coagulation factor X · Factor X activator · Metalloprotease · Serine protease

Abstract

Activators of blood coagulation factor X have been described in the venom of many snake species belonging to the genus Viperidae and Crotalidae as well as from a few Elapid species. Based on the structural and functional properties of purified activating principles, factor X activators are either metalloproteases or serine proteases. The best known activator is RVV-X from Russell's viper *(Daboia russelli)*, a metalloprotease consisting of a heavy chain containing the catalytic domain and two light chains which share homology with C-type lectins and which are thought to exert a regulatory function in the Ca^{2+}-dependent activation of factor X. This activator is also one of the best examples of the use of exogenous activators in coagulation research and in addition it is used in many diagnostic research kits. In this paper, an overview is given of the structural and functional properties of snake venom factor X activators thus far described in the literature.

Introduction

Blood coagulation factor X is one of the key components of the haemostatic system. Factor X is synthesized as a single-chain glycoprotein of 488 amino acids including a 40-amino-acid pre-pro leader peptide. Upon secretion, the leader sequence is cleaved off, but in addition a tripeptide (Arg-Lys-Arg) is also removed and, thus, the mature protein circulates in the blood as a two-chain protein (fig. 1). The light chain contains 11 γ-carboxyglutamic acid (Gla) residues which are the

KARGER

Fax + 41 61 306 12 34
E-Mail karger@karger.ch
www.karger.com

0301–0147/01/0316–0225$17.50/0

Accessible online at:
www.karger.com/journals/hae

G. Tans, PhD, Department of Biochemistry
Cardiovascular Research Institute Maastricht, Maastricht University
PO Box 616, NL–6200 MD Maastricht (The Netherlands)
Tel. +31 43 3881677, Fax +31 43 3884159
E-Mail g.tans@bioch.unimaas.nl

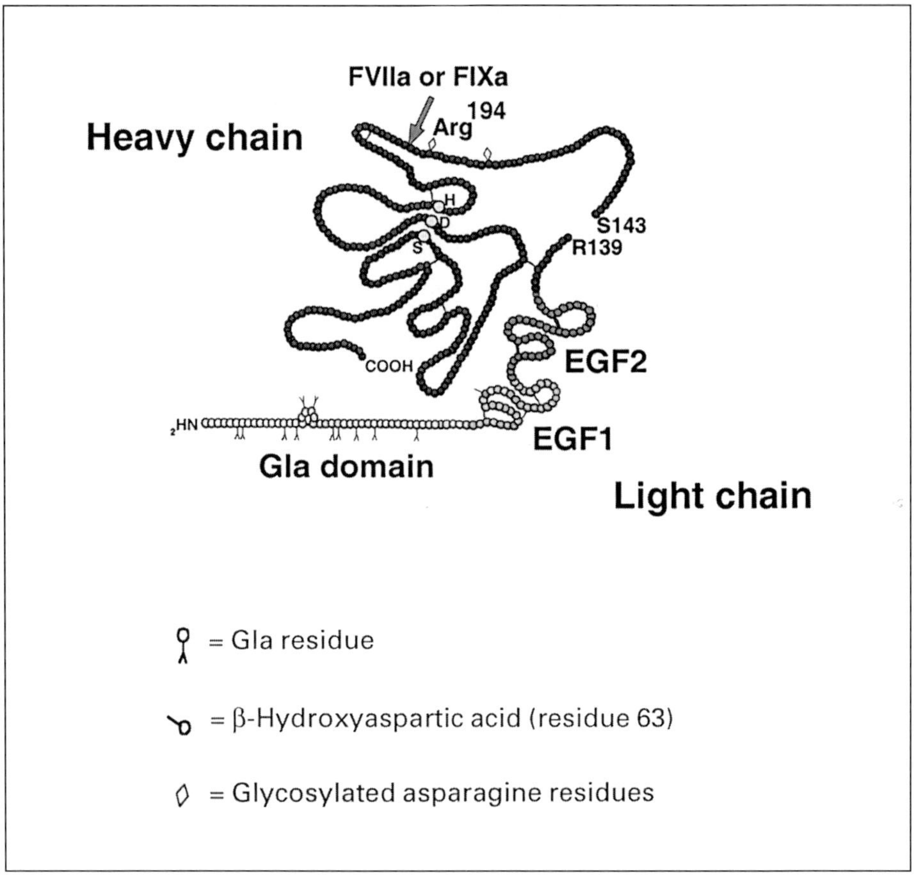

Fig. 1. Schematic representation of the primary sequence and domain structure of blood coagulation factor X. Each circle represents an amino acid residue. The structure is shown without the pre-pro leader sequence and without the tripeptide (Arg^{140}-Lys^{141}-Arg^{142}) which connects the heavy chain to the light chain. The catalytic triad (Ser195, His57 and Asp102) is indicated.

result of post-translational carboxylation of specific glutamic acid residues at the amino-terminal end (Gla domain), which are essential for the Ca^{2+}-dependent binding of factor X(a) to procoagulant (negatively charged) membranes. The Gla domain is followed by two EGF-like domains, the first of which contains a β-hydroxy aspartic acid residue at position 63 which is also introduced by post-translational modification. The heavy chain of factor X contains the catalytic triad Ser195-His57-Asp102 (chymotrypsin numbering [1]), typical of serine proteases. Activation of factor X is brought about by cleavage of a single peptide bond at Arg194-Ile195 (fig. 1) which removes the heavily glycosylated first 52 amino-terminal residues of the heavy chain and upon which the active site becomes fully exposed. Under physiological conditions factor X activation is a membrane-bound process in

Table 1. Snake species with factor-X-activating principles in the venom

Species		Reference
Bothrops	*atrox, brazili,columbiensis, insularis, marajoensis*	9, 11
	cotiara	10
	erythromelas	9, 10, 38
	alternatus, jararaca, jararacussu, moojeni	9, 10
	n. paranaensis, n. pauloensis, n. urutu	9, 10
	n. neuwiedi, n. diporus, n. matogrossensis	9
	n. pubescens, pirajai, pradoi, venezuelensis, megaera	9
Bungarus	*fasciatus*	12, 39
Cerastes	*cerastes, vipera*	8, 32, 34, 35
Calloselasma	*rhodostoma*	8
Daboia[a]	*russelli russelli, russelli siamensis*	7, 8
Echis	*carinatus, ocellatus*	8, 40, 41
Naja	*naja atra*	39
Ophiophagus	*hannah*	12, 39
Vipera	*berus berus, aspis aspis, ammodytes ammodytes lebetina*	8, 33, 36

[a] Only the initial original references are given.

which factor X is cleaved by the tissue factor-factor VIIa complex during the initiation phase of coagulation via the extrinsic pathway. This initial phase is efficiently attenuated by tissue factor pathway inhibitor (TFPI) but, depending upon the initial strength of the triggering event, the activation of factor X at the membrane can be sustained by the tenase complex (factor IXa together with factor VIIIa). However, even before the blood coagulation cascade hypothesis was formulated, it was already recognized that certain snake venoms were capable of very efficiently clotting blood [2]. Nowadays, it is recognized that such venoms can contain combinations of activating principles that can initiate and interfere with many steps of the blood coagulation cascade. Indeed, many of the activating principles have greatly facilitated research aimed at understanding and unravelling the mechanism of blood coagulation reactions. One of the targets to initiate or interfere with blood coagulation appears to be at the level of factor X or factor Xa since many proteins have been identified in snake venoms that are capable of activating factor X or inhibiting factor Xa. Some of the inhibitors show a remarkable specificity and have been found in blood-sucking animals such as vampire bats [3], ticks [4] or leeches [5, 6].

Snake Venom Activators of Factor X

Since the publication of McFarlane, who was the first to identify a factor X activating principle in the venom of Russell's viper [7], many other snake venoms were found to contain a factor X activator. Table 1 gives a comprehensive list of the snake species reported. Factor X activators appear widespread in Viperid and Crotalid venoms and, more recently, a few Elapid venoms were also found to contain factor-X-activating principles. Colu-

Table 2. Structural and functional properties of purified factor-X-activating principles

Species	Enzyme class	M_r	Subunits	Cofactor(s)	Inhibitors
D. russelli	metalloprotease	92,880	heavy chain two light chains	Ca^{2+}	chelating agents/PL
C. cerastes	metalloprotease	75,000	heavy chain[a] two light chains	Ca^{2+}	chelating agents/PL
	serine protease	40,000	two subunits		DFP
	serine protease	14,000	single chain		
C. vipera	serine protease	12,500	single chain		PMSF
B. atrox	metalloprotease	75,000	heavy chain[a] two light chains	Ca^{2+}	chelating agents
V. berus berus	metalloprotease	38,000		Ca^{2+}	chelating agents
V. aspis aspis	metalloprotease	75,000	multiple chains	Ca^{2+}	
V. lebetina		102,000[b]	heavy chain two light chains		
O. hannah	serine protease	64,500		Ca^{2+}	PMSF, TPCK, DFP
B. fasciatus	serine protease	70,000		Ca^{2+}	PMSF, DFP

[a] A heavy chain and two light chains were reported which makes it likely that these activators consist of three subunits.

[b] Calculated from the M_r reported for the heavy chain (67,000) and the two light chains (16,200 and 18,800).

brid venoms do not appear to contain factor X activators because thus far none have been reported and efforts to identify them in venom from Rhabdophis and Dispholidus genera failed [8]. Nahas et al. [9] screened the venom of 26 Bothrops species and found that only six venoms *(B. atrox asper, B. cotiara, B. fonsecai, B. ittapetiningae, B. castelnaudi)* were not able to activate factor X [9]. However, in a later screening Furtado et al. [10] did find a factor X activator in the venom of *B. cotiara* (especially in young snakes), but not in that of *B. jararacussu* (table 1). Thus, apart from some minor differences reported, the presence of factor X activators is rather widespread in snake venoms and apparently offers some advantage to the snake in captivating and/or killing its prey. From these many venoms, only a limited number of activating principles have been purified to homogeneity and characterized well. Based on what is known about their functional and physical properties, snake venom factor X activators are either metalloproteases or serine proteases (table 2). With only a few exceptions, efficient activation of factor X critically depends upon the presence of $CaCl_2$. Dose-response curves for $CaCl_2$ typically show a sigmoidal dependency with Hill coefficients ranging from 2.4 to 7.9 [11–13]. Since the mechanism of factor X activation by the activator present in Russell's viper venom (RVV-X) is quite well understood and can be regarded as an example for the other venom activators belonging to the group of metalloproteases, we will describe this activator in more detail.

The Activator from Russell's Viper Venom

By far the strongest and most important factor X activator is the one found in the venom of Russell's viper *(Daboia russelli,* formerly known as *Vipera russelli)*. Venom preparations of the major two subspecies, i.e *Daboia russelli russelli* and *Daboia russelli siamensis,* show potencies that are approximately the same and that are at least ten- to a hundred-fold higher than in any other venom [8]. The purified factor X activator (RVV-X) contains some 13% carbohydrate and runs as a single band with 79,000 apparent M_r on SDS-PAGE [14]. RVV-X activates factor X by cleaving the Arg-Ile bond at position 194 (cf. fig. 1) and is also able to activate factor IX and protein C by specific cleavage of Arg-Ile and Arg-Val bonds [14, 15]. The exact molecular weight, sugar composition as well as the composition of the subunit structure following reduction was clarified by Gowda et al. [16] who showed that RVV-X (M_r = 92,880) consists of a heavy chain of 57,600 and two light chains of 19,400 and 16,400 M_r, respectively, held together by disulfide bond(s). The molecule contains six N-linked oligosaccharides, four of which are located in the heavy chain and one in each light chain. Removal of easily accessible sugars does not appear to change the functional capacity of RVV-X, but removal of the core sugars with N-glycanase causes a virtually complete loss of factor-X-activating capacity, apparently as a result of major conformational changes in the molecule [17].

The complete amino acid sequence of the heavy chain and the smallest of the two light chains have been determined [18]. The 427-residue heavy chain was found to contain three distinct domains: a metalloproteinase domain containing the conserved HExxH metalloprotease catalytic site sequence, a disintegrin (platelet aggregation inhibitor) domain and a cysteine-rich region, the function of which is as yet not completely clear. The disintegrin domain is presumably responsible for the strong antiplatelet-aggregating effect of RVV-X, although the specific RGD sequence is missing [18]. The light chain (123 residues) shares homology with C-type (Ca^{2+}-dependent) lectins and with the anticoagulant proteins factor IX/X binding protein of *Trimeresurus flavoviridis* [18, 19] and factor-X-binding protein of *Deinagkistrodon acutus* [19]. These latter findings are of great interest and can explain some of the mechanistic properties of RVV-X described earlier. The factor-IX/X- and X- binding proteins consist of two C-type lectin subunits held together by a single disulfide bond, probably as the result of domain swapping between the two subunits [20, 21]. These C-type lectin heterodimers specifically bind to the Gla domain of factor IX and factor X in a Ca^{2+}-dependent manner [19, 22] as a result of which the binding of these clotting factors to negatively charged membranes is likely blocked [23]. Mizuno et al. [23] recently published the crystal structure of the factor-X-binding protein of *D. acutus* in complex with the Gla domain (residues 1–44) of factor X and showed that the two C-type lectin domains form a concave region into which the Gla domain of factor X fits by interactions which involve a number of Gla residues and Ca^{2+} ions that form salt bridges with a cluster of positively charged residues in the factor-X-binding protein.

These data fit well with the model that Morita [24] proposed for RVV-X in which it is supposed that the two C-type lectin light chains of RVV-X serve as an exosite by which RVV-X recognizes and binds to the Gla domain of factor X (fig. 2). From the crystal structure of factor-X-binding protein from *D. acutus* with the Gla domain of factor X, a critical role for Ca^{2+} ions and Gla residues can be

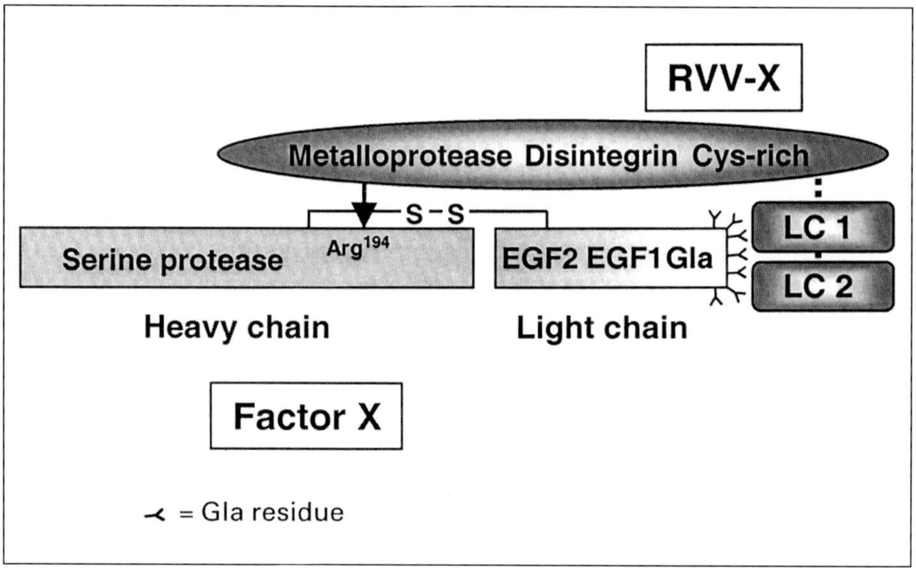

Fig. 2. Mechanism proposed by Morita et al. [24] for the activation of factor X by RVV-X. The heavy chain of RVV-X contains the metalloprotease domain responsible for the proteolytic cleavage of the target bond in factor X. The two C-type lectin light chains (LC1 and LC2) together form a secondary binding site specific for interaction with the Gla domain of factor X. Courtesy of Prof. T. Morita.

inferred which explains why Gla-domainless factor X or non-carboxylated factor X are poor substrates for RVV-X [25]. Together with the primary specificity of the metalloprotease domain for hydrophobic and bulky amino acid residues at P1′, which may be extending towards P2′ and P3′, the presence of such an exosite can further explain the substrate specificity and catalytic efficiency exhibited by RVV-X. Thus, factor X activation by RVV-X is effectively inhibited by the Gla domain of factor X (residues 1–44) [26] an inhibition that would be unlikely without direct interaction between the Gla domain and an exosite on RVV-X. In addition, RVV-X is capable of efficient activation not only of factor X but also of factor IX and protein C [14, 15], probably as a result of the fact that the Gla domains of these three proteins are very similar [27]. Such a model can also explain earlier findings in which it was reported that activation of bovine factor X by RVV-X was inhibited by the binding of factor X to negatively charged lipids [28]. Binding of factor X to negatively charged lipids involves the Gla domain and will, therefore, by competition diminish the interaction of RVV-X with factor X resulting in an inhibition of activation.

Finally, the similarity with prothrombin activators such as carinactivase and multactivase should be noted. The properties of these activators in prothrombin activation are similar to those of RVV-X in factor X activation. Thus, carinactivase and multactivase are also comprised of three subunits (a heavy chain and two light chains) and prothrombin activation by these activators is also critically dependent upon the presence of Ca^{2+} ions and a normally carboxylated prothrombin Gla domain [29, 30]. Moreover, the isolated regulatory subunits of carinactivase recognize the Ca^{2+}-bound conformation of the prothrombin Gla domain [29] and binding of prothrombin to negatively charged lipids results in inhibition of prothrombin activation [29, 31].

Metalloprotease Activators

From the snake venom activators that were purified to homogeneity, a number were found to be both structurally and functionally similar to RVV-X (table 2). Although in the original publications concerning the activators purified from *B. atrox* and from *Cerastes cerastes* venom, it was proposed that the purified activators showed heterogeneity because two light chains were observed on SDS-PAGE [11, 32], it is very likely that these activators, like RVV-X, consist of a heavy chain containing the metalloprotease domain and two C-type lectin light chains held together by disulfide bond(s). Hoffmann and Bon [11] isolated two factor X activators from *B. atrox* venom, but apart from differences in the purification both activators show similar functional characteristics. In addition, it was noted that these activators cleave the heavy chain of factor X at two additional positions, an activity not seen with RVV-X. Very recently, an RVV-X like activator was purified from the venom of *Vipera lebetina* which, like RVV-X, was found capable of activating factor IX [33].

The activators from *Cerastes* venom deserve special attention. Franssen et al. [32] purified an activator which showed typically the same characteristics as RVV-X, but these authors failed to find other activators. In later publications, however, several other activators were identified which appear to be serine proteases (see below) [34, 35].

Vipera berus berus venom may contain several activators and one of these (VBFXAE) was purified to homogeneity and found to be a single-chain glycoprotein of 38,000 M_r [36]. This enzyme, which contains two Ca^{2+} ions, specifically activates factor X, although a weak hydrolysis of the insulin B chain was noted [36]. The latter activity was inhibited by EDTA, which indicates that this preparation contains a metalloprotease.

Komori et al. [37] purified an activator of 75,000 M_r from *Vipera aspis aspis* venom which upon reduction showed bands only at 14,000 and 16,000 M_r. These authors, therefore, suggested that the activator likely comprises a multitude of these subunits held together by disulfide bonds and the exact subunit composition awaits clarification [37].

Serine Protease Activators

Until the early 90s, factor X activators had only been identified in Crotalid and Viperid venoms. However, since then specific factor X activators have also been found in Elapid venoms (table 1, 2). These venom activators are enzymes that are quite different from the one found in *D. russelli.* They appear to be serine proteases since they are inhibited by phenylmethylsulphonyl fluoride (PMSF), diisopropylfluorophosphate (DPF) and chloromethylketones [12, 13]. Moreover, these enzymes are single-chain proteins as judged by SDS-PAGE in the absence and presence of reducing agents [12, 13]. However, they share with RVV-X an absolute dependence on Ca^{2+} ions and the effect of Ca^{2+} on the activation of factor X is strongly cooperative with Hill coefficients of 6.83 and 7.9 for the activator of *Ophiophagus hannah* and *Bungarus fasciatus*, respectively [12, 13]. Similar activators have been found in venoms from other *Bungarus* species as well [Kini, R.M., pers. commun., July 12, 2001].

The activators that were reported in *C. cerastes* and *Cerastes vipera* venom are somewhat atypical. Thus far, this has been the only species for which both an RVV-like metalloproteinase and serine protease factor X activators have been reported. Moreover, a number of the serine proteases reported are rather small (14,000 and 12,500 M_r [35]) when compared to the other factor X activators described. Therefore, final sequence analysis of the various activators present in venom from *Cerastes* species will be necessary to reveal whether these low M_r enzymes reflect partially degraded products or whether these truly represent another class of activators.

References

1 Bode W, Turk D, Karshikov A: The refined 1.9-Å X-ray crystal structure of *D*-Phe-Pro-Arg chloromethylketone-inhibited human alpha-thrombin: Structure analysis, overall structure, electrostatic properties, detailed active-site geometry, and structure-function relationships. Protein Sci 1992;1:426–471.

2 Barnett B, Macfarlane RG: On the relative potency of certain snake venoms to coagulate haemophilic blood. Proc Zool Soc Lond 1934;pt 4:977.

3 Fernandez AZ, Tablante A, Beguin S, Hemker HC, Apitz-Castro R: Draculin, the anticoagulant factor in vampire bat saliva, is a tight-binding, noncompetitive inhibitor of activated factor X. Biochim Biophys Acta 1999;1434:135–142.

4 Waxman L, Smith DE, Arcuri KE, Vlasuk GP: Tick anticoagulant peptide (TAP) is a novel inhibitor of blood coagulation factor Xa. Science 1990;248:593–596.

5 Tuszynski GP, Gasic TB, Gasic GJ: Isolation and characterization of antistasin. An inhibitor of metastasis and coagulation. J Biol Chem 1987; 262:9718–9723.

6 Nutt E, Gasic T, Rodkey J, Gasic GJ, Jacobs JW, Friedman PA, Simpson E: The amino acid sequence of antistasin. A potent inhibitor of factor Xa reveals a repeated internal structure. J Biol Chem 1988;263:10162–10167.

7 MacFarlane RG: The coagulant action of Russell's viper venom; the use of antivenom in defining its reaction with a serum factor. Br J Haematol 1961;7:496–511.

8 Yamada D, Sekiya F, Morita T: Prothrombin and factor X activator activities in the venoms of *Viperidae* snakes. Toxicon 1997;35:1581–1589.

9 Nahas L, Kamiguti AS, Barros MA: Thrombin-like and factor X-activator components of *Bothrops* snake venoms. Thromb Haemost 1979;41: 314–328.

10 Furtado MF, Maruyama M, Kamiguti AS, Antonio LC: Comparative study of nine *Bothrops* snake venoms from adult female snakes and their offspring. Toxicon 1991;29: 219–226.

11 Hofmann H, Bon C: Blood coagulation induced by the venom of *Bothrops* atrox. 2. Identification, purification, and properties of two factor X activators. Biochemistry 1987;26: 780–787.

12 Zhang Y, Xiong YL, Bon C: An activator of blood coagulation factor X from the venom of *Bungarus fasciatus*. Toxicon 1995;33:1277–1288.

13 Lee WH, Zhang Y, Wang WY, Xiong YL, Gao R: Isolation and properties of a blood coagulation factor X activator from the venom of king cobra *(Ophiophagus hannah)*. Toxicon 1995;33:1263–1276.

14 Kisiel W, Hermodson MA, Davie EW: Factor X activating enzyme from Russell's viper venom: Isolation and characterization. Biochemistry 1976;15:4901–4906.

15 Kisiel W, Ericsson LH, Davie EW: Proteolytic activation of protein C from bovine plasma. Biochemistry 1976;15:4893–4900.

16 Gowda DC, Jackson CM, Hensley P, Davidson EA: Factor X-activating glycoprotein of Russell's viper venom. Polypeptide composition and characterization of the carbohydrate moieties. J Biol Chem 1994; 269:10644–10650.

17 Gowda DC, Jackson CM, Kurzban GP, McPhie P, Davidson EA: Core sugar residues of the N-linked oligosaccharides of Russell's viper venom factor X-activator maintain functionally active polypeptide structure. Biochemistry 1996;35: 5833–5837.

18 Takeya H, Nishida S, Miyata T, Kawada S, Saisaka Y, Morita T, Iwanaga S: Coagulation factor X activating enzyme from Russell's viper venom (RVV- X). A novel metalloproteinase with disintegrin (platelet aggregation inhibitor)-like and C-type lectin-like domains. J Biol Chem 1992;267:14109–14117.

19 Atoda H, Yoshida N, Ishikawa M, Morita T: Binding properties of the coagulation factor IX/factor X-binding protein isolated from the venom of *Trimeresurus flavoviridis*. Eur J Biochem 1994;224:703–708.

20 Mizuno H, Fujimoto Z, Koizumi M, Kano H, Atoda H, Morita T: Structure of coagulation factors IX/X-binding protein, a heterodimer of C-type lectin domains. Nat Struct Biol 1997;4:438–441.

21 Mizuno H, Fujimoto Z, Koizumi M, Kano H, Atoda H, Morita T: Crystal structure of coagulation factor IX-binding protein from habu snake venom at 2.6 Å: Implication of central loop swapping based on deletion in the linker region. J Mol Biol 1999; 289:103–112.
22 Sekiya F, Yamashita T, Morita T: Role of calcium(II) ions in the recognition of coagulation factors IX and X by IX/X-bp, an anticoagulant from snake venom. Biochemistry 1995;34:10043–10047.
23 Mizuno H, Fujimoto Z, Atoda H, Morita T: Crystal structure of an anticoagulant protein in complex with the Gla domain of factor X. Proc Natl Acad Sci USA 2001;98:7230–7234.
24 Morita T: Proteases which activate factor X; in Bailey G (ed): Enzymes from Snake Venoms. Fort Collins, Alaken, 1998, pp 179–209.
25 Skogen WF, Bushong DS, Johnson AE, Cox AC: The role of the Gla domain in the activation of bovine coagulation factor X by the snake venom protein XCP. Biochem Biophys Res Commun 1983;111:14–20.
26 Morita T, Atoda H, Sekiya F: Structure and functions of coagulation factor IX/factor X-binding protein isolated from the venom of *Trimeresurus flavoviridis*. Adv Exp Med Biol 1996;391:187–196.
27 Furie B, Bouchard BA, Furie BC: Vitamin K-dependent biosynthesis of gamma-carboxyglutamic acid. Blood 1999;93:1798–808.
28 van Dieijen G, Tans G, van Rijn J, Zwaal RF, Rosing J: Simple and rapid method to determine the binding of blood clotting factor X to phospholipid vesicles. Biochemistry 1981;20:7096–7101.
29 Yamada D, Sekiya F, Morita T: Isolation and characterization of carinactivase, a novel prothrombin activator in *Echis carinatus* venom with a unique catalytic mechanism. J Biol Chem 1996;271:5200–5207.
30 Petrovan RJ, Govers-Riemslag JW, Nowak G, Hemker HC, Rosing J, Tans G: Purification and characterization of multisquamase, the prothrombin activator present in *Echis multisquamatus* venom. Thromb Res 1997;88:309–316.
31 Govers-Riemslag JW, Johnsen L, Petrovan RJ, Rosing J, Tans G: A kinetic assay to determine prothrombin binding to membranes. Thromb Res 1998;92:239–247.
32 Franssen JH, Janssen-Claessen T, Van Dieijen G: Purification and properties of an activating enzyme of blood clotting factor X from the venom of *Cerastes cerastes*. Biochim Biophys Acta 1983;747:186–190.
33 Siigur E, Tõnismägi K, Trummal K, Aaspõllu A, Vija H, Subbi J, Siigur J: Factor X activator from *Vipera lebetina* snake venom with factor IX activating activity. Thromb Haemost 2001;suppl:P2108.
34 el-Asmar MF, Shaban E, Hagag M, Swelam N, Tu A: Coagulant component in *Cerastes cerastes* (Egyptian sand viper) venom. Toxicon 1986; 24:1037–1044.
35 Farid T, Nasser H, Zaki K, el-Asmar MF: Low molecular weight factor X activator from *Cerastes vipera* (Sahara sand viper) venom. Toxicon 1993;31:1007–1017.
36 Samel M, Siigur J: Medium molecular weight factor X activating enzyme from *Vipera berus berus* venom. Toxicon 1995;33:41–52.
37 Komori Y, Nikai T, Sugihara H: Isolation and characterization of factor X activator from the venom of *Vipera aspis aspis*. Int J Biochem 1990; 22:1053–1060.
38 Maruyama M, Kamiguti AS, Tomy SC, Antonio LC, Sugiki M, Mihara H: Prothrombin and factor X activating properties of *Bothrops erythromelas* venom. Ann Trop Med Parasitol 1992;86:549–556.
39 Zhang Y, Xiong YL, Bon C: Effects of Chinese snake venoms on blood coagulation, coagulation factors and synthetic substrates. Asiat Herpetol Res 1993;5:117–126.
40 Denson KW: Clot-inducing substances present in snake venoms with particular reference to *Echis carinatus*. Thromb Res 1976;8:351–360.
41 Hemker HC, van Dam-Mieras MC, Devilee PP: The action of *Echis carinatus* venom on the blood coagulation system. Demonstration of an activator of factor X. Thromb Res 1984;35:1–9.

Procoagulant Proteins and Diagnostic Agents

Hæmostasis

Haemostasis 2001;31:234–240

Snake Venom Prothrombin Activators Homologous to Blood Coagulation Factor Xa

Jeremiah S. Joseph[a,d] R. Manjunatha Kini[b,c]

[a]Department of Biochemistry, Faculty of Medicine, [b]Department of Biological Sciences, Faculty of Science, National University of Singapore, Singapore;
[c]Department of Biochemistry and Molecular Biophysics, Medical College of Virginia, Virginia Commonwealth University, Richmond, Va., USA; [d]Bioinformatics Centre, Institute of Molecular and Cell Biology, Singapore

Key Words

Snake venom · Procoagulants · Prothrombin activators · Factor Xa

Abstract

We have recently determined the complete amino acid sequence of trocarin, a group D prothrombin activator from the venom of *Tropidechis carinatus* (Australian rough-scaled snake). This proteinase is both functionally and structurally similar to mammalian blood coagulation factor Xa. It shows ~70% homology and possesses the characteristic Gla domain, two EGF domains and serine proteinase domain. To examine structure-function relationships, we generated a molecular model of trocarin based on a human factor Xa des-Gla crystal structure (1xka) as template. Based on known sites of interaction between mammalian factor Xa, factor Va and prothrombin, structure-function relationships of trocarin were explored. Unlike factor Xa, trocarin is glycosylated and has a large carbohydrate moiety at the entrance to its active site pocket. This might contribute to differences observed in the kinetics of hydrolysis of synthetic substrates by trocarin as compared to human factor Xa. A Ca^{2+}-binding loop present in the heavy chain of factor Xa also seems to be lost in trocarin. In addition to its role in hemostasis, factor Xa shows other biological effects, including inflammation via its interaction with effector protease receptor-1 (EPR-1). Interestingly, the EPR-1 recognition site is distinctly different in trocarin, the functional consequences of which are being investigated.

KARGER

Fax + 41 61 306 12 34
E-Mail karger@karger.ch
www.karger.com

0301–0147/01/0316–0234$17.50/0

Accessible online at:
www.karger.com/journals/hae

R. Manjunatha Kini
Department of Biological Sciences, Faculty of Science
Block S2, Room 04-11, 14 Science Drive
National University of Singapore, Singapore 117543
Tel. +65 874 5235, Fax +65 779 2486, E-Mail dbskinim@nus.edu.sg

Introduction

The activation of prothrombin to thrombin is accomplished in vivo by the prothrombinase complex, consisting of a serine proteinase factor Xa, and a cofactor factor Va, assembled on negatively charged phospholipid membranes in the presence of Ca^{2+} ions [1]. Various exogenous hemostatic factors have also been isolated that specifically act on prothrombin [2]. Snake venom is a rich source of prothrombin-activating proteinases [2–4]. Based on cofactor requirements in prothrombin activation, these venom proteinases have been classified into four groups: Groups A, B, C and D [5]. Group A activators (example, ecarin, *Echis carinatus*) are metalloproteinases that require no cofactors and efficiently convert prothrombin into meizothrombin [6]. Group B activators are metalloproteinases – C-type lectin complexes (example, carinactivase-1, *Echis carinatus*) that also convert prothrombin to meizothrombin, but require Ca^{2+} ions for activity [7]. Group C prothrombin activators (example, pseutarin, *Pseudonaja textilis*) are large multi-subunit proteinases that activate prothrombin to mature thrombin, cleaving both the necessary peptide bonds [8]. Prothrombin conversion by these proteins is greatly enhanced by Ca^{2+} ions and phospholipids. Group D activators (example, trocarin, *Tropidechis carinatus*) are found exclusively in Australian elapids [9]. They also convert prothrombin to mature thrombin, a reaction greatly enhanced by blood coagulation factor Va, negatively charged phospholipids and Ca^{2+} ions. In this respect, group D prothrombin activators strongly resemble blood coagulation factor Xa.

Structural characterization of these proteinases was expected to contribute significantly towards understanding structure-function relationships and mechanisms of factor Xa-mediated prothrombin activation. Moreover, the phylogenetic distance and functional similarities between these proteinases and factor Xa would help define molecular details of interactions among the protein components of the prothrombinase complex. Therefore, we initiated studies on trocarin, the group D prothrombin activator from the venoms of the Australian rough-scaled snake, *T. carinatus*.

Functional Characterization

Trocarin was purified from crude venom by reversed-phase HPLC [9]. It has similar dose-dependent procoagulant effects to human factor Xa on human plasma [9]. Similar to factor Xa, it cleaves prothrombin at both peptide bonds, Arg_{274}–Thr_{275} and Arg_{323}–Ile_{324}, required to convert it to mature thrombin [9]. Characteristic of group D activators, its prothrombin-converting activity is greatly enhanced by factor Va, Ca^{2+} ions and phospholipids [9]. Though the activity of the entire complex is comparable to that of the mammalian prothrombinase complex, trocarin has a higher basal activity in the absence of all cofactors compared to bovine and human factor Xa. Hence the stimulation of trocarin by the cofactors is generally ~ 10-fold lower than that of bovine and human factor Xa [9]. Trocarin cleaves the factor Xa-specific chromogenic substrates S-2222 and S-2765. Interestingly, while the K_ms of trocarin and factor Xa are comparable, the V_{max} for trocarin is about ~ 1,000-fold lower than that of mammalian factor Xa. This indicates that while the affinity of the enzymes for the peptide substrate is very similar, product release is likely impaired only in the venom enzyme. Similar low amidolytic activity was also reported for notecarin *(Notechis scutatus scutatus)* [10]. Finally, trocarin was lethal to mice at 1 mg/kg body weight i.p., likely causing disseminated intra-

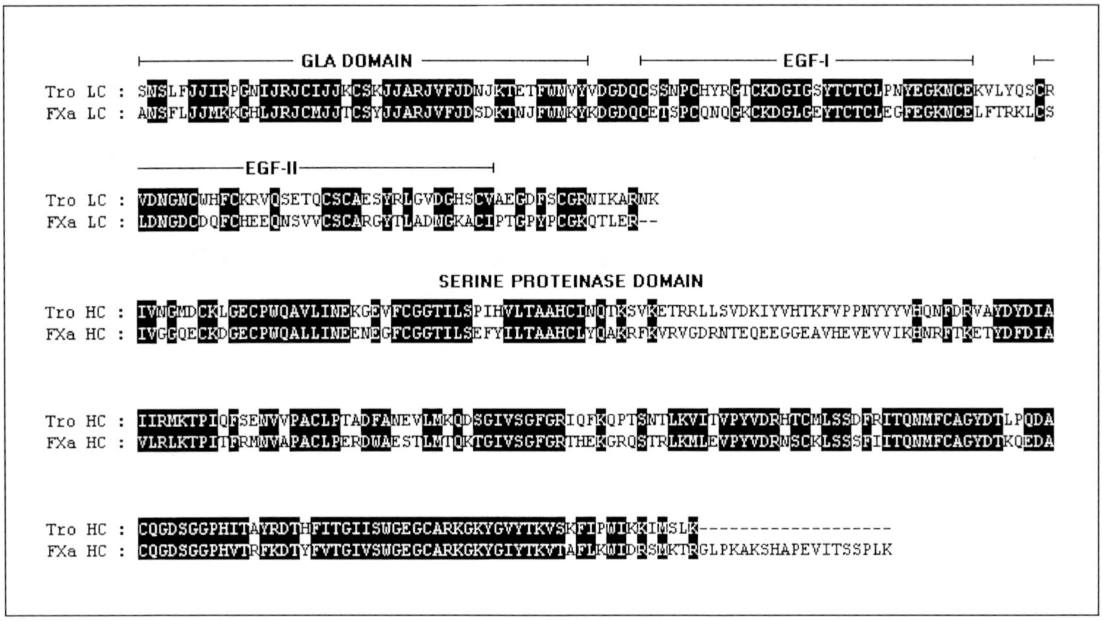

Fig. 1. Alignment of amino acid sequence of trocarin (SWISS-PROT #P81428) with human factor Xa sequence (#P00742). Identical residues are shaded. Gla, EGF-I, EGF-II and serine proteinase domain boundaries are indicated above light chain sequences. J = Gla residue.

vascular coagulopathy and is therefore expected to play the role of a toxin [9]. Hence a novel function has been ascribed to factor Xa, in addition to its well-known roles in coagulation, inflammation [for example, see ref. 11] and mitogenesis [for example, see ref. 12, 13].

Amino Acid Sequencing and Determination of Posttranslational Modifications

To ascertain its molecular identity, the N-terminal sequence of trocarin was determined. The protein consists of a light and a heavy chain; their N-terminal sequences have high homology to the corresponding chains of factor Xa. The complete amino acid sequence of trocarin was subsequently determined (fig. 1) [9]. Trocarin has 141 residues in its light chain and 235 residues in its heavy chain. All Cys residues have the same positions as those in factor Xa. Thus, based on sequence homology, trocarin has a domain structure identical to that of factor Xa (fig. 1, 2). The light chain consists of an N-terminal Gla domain (see below), followed by two EGF-like domains. The heavy chain is the serine proteinase catalytic domain (fig. 2).

During sequencing, several posttranslational modifications were observed. Eleven Gla residues in the N-terminal portion of their light chains were identified by mass spectrometry and resequencing the decarboxylated light chain. Comparison of trocarin and factor Xa sequences suggest that Gla residues in positions 6, 7, 14, 16, 19, 20, 25, 26,

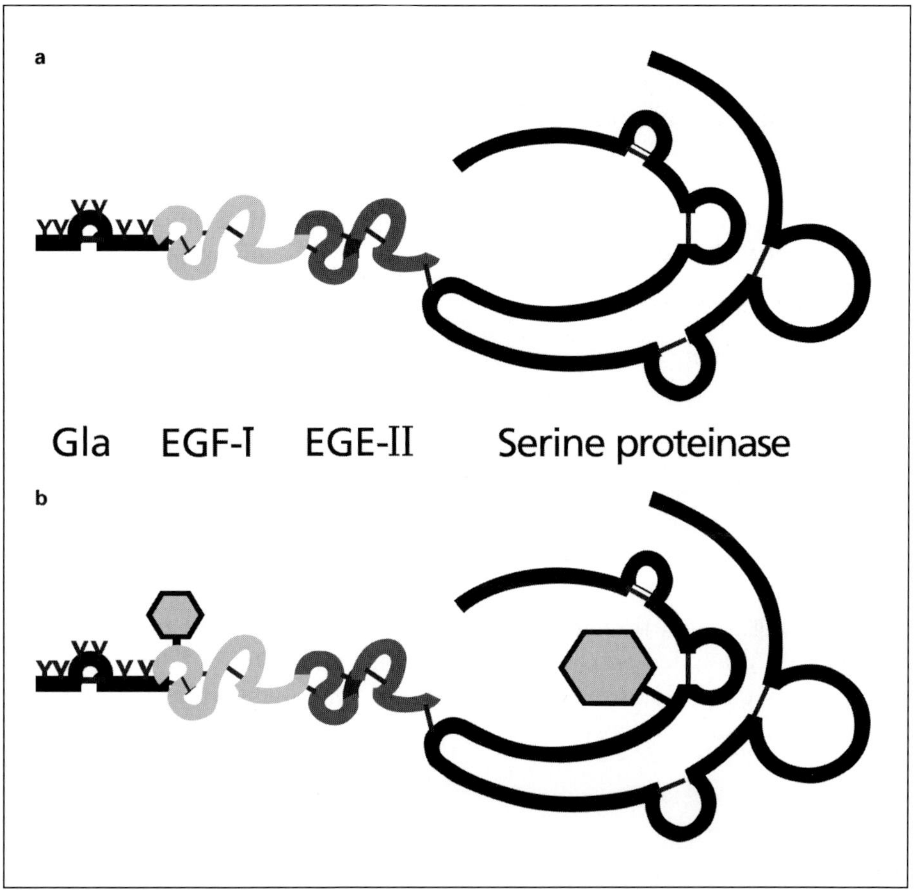

Fig. 2. Schematic representation of domain structure of factor Xa (**a**) and trocarin (**b**). Domains, disulfide linkages and post-translational modifications are depicted. Y = γ-Carboxylation; hexagons = carbohydrate moiety.

29 and 32 are critical for activity, while either one of the additional Glas in positions 35 and 39 could perhaps be dispensed with, without severely compromising activity [9]. Mass spectrometry and/or sugar compositional analysis indicated that trocarin is O-glycosylated at light chain Ser 52 (GlcNAc-Xyl or Gal-GalNAc) and N-glycosylated at heavy chain Asn 45 [9]. It is hence the first factor Xa-like protein known to be glycosylated. However, in factor VIIa, Ser 52 is glycosylated with disaccharide (-Glc-Xyl) or trisaccharide [-Glc-$(Xyl)_2$] moieties [14, 15]. Recombinant mutants in which Ser 52 was replaced by Ala had lowered (~60%) clotting activity, indicating that the glycosylation has significant functional implications [16, 17]. However, in trocarin the role of glycosylated Ser 52 is unknown. To the best of our knowledge, the GlcNAc-Xyl linkage has not been reported so far in proteins; it also represents the first known occurrence of O-linked

GlcNAc in secreted proteins. Structural heterogeneity of the N-linked carbohydrate in trocarin precluded unequivocal assignment of its structure. Finally, unlike factor Xa, trocarin is not β-hydroxylated at light chain Asp 63 [9].

Structural Comparison with Factor Xa

Trocarin is highly homologous to other factors Xa (62–70%) [9] (see figure 1 for sequence comparison with human factor Xa). By homology, it has a domain structure identical to factor Xa (fig. 2). Hence it is the first snake venom procoagulant that is structurally and functionally homologous to a blood coagulation factor. To complement 'linear' sequence analysis with three-dimensional perspective, we generated a molecular model of trocarin based on the human des-Gla factor Xa crystal structure, 1xka [18].

Ca^{2+} Binding

As mentioned above, the Ca^{2+}-dependent membrane-binding Gla domain is well conserved in trocarin. The EGF-I domain of factor Xa is also involved in Ca^{2+} binding. In factor Xa, the side chains of Asp 63, Asp 46 and Gln 49 and the backbone carbonyl groups of Gly 47 and Gly 64 serve as Ca^{2+} ligands [19, 20] (fig. 1). All these residues are completely conserved in trocarin. Though Asp 63 is not modified in trocarin, it is unlikely to be involved in Ca^{2+} binding, as in factor Xa, the β-hydroxyl group of hydroxyaspartate 63 points away from the metal ion [21, 22]. Another Ca^{2+}-binding site is seen in the factor Xa crystal structure, bound in a loop defined by heavy chain residues 56–66 [18]. Interestingly, in trocarin, except for one residue (Asp 54, conservative substitution), all of the other residues in this loop are not conserved (fig. 1). Loss of residues that are potential ligands for the Ca^{2+} ion (Asp 56, Asn 58, Thr 59, Gln 61, Glu 63 and/or Glu 66) seems to indicate loss of Ca^{2+}-binding ability at this site in the heavy chain. Functional consequences of this are not yet known.

Regions Important for Catalysis

The heavy chain of trocarin is homologous to serine proteinases and the heavy chain of factor Xa and is evidently the catalytic subunit (fig. 1). All residues required for the specific catalytic activity of factor Xa are completely conserved: the active site catalytic triad (His 42, Asp 88 and Ser 185), the substrate-binding pocket (Asp 179, Ala 180, Cys 181, Gln 182) and residues involved in forming the hydrogen-bonded structure antiparallel to the substrate (Ser 204, Trp 205, Gly 206, Gly 216) [9]. The regions of factor Xa (heavy chain residues 81–93 and 221–231) found to be involved in the recognition of the substrate prothrombin [23] are also fairly well conserved in trocarin. However, as mentioned above, significant differences were observed in the ability of trocarin to hydrolyze factor Xa-specific peptide substrates. Examination of the vicinity of the active site in the model suggests a structural basis for this difference in activity. As mentioned, trocarin is N-glycosylated at heavy chain Asn 45 (only three residues removed from active site His 42). The N-glycosylation (~ 3,300 D) is, intriguingly, positioned on the lip of the entrance to the active site. Hence it is possible that this large carbohydrate near the active site hinders the release of the hydrolysis products of synthetic peptide substrates, consequently affecting the Vmax of the reaction.

Regions Important for Interaction with Factor Va

The EGF-II domain of trocarin is much less homologous to factor Xa [9] (fig. 1). The precise function of this domain in factor Xa is still not well understood. Domain-swapping experiments by Hertzberg et al. [24], involving factor IX Gla and EGF-I domains 'spliced' to factor X EGF-II and serine proteinase domains, indicated that the EGF-II domain may mediate factor Va binding, the affinity of which could be enhanced by the EGF-I domain [24]. If this domain is indeed involved in factor Va binding, its low homology in trocarin is especially intriguing, since trocarin binds strongly to bovine factor Va [unpubl. obs.]. Chattopadhyay et al. [23] also identified regions corresponding to heavy-chain residues 17–28 and 60–82 as being in important for interaction with factor Va. While, the segment from residues 17–28 is quite conserved in trocarin, the heavy-chain residues 48–84 in trocarin show surprisingly little similarity to the corresponding region of factor Xa (fig. 1). The consequences of this would be interesting to examine.

Region Involved in Binding to Effector Cell Protease Receptor-1

The inter-EGF domain region (^{83}LFTK-RL^{88}) in factor Xa, is implicated in interaction with its specific physiological receptor, effector cell protease receptor-1 (EPR-1) [25]. This interaction mediates inflammation [11], as well as mitogenesis in several cell types [26, 27]. Interestingly, this region is completely different in trocarin (^{83}KVLYQS^{88}) (fig. 1). It is important to determine the functional consequences of these differences. We are currently investigating the ability of trocarin to induce inflammation and mitogenesis and the EPR-1 dependence of these effects, if any.

Acknowledgement

We acknowledge a National University of Singapore Academic Research Grants (R-154-000-093-112 and R-368-000-004-112) to RMK.

References

1 Suttie JW, Jackson CM: Prothrombin structure, activation and biosynthesis. Physiol Rev 1977;57:1–70.
2 Rosing J, Tans G: Inventory of exogenous prothrombin activators. Thromb Haemost 1991;65:627–630.
3 Markland FS Jr: Snake venoms. Drugs 1997;54(suppl 3):1–10.
4 Hutton RA, Warrell DA: Action of snake venom components on the haemostatic system. Blood Rev 1993;7:176–189.
5 Kini RM, Morita T, Rosing J: Classification and nomenclature of prothrombin activators isolated from snake venoms. Thromb Haemost, in press.
6 Morita T, Iwanaga S, Suzuki T: Activation of bovine prothrombin by an activator isolated from *Echis carinatus* venom. Thromb Res 1976; 8(suppl 2):59–65.
7 Yamada D, Sekiya F, Morita T: Isolation and characterization of carinactivase, a novel prothrombin activator in *Echis carinatus* venom with a unique catalytic mechanism. J Biol Chem 1996;271:5200–5207.
8 Masci PP, Whitaker AN, de Jersey J: Purification and characterization of the prothrombin activator of the venom of *Pseudonaja textiles*; in Gopalakrishnakone P, Tan CK (eds): Progress in Venom and Toxin Research. Proceedings of the First Asia-Pacific Congress on Animal, Plant and Microbial Toxins, Singapore 1987, pp 209–219.
9 Joseph JS, Chung MC, Jeyaseelan K, Kini RM: Amino acid sequence of trocarin, a prothrombin activator from *Tropidechis carinatus* venom: Its structural similarity to coagulation factor Xa. Blood 1999;94:621–631.

10 Tans G, Govers-Riemslag JWP, van Rihn JLML, Rosing J: Purification and properties of a prothrombin activator from the venom of *Notechis scutatus scutatus*. J Biol Chem 1985; 260:9366–9372.
11 Cirino G, Cicala C, Bucci M, Sorrentino L, Ambrosini G, DeDominicis G, Altieri DC: Factor Xa as an interface between coagulation and inflammation. Molecular mimicry of factor Xa association with effector cell protease receptor-1 induces acute inflammation in vivo. J Clin Invest 1997;99:2446–2451.
12 Gasic GP, Arenas CP, Gasic TB, Gasic GJ: Coagulation factors X, Xa, and protein S as potent mitogens of cultured aortic smooth muscle cells. Proc Natl Acad Sci USA 1992;89:2317–2320.
13 Ko FN, Yang YC, Huang SC, Ou JT: Coagulation factor Xa stimulates platelet-derived growth factor release and mitogenesis in cultured vascular smooth muscle cells of rat. J Clin Invest 1996;98:1493–1501.
14 Hase S, Kawabata S, Nishimura H, Takeya H, Sueyoshi T, Miyata T, Iwanaga S, Takao T, Shimonishi Y, Ikenaka T: A new trisaccharide sugar chain linked to a serine residue in bovine blood coagulation factors VII and IX. J Biochem 1988;104: 867–868.
15 Iwanaga S, Nishimura H, Kawabata S, Kisiel W, Hase S, Ikenaka T: A new trisaccharide sugar chain linked to a serine residue in the first EGF-like domain of clotting factors VII and IX and protein Z. Adv Exp Med Biol 1990;281:121–131.
16 Bjoern S, Foster DC, Thim L, Wiberg FC, Christensen M, Komiyama Y, Pederson AH, Kisiel W: Human plasma and recombinant factor VII. Characterization of O-glycosylations at serine residues 52 and 60 and effects of site directed mutagenesis at serine 52 to alanine. J Biol Chem 1991;266:11051–11057.
17 Iino M, Foster DC, Kisiel W: Functional consequences of mutations in Ser-52 and Ser-60 in human blood coagulation factor VII. Arch Biochem Biophys 1998;352:182–192.
18 Kamata K, Kawamoto H, Honma T, Iwama T, Kim SH: Structural basis for the chemical inhibition of human blood coagulation factor Xa. Proc Natl Acad Sci USA 1998;95: 6630–6635.
19 Stenflo J: Structure-function relationships of epidermal growth factor modules in vitamin K-dependent clotting factors. Blood 1991;78: 1637–1651.
20 Selander-Sunnerhagen M, Ullner M, Persson E, Teleman O, Stenflo J, Drakenberg T: How an epidermal growth factor (EGF)-like domain binds calcium. High resolution NMR structure of the calcium form of the NH_2-terminal EGF-like domain in coagulation factor X. J Biol Chem 1992;267:19642–19649.
21 Valcarce C, Selander-Sunnerhagen M, Tamlitz A-M, Drakenberg T, Bjork I, Stenflo J: Calcium affinity of the NH_2-terminal epidermal growth factor-like module of factor X. J Biol Chem 1993;268:26673–26678.
22 Selander-Sunnerhagen M, Persson E, Dahlqvist I, Drakenberg T, Stenflo J, Mayhew M, Robin M, Handford P, Tilley JW, Campbell ID, Brownlee GG: The effect of aspartate hydroxylation on calcium binding to epidermal growth factor-like modules in coagulation factors IX and X. J Biol Chem 1993;268: 23339–23344.
23 Chattopadhyay A, James HL, Fair DS: Molecular recognition sites on factor Xa which participate in the prothrombinase complex. J Biol Chem 1992;267:12323–12329.
24 Hertzberg MS, Ben-Tal O, Furie B, Furie BC: Construction, expression, and characterization of a chimera of factor IX and factor X. The role of the second epidermal growth factor domain and serine protease domain in factor Va binding. J Biol Chem 1992;267:14759–14766.
25 Ambrosini G, Plescia J, Chu KC, High KA, Altieri DC: Activation-dependent exposure of the inter-EGF sequence Leu^{83}-Leu^{88} in factor Xa mediates ligand binding to effector cell protease receptor-1. J Biol Chem 1997;272:8340–8345.
26 Herbert JM, Bono F, Herault JP, Avril C, Dol F, Mares AM, Schaeffer P: Effector protease receptor 1 mediates the mitogenic activity if factor Xa for vascular smooth muscle cells in vitro and in vivo. J Clin Invest 1998;10:993–1000.
27 Nicholson AC, Nachman RL, Altieri DC, Summers BD, Ruf, W, Edgington TS, Hajjar DP: Effector cell protease receptor-1 is a vascular receptor for coagulation factor Xa. J Biol Chem 1996;271:28407–28413.

Procoagulant Proteins and Diagnostic Agents

Hæmostasis

Haemostasis 2001;31:241–246

Factor V Activation and Inactivation by Venom Proteases

Jan Rosing José W.P. Govers-Riemslag Lev Yukelson
Guido Tans

Department of Biochemistry, Cardiovascular Research Institute Maastricht,
University of Maastricht, Maastricht, The Netherlands

Key Words
Venom · Snake venom · Caterpillar · Blood coagulation factor V · Factor V activation · Factor V inactivation

Abstract
Blood coagulation factor V is a single-chain glycoprotein with M_r = 330,000 which plays an important role in the procoagulant and anticoagulant pathways. Thrombin activates factor V into factor Va, a two-chain molecule which is composed of a heavy (M_r = 105,000) and a light chain (M_r = 71,000/74,000). Factor Va accelerates factor Xa-catalysed prothrombin activation more than 1,000-fold and under physiological conditions the cofactor activity of factor Va in prothrombin activation is down-regulated by activated protein C. Factor V can also be activated by a wide variety of snake venoms (e.g. from Vipera species, *Naja naja oxiana, Bothrops atrox*) and by proteases present in the bristles of a South American caterpillar *(Lonomia achelous)*. Some venoms, notably of *Vipera lebetina turanica* and *Lonomia achelous,* contain proteases that are able to inactivate factor V or factor Va. Venom factor V activators are excellent tools in studying the structure-function relationship of factor V(a) and they are also used in diagnostic tests for quantification of plasma factor V levels and for the screening of defects in the protein C pathway. In this review, the structural and functional properties of animal venom factor V activators and inactivators is described.

Introduction

For almost every clotting factor in the coagulation cascade there is a venom protease that can activate it. This paper focuses on ani-

KARGER

Fax + 41 61 306 12 34
E-Mail karger@karger.ch
www.karger.com

0301–0147/01/0316–0241$17.50/0

Accessible online at:
www.karger.com/journals/hae

Prof. Dr. J. Rosing, Department of Biochemistry
Cardiovascular Research Institute Maastricht, University of Maastricht
PO Box 616, NL–6200 MD Maastricht (The Netherlands)
Tel. +31 43 388 1678, Fax +31 43 388 4159
E-Mail j.rosing@bioch.unimaas.nl

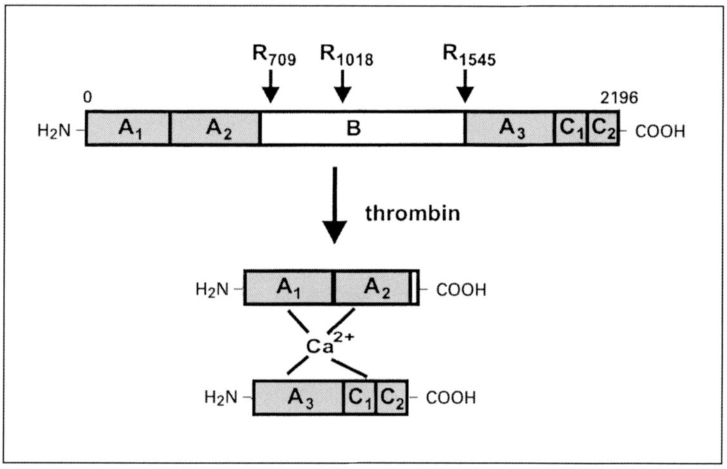

Fig. 1. Schematic diagram of factor V activation.

mal venom components that can activate or inactivate blood coagulation factor V, a clotting factor which, considering its central position in the coagulation cascade, has a key function in haemostasis.

Factor V is a multi-functional protein which plays an important role in both the procoagulant and anticoagulant pathways [for reviews, see ref. 1, 2]. Factor V (fig. 1) circulates in plasma as a single-chain glycoprotein with a molecular mass of 330 kD. However, factor V is not active as a coagulation factor and in order to express procoagulant activity it has to be first activated by thrombin or factor Xa. Activation proceeds via specific cleavage of peptide bonds at positions 709, 1018 and 1545, and yields factor Va, a heterodimer consisting of a heavy chain of 105 kD and a light-chain doublet of about 71/74 kD.

Factor Va acts as cofactor in factor Xa-catalysed prothrombin activation and it enhances the rate of thrombin formation more than 1,000-fold. Kinetic analysis has shown that factor Va accelerates prothrombin activation by: (1) acting as a receptor that promotes the binding of factor Xa and prothrombin to procoagulant membranes and (2) by enhancing the catalytic activity of factor Xa [2]. Factor V also plays a role in the anticoagulant pathway. Its cofactor activity in prothrombin activation is down-regulated by activated protein C and finally it has been shown that factor V acts as cofactor in activated protein C (APC)-mediated factor VIII inactivation [1, 2].

Animal venom proteins have been used as a tool in investigations of the structure-function relationship of factor V and in the development of diagnostic tests that probe the properties of factor V in both the pro- and anticoagulant pathways.

Animal Venom Factor V Activators and Inactivators

Table 1 summarises the animal venoms which have been reported to contain factor V activators. The venoms of Viperidae are historically linked with blood coagulation. Already around 1960 it was reported that Russel's viper venom contains both a potent factor X [3] and a factor V activator [4]. In 1978 it was shown that thrombocytin, the throm-

Table 1. Animal venom factor V activators

Venom source	Species
Bristles of larvae (caterpillar) of Lepidoptera	*L. achelous*
Venoms of Crotalidae	*B. atrox* (thrombocytin)
Venoms of Elapidae	*N. n. oxiana*
Venoms of Viperidae	*V. russelli, V. lebetina, V. ursini*

bin-like enzyme from *Bothrops atrox* was able to activate factor V [5] and in the early 1990s our laboratory in Maastricht reported that the venom of several *Naja* species contained a factor V activator [6]. Recently, it was shown that the bristles from certain South American caterpillars contain two different proteases that can activate and inactivate factor V, respectively [7]. The venom factor V activator from *Naja naja oxiana* partially inactivates thrombin-activated factor V [6], and unpublished observations reported in the present paper indicate that the venom from *Vipera lebetina turanica* is also able to inactivate factor Va. In this paper we will provide further information on the structural and functional properties of the factor V activators and inactivators.

Factor V Activators in Venoms from Viperidae

Russel's viper venom contains a powerful factor V activator, which is called RVV-V. The ability to activate factor V was first recognized by Hjort [4] in 1957 and RVV-V was purified to homogeneity by Kisiel [8]. RVV-V is a single-chain serine protease which consists of 236 amino acids [9]. This enzyme activates factor V [10, 11] by a specific cleavage of a single peptide bond at Arg^{1545} [12] which distinguishes it from thrombin which also cleaves factor V at positions 709 and 1018. The product obtained after activation by RVV-V does however have the same procoagulant activity as thrombin-activated factor V [10, 11]. RVV-V is unique in that no other substrate has as yet been identified and it is widely used in research in the field of haemostasis and thrombosis.

The RVV-V purified by Kisiel [8] appeared to contain 3 isoproteins that could be separated by HPLC [9]. The three isoproteins called RVV-Vα, RVV-Vβ and RVV-Vγ, which are present in the purified preparation in a weight ratio 2:1:6, have molecular masses of 29, 27.5 and 29 kD, respectively [9]. The amino acid composition of RVV-Vβ differed from that of RVV-Vα and RVV-Vγ. Since only limited amounts of protein were available, RVV-Vβ was not further characterised. The complete amino acid sequences of RVV-Vα and RVV-Vγ were determined and both contained 236 amino acids and they differed in 6 amino acid residues [9]. It is not known whether RVV-Vα, RVV-Vβ and RVV-Vγ exhibit the same activity in factor V activation since it was not possible to study their functional properties due to the fact that they were only separable by HPLC under denaturing conditions.

Other *Vipera* species also contain factor V activators. Siigur et al. [13, 14] characterised the activator from *Vipera lebetina*. In our laboratory [unpubl. obs.] we compared the factor V activators from *Vipera russelli, V. lebetina* and *Vipera ursini* called RVV-V, LVV-V and

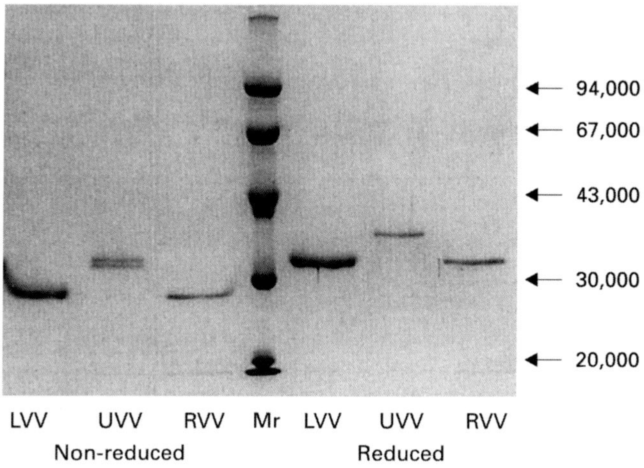

Fig. 2. SDS-PAGE analysis of factor V activators purified from the venoms of *V. russelli* (RVV), *V. lebetina* (LVV) and *V. ursini* (UVV). Molecular weight markers are indicated M_r.

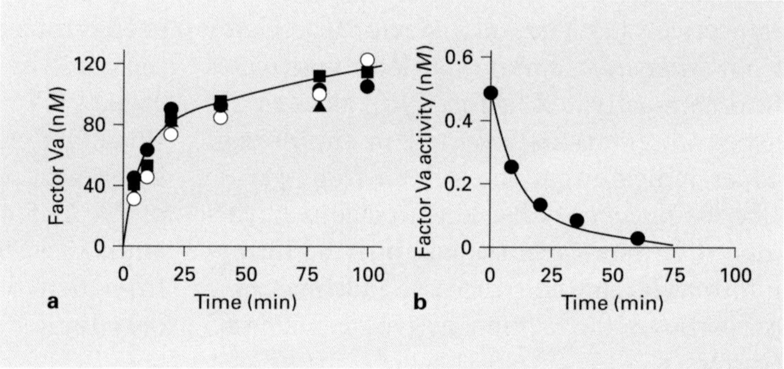

Fig. 3. a, **b** Time courses of activation and inactivation of factor V by vipera activators. **a** Concentrations of activators were: ● = 10 n*M* RVV-V, ■ = 10 n*M* LVV-V, ▲ = 9 n*M* UVV-V and ○ = 0.25 n*M* human α-thrombin. **b** Inactivation of factor Va by 250 n*M* LVV-V.

UVV-V, respectively. The activators were purified to homogeneity and had molecular weights of 29,000 in the case of *V. lebetina* and *V. russelli* and 34,000 in the case of *V. ursini* (fig. 2).

The three venom activators were equally active in activating factor V (fig. 3a). For comparison, the time course of factor V activation by thrombin is also given in this figure. It should be emphasised that the thrombin concentration in the activation mixture was some 40-fold lower than the concentration of the venom activators. This means that thrombin is about 40 times more effective in activating factor V than the *Vipera* activators.

Finally, it was observed that the purified activator from *V. lebetina* (LVV-V) not only activated factor V, but also inactivated factor Va. When thrombin- or venom-activated factor V was incubated with 250 n*M* LVV-V, the cofactor activity of factor Va in prothrombin activation was lost with a half-life of about 10 min (fig. 3b). Since the concentration of LVV-V used in this experiment was about 25 times higher than in the activation experiments, this means that under normal activation conditions there is only a minor degradation of factor Va by LVV-V. Currently, we are investigating which peptide bond cleavages in factor Va are responsible for LVV-V-catalysed factor Va inactivation.

Factor V Activator from the Venom of *Naja naja oxiana*

During a screening of the pro- and anticoagulant properties of snake venoms from the Elapidae family it was observed that the venoms from several *Naja* species were able to activate factor V [6, 15]. The factor V activator present in the venom of *N. n. oxiana,* which contained the largest amount of activator, was purified by repeated chromatography on a Mono S column [6]. The purified activator appears to be a single-chain protein which, as judged by polyacrylamide gel electrophoresis in the presence of SDS, has an apparent molecular mass of 48 kD.

The *N. n. oxiana* protease activates factor V by cleaving two peptide bonds yielding a factor Va molecule which, compared with thrombin-activated factor V, has a somewhat smaller heavy and larger light chain [6]. These small differences have a rather large effect on the cofactor activity of factor Va in prothrombin activation. *N. n. oxiana*-activated factor V had an 8-fold lower cofactor activity than thrombin-activated factor V. Addition of *N. n. oxiana* protease to thrombin activated-factor V resulted in some 90% loss of the cofactor activity of factor Va in prothrombin activation, indicating that the factor V activator from *N. n. oxiana* cleaves a peptide bond in both factor V and factor Va that results in the formation of a factor Va derivative which has a strongly diminished cofactor activity. SDS-PAGE in combination with HPLC and amino acid sequencing of a peptide isolated from a venom-treated factor Va preparation indicated that the venom activator removed the last 27 amino acids from the carboxy terminal domain of the heavy chain of factor Va [16]. This indicates that the venom protease from *Naja oxiana* cleaves the His^{682}-Asp^{683} peptide bond in factor V. Kinetic analysis showed that *Naja oxiana*-activated factor V had a diminished affinity for factor Xa and prothrombin [16], which indicates that the Asp^{683}-Arg^{709} domain of factor Va is essential for optimal interaction with prothrombin and factor Xa.

Factor V Activators Present in the Venoms from *Bothrops atrox* and *Lonomia achelous*

Thrombocytin, a protease isolated from the venom of *Bothrops atrox* by Niewiarowski et al. [17] and Kirby et al. [18], is also able to activate factor V [5]. Purified thrombocytin is a glycoprotein with a molecular mass of 36 kD which is a thrombin-like serine protease which not only activates factor V, but also platelets, factor XIII and factor VIII [17].

Activation of factor V by thrombocytin proceeds via the cleavage of two peptide bonds yielding a product which on the basis of SDS-PAGE is similar to thrombin-activated factor V [5]. The exact position of the peptide bonds in factor V that are cleaved by thrombocytin during activation is not yet known and there is also no information on the cofactor activity of thrombocytin-activated factor V.

It is well known that contact with caterpillars of moths belonging to the genus *Lonomia* may cause a haemorrhagic syndrome that likely results from the presence of proteolytic proteins in the bristles of these caterpillars which affect coagulation and fibrinolysis [19–21]. Lopez and Arocha-Pinango [7] identified a factor V activator in the hemolymph of *L. achelous*. The partially purified activator appears to be a thermostable metalloprotease which has a molecular mass less than 97 kD. The crude venom also contains a high molecular mass serine or cysteine protease that inactivates factor V. Unfortunately, there is as yet no information on the peptide bonds that are cleaved in factor V during activation and inactivation by the proteases present in the venom of *L. achelous.*

References

1 Dahlback B: Procoagulant and anticoagulant properties of coagulation factor V: Factor V Leiden (APC resistance) causes hypercoagulability by dual mechanisms. J Lab Clin Med 1999;133:415–422.

2 Rosing J, Tans G: Coagulation factor V: An old star shines again. Thromb Haemost 1997;78:427–433.

3 MacFarlane RG: The coagulant action of Russell's viper venom; the use of antivenom in defining its reaction with a serum factor. Br J Haematol 1961;7:496–511.

4 Hjort PF: Intermediate reactions in the coagulation of blood with tissue thromboplastin. Scand J Clin Lab Invest 1957;9:1–18.

5 Rawala R, Saraswati S, Niewiarowski S, Colman R: Molecular changes during the activation of bovine factor V by snake venom proteases (abstract). Circulation 1978;58:209.

6 Gerads I, Tans G, Yukelson L, Zwaal RF, Rosing J: Activation of bovine factor V by an activator purified from the venom of *Naja naja oxiana*. Toxicon 1992;30:1065–1079.

7 Lopez M, Gil A, Arocha-Pinango CL: The action of *Lonomia achelous* caterpillars venom on human factor V. Thromb Res 2000;98:103–110.

8 Kisiel W: Molecular properties of the factor V-activating enzyme from Russell's viper venom. J Biol Chem 1979;254:12230–12234.

9 Tokunaga F, Nagasawa K, Tamura S, Miyata T, Iwanaga S, Kisiel W: The factor V-activating enzyme (RVV-V) from Russell's viper venom. Identification of isoproteins RVV-V alpha, -V beta, and -V gamma and their complete amino acid sequences. J Biol Chem 1988;263: 17471–17481.

10 Suzuki K, Dahlback B, Stenflo J: Thrombin-catalyzed activation of human coagulation factor V. J Biol Chem 1982;257:6556–6564.

11 Kane WH, Majerus PW: Purification and characterization of human coagulation factor V. J Biol Chem 1981;256:1002–1007.

12 Kane WH, Davie EW: Blood coagulation factors V and VIII: Structural and functional similarities and their relationship to hemorrhagic and thrombotic disorders. Blood 1988; 71:539–555.

13 Siigur E, Aaspollu A, Siigur J: Molecular cloning and sequence analysis of a cDNA for factor V activating enzyme. Biochem Biophys Res Commun 1999;262:328–332.

14 Siigur E, Samel M, Tonismagi K, Subbi J, Reintamm T, Siigur J: Isolation, properties and N-terminal amino acid sequence of a factor V activator from *Vipera lebetina* (Levantine viper) snake venom. Biochim Biophys Acta 1998;1429:239–248.

15 Yukelson LY, Tans G, Thomassen MC, Hemker HC, Rosing J: Procoagulant activities in venoms from central Asian snakes. Toxicon 1991; 29:491–502.

16 Bakker HM, Tans G, Thomassen MC, Yukelson LY, Ebberink R, Hemker HC, Rosing J: Functional properties of human factor Va lacking the Asp683-Arg709 domain of the heavy chain. J Biol Chem 1994; 269:20662–20667.

17 Niewiarowski S, Kirby EP, Brudzynski TM, Stocker K: Thrombocytin, a serine protease from *Bothrops atrox* venom. 2. Interaction with platelets and plasma-clotting factors. Biochemistry 1979;18:3570–3577.

18 Kirby EP, Niewiarowski S, Stocker K, Kettner C, Shaw E, Brudzynski TM: Thrombocytin, a serine protease from *Bothrops atrox* venom. 1. Purification and characterization of the enzyme. Biochemistry 1979;18: 3564–3570.

19 Arocha-Pinango CL, Marval E, Guerrero B: *Lonomia* genus caterpillar toxins: Biochemical aspects. Biochimie 2000;82:937–942.

20 Amarant T, Burkhart W, LeVine H, 3rd, Arocha-Pinango CL, Parikh I: Isolation and complete amino acid sequence of two fibrinolytic proteinases from the toxic Saturnid caterpillar *Lonomia achelous*. Biochim Biophys Acta 1991;1079:214–221.

21 Donato JL, Moreno RA, Hyslop S, Duarte A, Antunes E, Le Bonniec BF, Rendu F, de Nucci G: *Lonomia obliqua* caterpillar spicules trigger human blood coagulation via activation of factor X and prothrombin. Thromb Haemost 1998;79:539–542.

Procoagulant Proteins and Diagnostic Agents

Hæmostasis

Haemostasis 2001;31:247–256

Molecular Basis for the Partition of the Essential Functions of Thrombin among Snake Venom Serine Proteinases

The Case of Thrombin-Like Enzymes

R.C. Maroun

Unité des Venins, Institut Pasteur, Paris, France

Key Words

Thrombin · Snake venoms · Thrombin-like serine proteinases · Molecular modeling · Molecular recognition

Abstract

Thrombin is a mammalian serine proteinase that plays a prominent role in the maintenance and regulation of hemostasis through its interaction with various substrates and/or ligands. The venoms of several snakes contain glycosylated serine proteinases that have been recognized to possess one or more of the essential activities of thrombin on fibrinogen (Fg) and/or platelets. These proteinases share about 60% sequence identity. One class of snake venom serine proteinases are those known as thrombin-like (TLE), named after their ability to directly clot Fg in order to preferentially produce fibrinopeptide A, fibrinopeptide B or both. To understand the molecular basis of this phenomenon, the corresponding amino acid sequences and molecular structures need to be analyzed. Given the absence of experimentally determined tertiary structures of snake venom, TLEs, three-dimensional molecular models should prove useful in this context. Towards this goal, we obtained models of snake venom TLEs that used TSV-PA as template, TSV-PA being the only snake venom serine proteinase whose crystal structure is known to date. Along with a comparative sequence analysis the models contribute to the identification and description of thrombin-homologous or alternative binding sites, helping thus to understand differences in specificity.

Introduction

Serine proteinases of a given fold share an identical catalytic mechanism, but markedly differ in substrate specificity [1–3], a property that is the outcome of evolutionary diver-

KARGER

Fax + 41 61 306 12 34
E-Mail karger@karger.ch
www.karger.com

0301–0147/01/0316–0247$17.50/0

Accessible online at:
www.karger.com/journals/hae

R.C. Maroun
Unité des Venins, Institut Pasteur
25, rue du Docteur-Roux, F–75724 Paris Cedex 15 (France)
Tel. +33 1 40 61 34 97, Fax +33 1 40 61 30 57
E-Mail rmaroun@pasteur.fr

gence [3, 4]. This has led to a large diversity of the structure of surface loops that surround the extended substrate-binding site.

Thrombin is a multifunctional mammalian serine proteinase that plays a major role in the maintenance and regulation of hemostasis. It stimulates plasma coagulation and platelet aggregation and thus possesses a procoagulant action. It acts on fibrinogen (Fg) by cleaving two peptide bonds, producing fibrin monomers that polymerize to form a clot. Thrombin is also the most potent activator of platelets [5]. It mediates platelet activation and aggregation by hydrolysis of a platelet thrombin receptor [5, 6]. It also has anticoagulant properties, since after its interaction with thrombomodulin it activates protein C, which is a major anticoagulant factor [7]. The crystal structure of human α-thrombin bound to specific inhibitors revealed prominent structural features of the thrombin molecule [8–10]. These features are the location of the catalytic triad within a deep canyon-like active-site cleft, and the presence of two extended surfaces. One surface, composed of positively charged and hydrophobic residues, is referred to as exosite I (ESI) or Fg recognition exosite. The other surface (exosite ESII) is solely composed of positively charged residues. ESI is required for thrombin binding to several substrates (Fg, proteinase activated thrombin receptor, and heparin cofactor II), ligands (thrombomodulin and glycoprotein 1b) [11] or the inhibitor hirudin [12]. For ESI, apolar as well as electrostatic interactions appear to be important determinants of complex formation [13]. ESII, which is located close to the carboxy-terminal B-chain helix, is involved in heparin [14], haemadin [15] and prothrombin fragment 2 binding [18].

Snake venom serine proteinases belong to the trypsin family of serine proteinases (family S1 of clan SA, [17]) and exhibit a high degree of sequence identity (60–68%). They have been characterized for their activity on a variety of substrates and it appears that each one is quite specific for a given macromolecular substrate, particularly the components of the blood coagulation cascade [18, 19]. The so-called snake venom thrombin-like serine proteinases (SVTLE) possess primarily one of the essential functions of thrombin: the clotting of Fg. The enzymes may thus clot Fg by splitting off preferentially either fibrinopeptide A, fibrinopeptide B or both. They are named venombin A, B or AB, respectively. Hirudin has no effect or an only negligible effect on their activity. Although to a much lower extent than thrombin, some may activate other blood coagulation factors.

The biological significance of the display of the essential Fg-clotting function of thrombin in SVTLEs is not clear. In an attempt to better understand the determinants of enzyme specificity, we obtained and analyzed their three-dimensional (3D) molecular models. We considered most appropriate to use the crystal structure of the one-chain nonzymogen plasminogen activator of *Trimeresurus stejnegeri* snake venom (TSV-PA) as the template molecule for the knowledge-based approach used. Indeed, TSV-PA, a snake venom serine proteinase, possesses functional, amino acid sequence (61–73% sequence identity) and structural features in common with the SVTLEs. Among the structural features are the pattern of Cys residues and, most of all, the C-terminal extension characteristic of snake venom serine proteinases. A comparative sequence analysis along with the resulting 3D models allows us to highlight and locate spatially alternative and/or equivalent residues that could be involved in the interaction of these enzymes with their natural substrates.

Materials and Methods

Sequence Analysis

The proteins used were those in table 1 of Pirkle [19] for which complete amino acid sequences exist, plus contortrixobin, whose sequence has been made available recently. The SVTLEs and their accession numbers are as follows: ancrod B (SwissProt P47797), batroxobin (SwissProt P04971), bilineobin (PIR S65621), bothrombin (PIR A54361), calobin (SwissProt Q91053), contortrixobin (TrEMBL P82981), crotalase [20], flavoxobin (SwissProt P05620), and the gyroxin analog from *Lachesis m. muta* (SwissProt P33589). The other two sequences are TSV-PA (SwissProt Q91516) and human α-thrombin heavy chain (PDB code 1ppb; since we deal only with the heavy chain, the letter H used to denote it will not be used in this text). The statistical analysis of the protein sequences was performed with the SAPS software [21].

Sequence Homology Modeling and Structure Prediction

3D structure visualization, modeling and analysis of the 9 SVTLEs were performed with the Homology module of the insightII 98.0 package (Accelrys, San Diego, Calif., USA) using as template TSV-PA complexed to the EGRcmk inhibitor [22]. The sequence alignment was that in figure 1 of Pirkle [19] with slight modifications in order to optimize the alignment to TSV-PA. The sequence alignment between TSV-PA and α-thrombin is structure based. The program ClustalW 1.6 was then applied to the alignment in figure 1 to obtain alignment scores and define more precisely, with the help of the crystal structures, the variable sequence regions that were subsequently mapped onto the structure of TSV-PA.

The putative secondary sites of the SVTLEs were searched through visual comparison to the structure of thrombin and its defined exosite residues [13]. The amino acid residue numbering is based on that of α-chymotrypsinogen.

Results

Sequence Analysis

The SAPS composition and distribution sequence analysis of the SVTLEs indicated that the composition in the major hydrophobic residues Leu, Val, Ile, Phe and Met is of 23.9–28.5%. The acidic residues significantly outnumber the basic residues for contortrixobin, flavoxobin and crotalase, just like TSV-PA. In analogy to α-thrombin, the basic residues outnumber the acidic ones in ancrod and bilineobin. For batroxobin, bothrombin, calobin and the *L. m. muta* gyroxin analog, both kinds of residues are in approximately the same proportions. SAPS detects significant mixed-charge linear clusters of amino acid residues for six of the enzymes. These clusters vary in length from 25 to 44 residues, as compared to TSV-PA with a cluster of 31 residues. Thrombin, batroxobin, calobin and crotalase contain no significant mixed-charge clusters according to SAPS. Weak mixed-charge clusters could nevertheless be detected by inspection. For example, for thrombin, the cluster spans Asp60e and Lys81, which encompasses the Fg-binding exosite (FBE) segment 67–80 [13].

The multiple-sequence alignment in figure 1 shows a number of variable regions (underlined) common to the SVTLEs. These are the 34–39, 60–67, 72–76, 90–101, 146–148, 165–167, 171–178 fragments (A–G), as well as the 245–245e C-terminal portion (H). With respect to the 7 venombins A (crotalase to calobin), and outside the variable regions, six residues (Val17, Phe34, Ser42, Arg111, His137 and Tyr159) are unique to contortrixobin (a venombin B), whereas only 3 are unique to bilineobin, a venombin AB (Ile16, Val158 and Ser229). It is also possible to detect the residues shared by all venombins and α-thrombin, excluding Cys, catalytic residues and conserved segments around molecular markers (see below; ref. [23]; green residues in the sequence of thrombin in figure 1). Those residues that are unique to all SVTLEs are indicated by an X. For example, at position 50 and 225 all SVTLEs possess a Glu and a Pro, respectively, as compared to an Arg and a Tyr for thrombin. Ancrod B contains a free sulfhydryl at Cys193.

```
TSV-PA:   VFGGDECNINEHRSLVVLF--NS--NGFLCGGTLINQDWVVTAAHCDS----------NNFQ

CROTAL:   VIGGDECNINEHRFLVALYDYWS--QLFLCGGTLINNEWVLTAAHCD-----------RTH-
BATROX:   VIGGDECDINEHPFLAFMY--YS--PRYFCGMTLINQEWVLTAAHCN-----------RRF-
ANCROD:   VIGGDECNINEHRFLVALYDSTT--RNFLCGGVLIHPEWVITAKHCN-----------KKS-
BOTHRO:   VIGGDECDINEHPFLAFMY--YS--PQYFCGMTLINQEWVLTAAHCD-----------KTY-
GLMUTA:   VIGGDECNINEHRFLVALYDGLS--GTFLCGGTLINQEWVLTAQHCN-----------RSL-
FLAVOX:   VIGGDECNINEHPFLVALYDAWS--GRFLCGGTLINPEWVLTAAHCDS----------KNFK
CALOBI:   VIGGDECNINEHRFLVALYNSRS--RTLFCGGTLINQEWVLTAAHCER----------KNF-
CONTOR:   VVGGDECNINEHRFLVAIF--NS--NGFVCSGTLINQEWVLTAAHCDS----------TDFQ
BILINE:   IIGGDECNINEHRFLVALYDVWS--GSFLCGGTLINQEWVLTAAHCN-----------MSN-
            X XXX  XXX XX    --------          X         ---------------
Region                          A                                 B
ATHROM:   IVEGSDAEIGMSPWQVMLF--RKSPQELLCGASLISDRWVLTAAHCLLYPPWDKNFTENDLL
          16 20        30     35    40        50      * 60
                               36a                      60abcdefghi   65


TSV-PA:   --LLFGVHSKKILN-EDEQTRDPKEKFF--CPNRKKDD---EVDKDIMLIKLDS

CROTAL:   ILIYVGVHDRSVQF-DKEQRRFPKEKYFFDCSNNFTKW-----DKDIMLIRLNK
BATROX:   MRIHLGKHAGSVAN-YDEVVRYPKEKFI--CPNKKKNV---ITDKDIMLIRLDR
ANCROD:   MVLYLGKHKQSVKF-DDEQERFPKEKHFIRCNKPRTRW-----GEDIMLIRLNK
BOTHRO:   MRIYLGIHTRSVAN-DDEVIRYPKEKFI--CPNKKKNV---ITDKDIMLIRLNR
GLMUTA:   MNIYLGMHNKNVKF-DDEQRRYPKKKYFFRCNKNFTKW-----DEDI---RLNR
FLAVOX:   MK--LGAHSKKVLN-EDEQIRNPKEKFI--CPNKKNDE---VLDKDIMLIKLDS
CALOBI:   -RIKLGIHSKKVPN-EDEQTRVPKEKFF--CLSSKNYT---LWDKDIMLIRLDS
CONTOR:   --IKLGAHSKKVLN-EDEQIRNPKEKFI--CPNKKNDE---VLDKDIMLIKLDS
BILINE:   IYIYLGMHNQSVQF-DDEERRYPKEKYLFRCSKNFTKW-----DKDIMLIRLNK
          ----  ---X-         X XX X  --X------------      X
Region               C                       D
ATHROM:   --VRIGKHSRTRYERNIEKISMLEKIYI--HPRYN-WRENLDRDIALMKLKK
               70         80        90         100   *     110
          ab          77a           90ab      97a


TSV-PA:   SVSNSEHIAPLSLPS-SPP----SVGSVCRIMGWGKTIPTKE-----IYPDVPHCANINI

CROTAL:   PVSYSEHIAPLSLPS-SPP----IVGSVCRAMGWGQTTSPQE-----TLPDVPHCANINL
BATROX:   PVKNSEHIAPLSLPS-NPP----SVGSVCRIMGWGAITTSED-----TYPDVPHCANINL
ANCROD:   PVNNSEHIAPLSLPS-NPP----IVGSVCRVMGWGSINKYID-----VLPDEPRCANINL
BOTHRO:   PVKNSTHIAPISLPS-NPP----SVGSVCRIMGWGAITTSED-----TYPDVPHCANINL
GLMUTA:   PVRFSAHIEPLSLPS-NPP----SEDSVCRVMGWGQITSPPE-----TLPDVPHCANINL
FLAVOX:   PVSYSEHIAPLSLPS-SPP----SVGSVCRIMGWGSITPVEE-----TFPDVPHCANINL
CALOBI:   PVSNSEHIAPLSLPS-SPP----SVGSVCRIMGWGRISPTKE-----TYPDVPHCANINL
CONTOR:   RVSNSEHIAPLSLPS-SPP----SVGSVCHIMGWGSITPIEV-----TFPDVPHCAYINL
BILINE:   PVRNSEHIAPLSLPS-SPP----IVGSVCRVMGWGTITSPNE-----TLPDVPRCVNINL
                X X  X  X  XX       XXX  X       ---        X X X  XXX
Region                                            E
ATHROM:   PVAFSDYIHPVCLPDRETAASLLQAGYKGRVTGWGNLKETWTTNVGKGQPSVLQVVNLPI
                 120          130       140            150       160
                          129abc                 149abcde
```

```
TSV-PA:   LDHAVCRTA--YSWRQVANTTLCAG-IL-QG-GRDTCHFDSGGPLICN------GIFQG

CROTAL:   LDYEVCRTAHPQFRLPATSRTLCAG-VL-EG-GIDTCNRDSGGPLICN------GQFQG
BATROX:   FNNTVCREAYNG--LPA--KTLCAG-VL-QG-GIDTCGGDSGGPLICN------GQFQG
ANCROD:   YNYTVCRGVFPR--IPKKSKILCAG-DL-QG-RLDSCHCDSGGPLICS------EEFHG
BOTHRO:   FNNTVCREAYNG--LPA--KTLCAG-VL-QG-GIDTCGGDSGGPLICN------GQFQG
GLMUTA:   FNYTVCRGAYPR--MP--TKVLCAG-VL-EG-GIDTCNRDSGGPLICN------GQFQG
FLAVOX:   LDDVECKPGYPEL-LPEY-RTLCAG-VL-QG-GIDTCGFDSGTPLICN------GQFQG
CALOBI:   LEYEMCRAPYPEFGLPATSRTLCAG-IL-EG-GKDTCRGDSGGPLICN------GQFQG
CONTOR:   LDDAACQPGYPEV-LPEY-RTLCAG-IL-EG-GKDTCNYDSGGPLICN------GQFQG
BILINE:   FNYTVCRGVFPR--LPERSRILCAG-VL-EG-GIDTCKRDSGGPLICN------GQFQG
            ---     ---------  X     X  X X  X        XX          X
Region       F          G
ATHROM:   VERPVCKDS--T-RIRITDNMFCAGYKPDEGKRGDACQGDSGGPFVMKSPFNNRWYQMG
               170          180            190    *  200         210
                171ab           184a                    204ab
                   172a            186abcd


TSV-PA:   IVSWGGHPCGQPGEPGVYTKVFDYLDWIKSIIAGNKDATCPP

CROTAL:   IVFWGPDPCAQPDKPGLYTKVFDHLDWIQSIIAGEKTVNCP
BATROX:   ILSWGSDPCAEPRKPAFYTKVFDYLPWIQSIIAGNKTATCP
ANCROD:   IVYRGPNPCAQPDKPALYTNIFDHLHWILSIMAGNAT--CYP
BOTHRO:   ILSWGSDPCAEPRKPAFYTKVFDYLPWIQSIIAGNKTATCPP
GLMUTA:   IVFWGPDPCAQPDKPGVYTKVFDYLDWIQSVIAGNTT--CS
FLAVOX:   IVSYGGHPCGQSRKPGIYTKVFDYNAWIQSIIAGNTAATCLP
CALOBI:   IASWGDDPCAQPHKPAAYTKVFDHLDWIQSIIAGNTDASCPP
CONTOR:   IVSYGAHPCGQSLKPGIYTKVFDYNDWIQSIIAGNTAATCPP
BILINE:   IVSWGPKRCAQPRKPALYSKVFDHLDWIQSIIAGNKTVNCP
                        X       X      XX    -------
Region                                           H
ATHROM:   IVSWGE-GCDRDGKYGFYTHVFRLKKWIQKVIDQF       GE
                220        230       240  245     246
                221A                      245abcdefg
                  221
```

Fig. 1. Multiple alignment for the SVTLEs, TSV-PA and α-thrombin. First, α-thrombin and TSV-PA were aligned topologically. Thereafter, the alignment in figure 1 of Pirkle [19] for SVTLEs was added and optimized with respect to TSV-PA. Exosite I basic, hydrophobic and extension residues are in blue, magenta and gold, respectively. ESII residues are in red. Residues common to all SVTLEs and thrombin are in green, excluding Cys, catalytic residues and sequence motifs around molecular markers. Variable regions A–H (see text) for all 9 SVTLEs together are underlined. PDB file 1PPB [14] was used as the crystallographic reference of thrombin for structure and sequence. The sequences are arranged as venombin A (crotalase to calobin), venombin B (contortrixobin) and venombin AB (bilineobin). Chymotrypsinogen numbering for thrombin is shown.

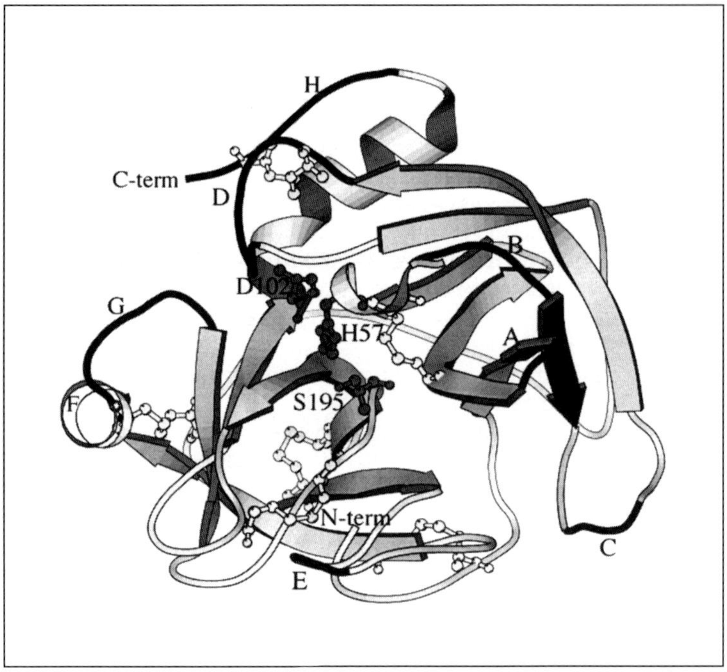

Fig. 2. MOLSCRIPT [36] representation of the TSV-PA crystallographic structure in which the detected variable regions in SVTLEs (underlined in fig. 1) are labeled A–H. Active site residues His57, Asp102 and Ser195 and disulfide bonds are in ball and stick representation. Arrows: β-strands; ribbons: α-helices.

Structure Analysis

Certain residues occupied positions in space equivalent to the corresponding thrombin residues, other residues were particular to the SVTLEs (not shown). As compared to thrombin, with 9 nonneutralized basic residues composing ESI, the SVTLEs contain up to 5 residues (fig. 1). Hydrophobic residues prone to forming a hydrophobic cluster are on the average less numerous than thrombin, except for bothrombin, 7. Calobin presents only 2 hydrophobic residues. In brief, it is possible to detect the presence of an exosite in SVTLEs analogous to ESI of thrombin and composed of basic and hydrophobic residues. However, ESII of thrombin is poorly represented in the SVTLEs, none of which presents the 'heparin-binding' motif of the ESII of thrombin [13], and whose basic residues are outnumbered by those of thrombin.

Figure 2 shows the variable regions underlined in figure 1 as they are mapped onto the 3D structure of TSV-PA. Some correspond to well-known regions or loops (36, 60, 99, 174) and to the segment on the N-terminal side of the autolysis loop of thrombin. The five regions are located east, northeast, northwest, west and south of the catalytic site in the canonical view. Segment 72–76 is southeast of the catalytic cavity and relatively far from it. Segment 165–167 is part of helix 2 of the trypsin fold, east of the active site. The remaining segment comprises the C-terminal extension unique to snake venom serine proteinases.

A number of sequence motifs surrounding serine proteinase evolutionary markers and active site residues for the S1 family of clan SA have been identified [23]. Thus, for the motif surrounding catalytic His57 (^{54}TAAHC^{58}), ancrod and *L. m. muta* gyroxin analog present a residue other than the expected Ala56. Interestingly enough, all SVTLEs present a DIML motif at Asp102, instead of the ^{102}DIAL^{105} motif. The gyroxin analog differs even further in the sense that it is missing the 2 last residues of this tetrad. As for the motif around Ser195 (^{193}GDSGGP^{198}), position 193 is occupied by a variety of residues, instead of the conserved Gly in mammalian serine proteinases. In addition, flavoxobin presents a Thr at 197. As concerns the Ser214 sequence pattern (^{211}GIVSW^{215}), apart from bilineobin, either position 213 or position 215 shows variations. Ser214 itself is not conserved in crotalase, ancrod and the *L. m. muta* gyroxin analog. Finally, for the triad following position 225, ^{225}(P/Y)GV(Y/F)^{228}, Gly226 and Val227 turn into Ala and (Leu/Phe/Ile/Ala), respectively.

It is interesting to note that, of the 6 thrombin single-point mutants that retain less than 5% of the wild-type Fg-clotting activity (Trp60d, Lys70, His71, Tyr76, Glu217, Arg221) [24] and are not located in the active site cleft (Trp60d), only Lys70 and Tyr76 belong to the canonical Fg recognition exosite. As for the remaining residues, amino acid changes in His71 and Arg221 may indirectly disturb Fg binding, given that they destabilize loops that are involved in the binding. On the other hand, Glu217 makes direct contacts with Fg. Of the 6 positions, only His71 is strictly conserved throughout the SVTLEs. Tyr76 becomes a Val and the positive charge at position 70 is conserved only in batroxobin and ancrod B, indicating a certain modification of the corresponding exosite. The charge at position 217 is preserved in five cases (positions 217 or 218, fig. 1); however, Arg221 becomes a Glu or a Gln.

Inspection of the structure of TSV-PA reveals that a cluster of hydrophobic residues is indeed present in the region corresponding to ESI, but that the basic residues number three and do not constitute a cluster. Five basic residues represent ESII. Four of them form a cluster, with one residue poorly exposed. None of the 5 residues are located on the C-terminal α-helix (fig. 1). TSV-PA has an acidic segment ^{96}DDE^{98} in the active site cavity. This tripeptide, and especially Asp97, has been shown to play a key role in plasminogen activation and thus substrate recognition [25]. Contortrixobin and flavoxobin contain an Asp-Glu peptide in the same region – the remaining SVTLEs show no acidic residues here.

Conclusion and Discussion

Overall, no quantitative correlation can be found between the specificity of venombin (A, B or AB) and either the hydrophobic or charged-residue content or the described sequence motifs around the molecular markers. In spite of these facts, it is possible to gather some useful information from figure 1 concerning those residues that are unique to each class of enzyme and that could contribute to the subfamily or specific enzyme specificity. Thus, certain residues are unique to each venombin B, venombin AB and all SVTLEs. Others are shared by thrombin and the SVTLEs. Furthermore, several of the variable-sequence segments are located around the active site, whereas one is localized relatively far away from it. Yet another comprises a C-terminal extension characteristic of snake venom serine proteinases. This region is part of the C-terminal portion of the sequence in the proteinase domain that encodes function in serine proteinases [26].

It is interesting to note that TSV-PA, which does not clot Fg, indeed lacks a patch of basic residues in the ESI region. The 3–4 basic residue cluster in the ESII region could present a binding surface to heparin.

The diversity in the functionality of thrombin is thus no doubt a reflection of the presence of specific residues or regions, as well as allosteric changes that may affect its functions [27]. Furthermore several residues, by overlapping other subsites, may fulfill more than one functional role. The modified or reduced functionality of some SVTLEs partially resides in the absence or variation of those specific residues or regions. Whereas in some cases the latter may be detected by sequence comparison, the allosteric changes during binding, if any, may not and would have to be modeled in order to study their influence. On the other hand, the SVTLEs are generally not purely monofunctional, i.e., do not act uniquely on a substrate, which makes it more difficult to precisely define the key residues. An example is crotalase, which in addition to being a SVTLE, displays kallikrein-like activity [28].

Position 193 plays a significant role in determining substrate specificity and resistance to inhibition [29]. As in thrombin, residue 193 is a Gly in calobin, bothrombin and batroxobin. These SVTLEs should be then sensitive to Kunitz-type inhibitors, such as bovine pancreatic trypsin inhibitor. The other SVTLEs contain either an Arg or the bulky residues Phe and Tyr. Because of the restricted accessibility of the S′2 pocket that these side chains cause, the remaining SVTLEs should not be sensitive to the inhibitor, just like TSV-PA, which has a Phe, or mesotrypsinogen with an Arg.

All the SVTLEs possess a Pro at position 225. This position has been shown to influence Na^+ binding near the primary specificity site and to contribute to its architecture. Binding of Na^+ allosterically enhances the catalytic activity of the enzyme. Pro at position 225 abrogates Na^+ binding through an effect on the architecture of the water channel around the primary specificity site. In addition, this Pro indicates that the SVTLEs evolved from the TCN/P225 lineage with a trypsin-like ancestor [30–32]. It follows then that the molecular mechanism of recognition and binding of ligands by SVTLEs is different from those of thrombin and not mediated by allosteric transitions. Some of the factors contributing then in a combined fashion to these differences include: (1) the absence of the cleft formed by the 60-loop of thrombin (making the active site more accessible); (2) the number of acidic residues lining the primary-specificity pocket; (3) the number and type of residues in the apolar binding site near the catalytic center; (4) the number and location of the basic residues constituting the ESI [33].

On the basis of a model, an alternative Fg recognition site was proposed for crotalase [20, 34]. Given that a number of basic and hydrophobic residues are present in SVTLEs at a region equivalent to the ESI in thrombin, the alternative site seems to be rather an extension of the ESI specific to SVTLEs, that includes Lys84a and Lys86. These residues are indeed conserved among SVTLEs and could interact with the acidic Fg sequence on the C-terminal side of the cleavage site. Surprisingly, the ESI extension residues are also present in TSV-PA.

Carbohydrates may play an important role in the structure and stability of a glycoenzyme, and sometimes lead to considerable loss of enzymatic activity. The presence of carbohydrates is usually not taken into account when modeling. Nevertheless, theoretical tools are now available that make it possible to model these moieties if their locations, nature and content are known. In spite of this, molecular models of SVTLEs obtained by sequence homology modeling, by keeping the overall serine proteinase fold and bringing together residues that are far in the sequence, ease the task of identifying regions involved in biomolecular recognition. Several structural features were thus reported and validated experimentally on the basis of a model of the gyroxin analog based on α-thrombin and trypsin [35]. Finally, the TSV-PA structure and the models derived from it should provide a basis for the study of structure-function relationships of snake venom serine proteinases. In a forthcoming paper we intend to develop in detail the structural features of the SVTLEs obtained here.

Acknowledgments

I am grateful to Prof. Wolfram Bode (Max-Planck Institut für Biochemie, Martinsried, Germany) for having encouraged me to start this work. I am a research scientist at the Institut National de la Santé et de la Recherche Médicale (INSERM), France.

References

1 Greer J: Comparative modeling methods: Application to the family of the mammalian serine proteases. Proteins 1990;7:317–334.
2 Stubbs MT, Bode W: A player of many parts: The spotlight falls on thrombin's structure. Thromb Res 1993;69:1–58.
3 Perona JJ, Craik CS: Structural basis of substrate specificity in the serine proteases. Protein Sci 1995;4: 337–360.
4 Perona JJ, Craik CS: Evolutionary divergence of substrate specificity within the chymotrypsin-like serine protease fold. J Biol Chem 1997; 272:29987–29990.
5 Coughlin SR: Thrombin signalling and protease-activated receptors. Nature 2000;407:258–264.
6 Vu TK, Wheaton VI, Hung DT, Charo I, Coughlin SR: Domains specifying thrombin-receptor interaction. Nature 1991;353:674–677.
7 Fenton JW 2nd, Ofosu FA, Moon DG, Maraganore JM: Thrombin structure and function: Why thrombin is the primary target for antithrombotics. Blood Coagul Fibrinolysis 1991;2:69–75.
8 Grutter MG, Priestle JP, Rahuel J, Grossenbacher H, Bode W, Hofsteenge J, Stone SR: Crystal structure of the thrombin-hirudin complex: A novel mode of serine protease inhibition. EMBO J 1990;9: 2361–2365.
9 Rydel TJ, Tulinsky A, Bode W, Huber R: Refined structure of the hirudin-thrombin complex. J Mol Biol 1991;221:583–601.
10 Bode W, Huber R: Natural protein proteinase inhibitors and their interaction with proteinases. Eur J Biochem 1992;204:433–451.
11 Guillin MC, Bezeaud A, Bouton MC, Jandrot-Perrus M: Thrombin specificity. Thromb Haemost 1995; 74:129–133.
12 Fenton JW 2nd, Villanueva GB, Ofosu FA, Maraganore JM: Thrombin inhibition by hirudin: How hirudin inhibits thrombin. Haemostasis 1991;21(suppl 1):27–31.
13 Bode W, Huber R, Rydel TJ, Tulinsky A: X-ray crystal structures of human α-thrombin and of the human thrombin-hirudin complex; in Berliner LJ (ed): Thrombin: Structure and Function. New York, Plenum Press, 1992, pp 3–61.
14 Gan ZR, Li Y, Chen Z, Lewis SD, Shafer JA: Identification of basic amino acid residues in thrombin essential for heparin-catalyzed inactivation by antithrombin III. J Biol Chem 1994;269:1301–1305.
15 Richardson JL, Kroger B, Hoeffken W, Sadler JE, Pereira P, Huber R, Bode W, Fuentes-Prior P: Crystal structure of the human alpha-thrombin-haemadin complex: An exosite II-binding inhibitor. EMBO J 2000;19:5650–5660.
16 Arni RK, Padmanabhan K, Padmanabhan KP, Wu TP, Tulinsky A: Structures of the noncovalent complexes of human and bovine prothrombin fragment 2 with human PPACK-thrombin. Biochemistry 1993;32:4727–4737.
17 Halfon S, Craik CS: Introduction: Serine peptidases and their clans; in Barrett AJ, Rawlings ND, Woessner JF (ed): Handbook of Proteolytic Enzymes. London, Academic Press, 1998, pp 3–4.
18 Markland FS: Snake venoms. Drugs 1997;54(suppl 3):1–10.
19 Pirkle H: Thrombin-like enzymes from snake venoms: An updated inventory. Scientific and Standardization Committee's Registry of Exogenous Hemostatic Factors. Thromb Haemost 1998;79:675–683.
20 Massova I, Pirkle H, Edwards BF, Mobashery S: Insights into the three-dimensional structure of crotalase: Implications for biological activity and substrate specificity. Bioorg Med Chem Lett 1997;7: 3139–3144.
21 Brendel V, Bucher P, Nourbakhsh IR, Blaisdell BE, Karlin S: Methods and algorithms for statistical analysis of protein sequences. Proc Natl Acad Sci USA 1992;89:2002–2006.
22 Parry MA, Jacob U, Huber R, Wisner A, Bon C, Bode W: The crystal structure of the novel snake venom plasminogen activator TSV-PA: A prototype structure for snake venom serine proteinases. Structure 1998;6: 1195–1206.

23 Krem MM, Di Cera E: Molecular markers of serine protease evolution. EMBO J 2001;20:3036–3045.
24 Tsiang M, Jain AK, Dunn KE, Rojas ME, Leung LL, Gibbs CS: Functional mapping of the surface residues of human thrombin. J Biol Chem 1995;270:16854–16863.
25 Zhang Y, Wisner A, Maroun RC, Choumet V, Xiong Y, Bon C: *Trimeresurus stejnegeri* snake venom plasminogen activator. Site-directed mutagenesis and molecular modeling. J Biol Chem 1997;272:20531–20537.
26 Krem MM, Rose T, Di Cera E: The C-terminal sequence encodes function in serine proteases. J Biol Chem 1999;274:28063–28066.
27 Dang OD, Vindigni A, Di Cera E: An allosteric switch controls the procoagulant and anticoagulant activities of thrombin. Proc Natl Acad Sci USA 1995;92:5977–5981.
28 Markland FS, Kettner C, Schiffman S, Shaw E, Bajwa SS, Reddy KN, Kirakossian H, Patkos GB, Theodor I, Pirkle H: Kallikrein-like activity of crotalase, a snake venom enzyme that clots fibrinogen. Proc Natl Acad Sci USA 1982;79:1688–1692.
29 Braud S, Parry MA, Maroun R, Bon C, Wisner A: The contribution of residues 192 and 193 to the specificity of snake venom serine proteinases. J Biol Chem 2000;275:1823–1828.
30 Guinto ER, Caccia S, Rose T, Futterer K, Waksman G, Di Cera E: Unexpected crucial role of residue 225 in serine proteases. Proc Natl Acad Sci USA 1999;96:1852–1857.
31 Dang QD, Guinto ER, di Cera E: Rational engineering of activity and specificity in a serine protease (see comments). Nat Biotechnol 1997; 15:146–149.
32 Dang QD, Di Cera E: Residue 225 determines the Na(+)-induced allosteric regulation of catalytic activity in serine proteases. Proc Natl Acad Sci USA 1996;93:10653–10656.
33 Amiconi G, Amoresano A, Boumis G, Brancaccio A, De Cristofaro R, De Pascalis A, Di Girolamo S, Maras B, Scaloni A: A novel venombin B from *Agkistrodon contortrix contortrix:* Evidence for recognition properties in the surface around the primary specificity pocket different from thrombin. Biochemistry 2000; 39:10294–10308.
34 Henschen-Edman AH, Theodor I, Edwards BF, Pirkle H: Crotalase, a fibrinogen-clotting snake venom enzyme: Primary structure and evidence for a fibrinogen recognition exosite different from thrombin. Thromb Haemost 1999;81:81–86.
35 Castro HC, Silva DM, Craik C, Zingali RB: Structural features of a snake venom thrombin-like enzyme: Thrombin and trypsin on a single catalytic platform? Biochim Biophys Acta 2001;1547:183–195.
36 Kraulis PJ: MOLSCRIPT: A program to produce both detailed and schematic plots of protein structures. J Appl Crystallogr 1991;24: 946–950.

Procoagulant Proteins and Diagnostic Agents

Hæmostasis

Haemostasis 2001;31:257–265

Effects of Lopap on Human Endothelial Cells and Platelets

A.M. Chudzinski-Tavassi[a] M. Schattner[b] M. Fritzen[a]
R.G. Pozner[b] C.V. Reis[a] D. Lourenço[c] M.A. Lazzari[b]

[a]Laboratory of Biochemistry and Biophysics, Butantan Institute, Center for Applied Toxinology, CEPID-FAPESP, São Paulo, Brazil; [b]Hematological Research Institute, National Academy of Medicine and CONICET, Buenos Aires, Argentina; [c]Hematology Department UNIFESP, São Paulo, Brazil

Key Words

Lopap · Prothrombin activator · *Lonomia obliqua* · Platelets · ICAM-1 · E-selectin · Endothelial cells

Abstract

Severe consumption coagulopathy has been detected in rats after Lopap (a prothrombin activator from *Lonomia obliqua* caterpillar bristles) infusion and in humans after accidental contact with *L. obliqua* bristles. However, platelet count and antithrombin (AT) levels were only modestly affected, suggesting that a different form of blood coagulation activation may be involved in this hemorrhagic syndrome. Here we describe that Lopap had no effect on aggregation of washed human platelets induced by several agonists, suggesting that it might not impair platelet function in vivo. AT was able to inhibit the amidolytic activity of thrombin generated by incubation of Lopap with prothrombin in a purified system, which may be different from that generated by the prothrombinase complex in vivo. The surface expression of both ICAM-1 and E-selectin but not of VCAM-1 was upregulated by Lopap in cultured HUVEC, suggesting that it may behave differently from other mediators, such as thrombin and tumor necrosis factor-α.

Introduction

Lopap (*Lonomia obliqua* prothrombin activator protease) is a prothrombin-activating serine protease of 69 kD isolated from the crude extract of *Lonomia obliqua* bristles [1, 2]. Lopap showed procoagulant activity since it has been able to shorten the recalcification time of plasma. Infusion of the purified protein provoked thrombus formation in the cremaster microcirculatory network in rats, followed by hemorrhage in ischemic areas. Plas-

KARGER

Fax + 41 61 306 12 34
E-Mail karger@karger.ch
www.karger.com

0301–0147/01/0316–0257$17.50/0

Accessible online at:
www.karger.com/journals/hae

Dr. A.M. Chudzinski-Tavassi
Laboratory of Biochemistry and Biophysics, Butantan Institute
Av. Vital Brazil 1500, 05503-900 São Paulo-SP (Brazil)
Tel. +55 11 3726 7222, ext. 2109, Fax +55 11 3726 1024
E-Mail amchudzinski@hotmail.com

ma obtained from these animals was unclottable whereas platelet count decreased moderately [3].

Patients who had experienced accidental contact with the bristles of *L. obliqua* developed a hemorrhagic diathesis characterized by intense fibrinogen depletion and consumption of coagulation factors, such as factors V, VIII and XIII. Thrombin generation was demonstrated in these patients with the production of fragment 1 + 2 and thrombin-antithrombin complexes. Fibrinolysis was also activated, as shown by extremely high D dimer levels. However, neither a significant fall in platelet count nor a significant reduction in antithrombin (AT) levels was observed in humans, suggesting that a different form of blood coagulation activation may be involved in this hemorrhagic syndrome [4, 5].

A prothrombin activator, such as Lopap, may be involved in the activation of the coagulation cascade in these patients. However, its direct effects on prothrombin may not be the only mechanism by which it causes the bleeding disorder in humans.

In order to understand the possible mechanisms involved in this process, we performed experiments to evaluate: (1) the action of Lopap on aggregation of washed human platelets induced by several agonists; (2) the effect of purified AT on the amidolytic activity of thrombin generated by incubation of Lopap with prothrombin, and (3) the effect of Lopap on the expression of cell adhesion molecules as well as on the production of von Willebrand factor (vWF) by cultured endothelial cells.

Materials and Methods

Materials

RPMI culture medium and fetal bovine serum (FBS) were from Gibco BRL, Grand Island, N.Y., USA. Endothelial cell growth supplement from bovine neural tissue, heparin, pyruvate, adenosine 5′-diphosphate (ADP), α-thrombin, human prothrombin, AT and prostaglandin E_1 (PGE_1) were from Sigma Chemical Co., St. Louis, Mo., USA. Tumor necrosis factor-α (TNF-α) and fluorescein-conjugated mouse antihuman monoclonal antibodies IgG1 anti-ICAM-1 (CD54), anti-VCAM-1 (CD106) and control IgG1 were purchased from Calbiochem, La Jolla, Calif., USA. Monoclonal antibodies anti-E-selectin (CD62E) and fluorescein goat antimouse IgG $F(ab')_2$ were from Coulter, Immunotech, Marseille, France. An enzyme immunoassay kit for von Willebrand factor (vWF) determination was from RDI, Inc. Flanders, N.J., USA. Equine collagen was from Hormon-Chemie, Germany.

Methods

Platelet Aggregation Studies. Venous blood was obtained from normal individuals using syringes containing ACD anticoagulant (85 m*M* trisodium citrate, 71.4 m*M* citric acid, 111 m*M* dextrose) in a proportion of 1:7 (v/v).

Preparation of Washed Platelets. Suspensions of washed platelets were prepared as described elsewhere [6] with slight modifications. Immediately after collection, blood samples were centrifuged at 190 *g* for 15 min at room temperature to obtain platelet-rich plasma (PRP). PGE_1 (200 µg/ml solution in ethanol) was added to PRP (final concentration 0.25 µg/ml), remaining at 37 °C for 15 min. Then PRP was centrifuged at 1,700 *g* for 15 min, and the platelet pellet was carefully resuspended in modified Tyrode's solution [137 m*M* NaCl, 2.7 m*M* KCl, 0.42 m*M* NaH_2PO_4, 0.35 g% (w/v) bovine albumin, 10 m*M* Hepes, 5.55 m*M* dextrose, 1 m*M* $MgCl_2$, 0.05 µg/ml PGE_1, pH 6.2]. After remaining at 37 °C for 10 min, the suspension was centrifuged at 1,700 *g* for 15 min. The procedure of washing platelets was repeated once again, and finally platelet pellets were resuspended in Tyrode's solution containing calcium [137 m*M* NaCl, 2.7 m*M* KCl, 0.42 m*M* NaH_2PO_4, 0.35 g% (w/v) bovine albumin, 10 m*M* Hepes, 5.55 m*M* dextrose, 1 m*M* $MgCl_2$, 2 m*M* $CaCl_2$, pH 7.4]. Platelet count was adjusted to 300 × 10^9/l and the suspension was maintained at room temperature until testing.

Aggregation Experiments. Platelet aggregation assays were determined turbidimetrically [7] using a Chronolog aggregometer (model 560). A 400 µl aliquot of the washed platelet suspension was placed in the aggregometer cuvettes, and allowed to warm to 37 °C for 5 min. Then, the agonist solution containing Lopap or buffer was added and the aggregation curve was recorded for 5 min. Final concentrations of the agonists used for platelet aggregation were 50 µ*M* ADP and 300 μg/ml fibrinogen, 2.5 μg/ml equine collagen, 0.19 U NIH/ml thrombin, 0.1 U/ml human prothrombin, 14.1 μg/ml Lopap.

Studies of Thrombin Generated by Lopap in a Purified System

The amidolytic activity of the thrombin generated by Lopap was determined using the chromogenic substrate S-2238 in a purified system, as well as by its inhibition by AT. The time courses of prothrombin activation were evaluated after preincubation for 10 min at 37 °C of 2.5 μ*M* purified Lopap with 90 n*M* prothrombin in the presence or absence of 4.1 μ*M* AT, with 2.5 m*M* $CaCl_2$ to a final volume of 100 μl. The reaction was carried out in 100 m*M* Tris-HCl, pH 8.3. The hydrolysis of 40 μ*M* S-2238 by thrombin formed by Lopap was followed spectrophotometrically in a Spectra MAX 190 (Molecular Devices) spectrophotometer at 405 nm during 20 min at 37 °C.

Endothelial Cell Studies

Cell Cultures. Endothelial cells were obtained from HUVEC by collagenase digestion according to the method of Jaffe et al. [8]. HUVEC were identified by their cobblestone morphology and the binding of vWF antibody. Cells were grown in RPMI-1640 medium supplemented with FBS (10%), heparin (90 μg/ml), endothelial supplement growth factor (50 μg/ml), pyruvate (1 m*M*), glutamine (2 m*M*), penicillin (100 U/ml) and streptomycin (100 μg/ml) at 37 °C in a humidified 5% CO_2 incubator. HUVEC used routinely for experiments were between the first and third passage.

To examine the effect of Lopap, HUVEC were plated on 12-well dishes coated with 2% gelatin and grown to confluence. In order to avoid the interference of heparin or growth factors, complete medium was removed and after three washes, cells were incubated for 1 h in RPMI containing glutamine, antibiotics and different Lopap concentrations. TNF-α was used as a positive control.

Expression of Cell Adhesion Molecules by Flow Cytometry. After treatment of HUVEC with Lopap, medium was removed and after washing, cells were further cultured for 4 h in RPMI containing FBS (10%). Cells were harvested by treatment with 0.25% trypsin-0.02% EDTA solution. After washing with Ca^{2+}/Mg^{2+}-free PBS containing 10% FBS (PBS/FBS), pellets were resuspended in 50 μl of PBS/FBS containing saturating concentrations of FITC-conjugated CD54, CD106, or equivalent concentrations of an isotypic control IgG1. E-selectin was detected by staining cells with CD62E for 1 h at 4 °C. The cells were then washed and incubated with goat antimouse IgG FITC-conjugated $F(ab')_2$ fragments (1:100). Cells were fixed with1% paraformaldehyde, and analyzed by flow cytometry in a FACScan cytofluorometer (Becton Dickinson, Mountain View, Calif., USA). Appropriate settings of forward and side scatter gates were used to examine 10,000 cells per experiment. The percentage of positive cells was determined by the thresholds set using isotypic controls. The numbers of fluorescent molecules per cell were indirectly measured by assessing the mean intensity of arbitrary units of fluorescence (AUF) of cells.

vWF Production. The release of vWF into the culture medium was measured quantitatively by an enzyme immunoassay. Microtiter plates were coated overnight at 4 °C with 100 μl purified goat antihuman vWF antibody (3 μg/ml in 50 m*M* Na carbonate, pH 9.6). The remaining protein-binding sites were blocked with 150 μl of blocking solution. After washing, samples and standards diluted in HEPES-Tween-BSA solution were added and incubated for 120 min at 22 °C. Then a second anti-vWF antibody conjugated with horseradish peroxidase was added for another 90 min. Binding of the secondary antibody was detected by the addition of 100 μl of [2,2′-azino-di-[3-ethyl-benzthiazcline sulfonate] and the absorbance was read at 405 nm. Results are expressed in ng/ml and were extrapolated from a standard curve constructed from serial dilutions of normal pooled plasma, assuming a vWF concentration of 10 μg/ml.

Statistical Analysis

All results are expressed as the mean ± SEM. Statistical analysis was performed using Student's t test for paired data. Differences with a p value <0.05 were considered significant.

Results

Platelet Aggregation

Washed platelet aggregation induced by different agonists was not affected by preincubation of platelets with Lopap. On the other hand, thrombin formation from prothrombin by Lopap was able to induce platelet aggregation similarly to that induced by purified thrombin used as control. These results are shown in figure 1.

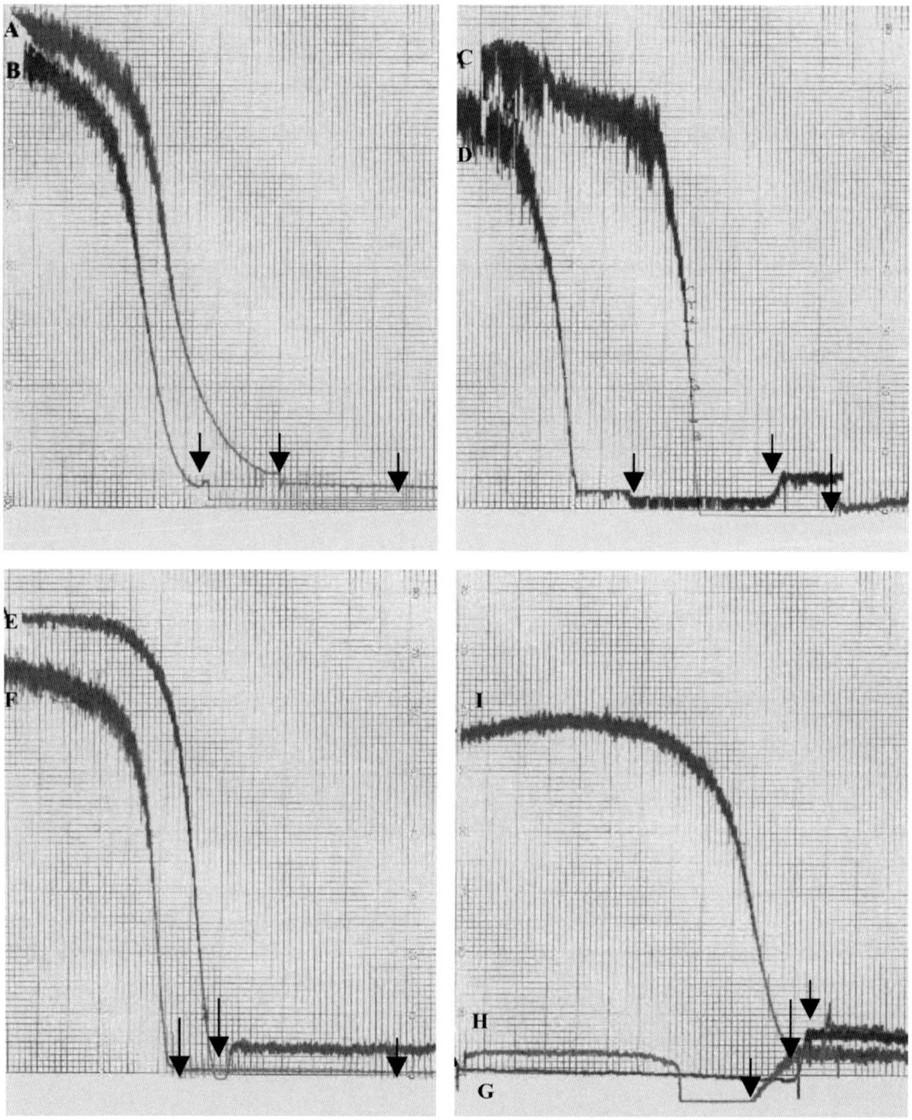

Fig. 1. Washed platelet aggregation after addition of Lopap and different agonists. Arrows indicate agonist addition. A = Thrombin; B = Lopap and thrombin; C = ADP; D = Lopap and ADP; E = collagen; F = Lopap and collagen; G = prothrombin; H = Lopap; I = Lopap and prothrombin.

Inhibition of Thrombin Generated by Lopap

Prothrombin (90 n*M*) incubated with Lopap (2.5 μ*M*) generated thrombin which was able to hydrolyze 40 μ*M* of S2238 chromogenic substrate. Adding AT blocked the amidolytic activity of α-thrombin as well as that of thrombin generated by Lopap (fig. 2).

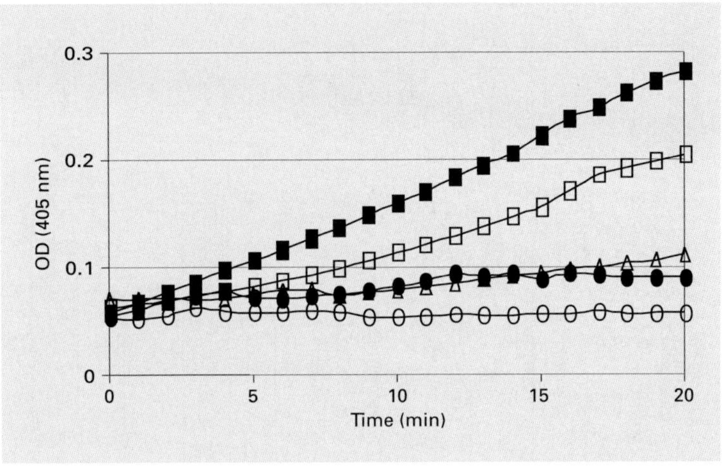

Fig. 2. Amidolytic activity of thrombin generated after Lopap action on prothrombin in a purified system, and its inhibition by AT. ○ = Prothrombin control; □ = prothrombin and Lopap; ■ = thrombin; △ = prothrombin, Lopap and antithrombin; ● = thrombin and AT.

Endothelial Cell Surface Adhesion Molecule Expression

Exposure of HUVEC to Lopap resulted in a concentration-dependent expression of ICAM-1 and E-selectin without detectable increases in VCAM-1 levels, as shown in figure 3. ICAM-1 was still detected 24 h later, whereas E-selectin could not be detected by this time.

Inhibition of Lopap-Induced Cellular Adhesion Molecules

The increased expression of adhesion molecules was specifically mediated by Lopap since preincubation of the purified protein with a serum against whole venom almost completely abrogated E-selectin increase induced by Lopap (fig. 4). Inhibition was not related to an unspecific effect of the serum or the immune complex formed, since the presence of serum did not modify constitutive or E-selectin expression induced by TNF-α (data not shown).

VWF Release Induced by Lopap

While incubation of HUVEC with α-thrombin for 1 h increased the amount of vWF in culture medium, Lopap was not able to modify vWF production (fig. 5). Moreover, levels of vWF after 24 h of endothelial cell treatment with Lopap for 1 h showed results similar (data not shown) to control samples, suggesting that Lopap was not able to induce either vWF release or de novo synthesis of vWF by endothelial cells.

Discussion

A remarkable finding in patients with envenomation by contact with *L. obliqua* is the contrast between a severe consumption coagulopathy, with fibrinogen depletion and production of high levels of prothrombin fragment 1 + 2, thrombin-antithrombin complexes, D dimer, but a modest fall in platelet count and AT concentration [4].

Similar results were observed after infusion of Lopap into rats, showing an almost complete fibrinogen depletion while platelet count was reduced only by 30%. In these animals, collagen-induced platelet aggregation was completely inhibited [3].

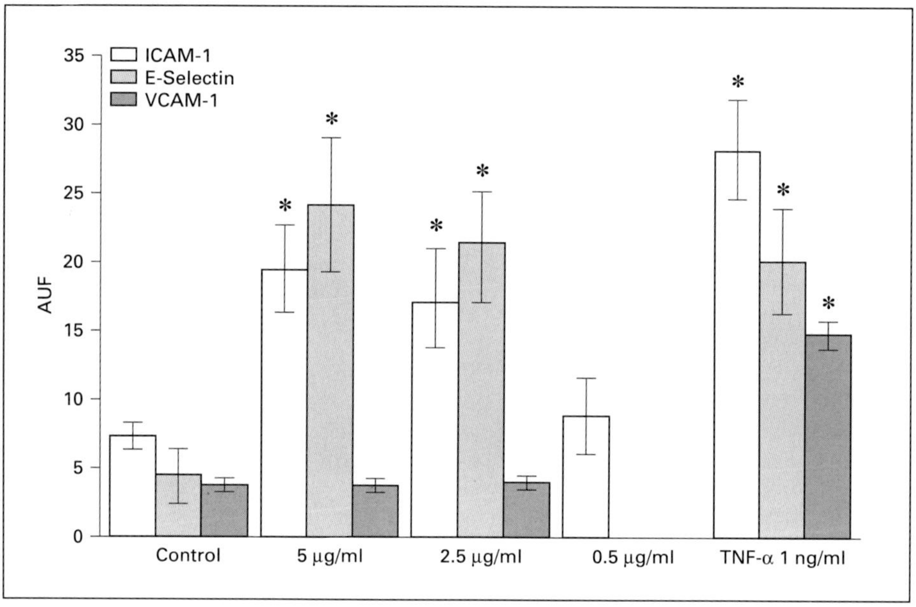

Fig. 3. Expression of ICAM-1, E-selectin and VCAM-1 by HUVECs exposed to different concentrations of Lopap.

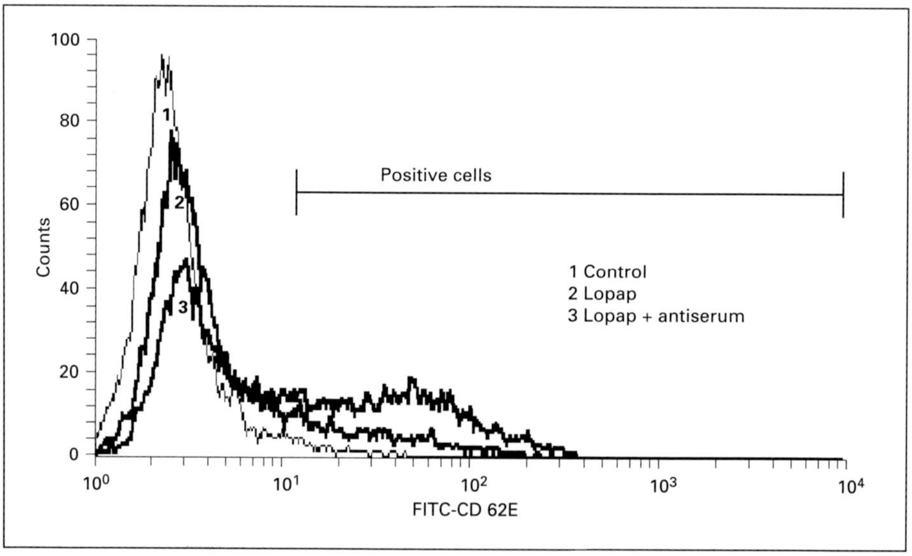

Fig. 4. Inhibition of Lopap-induced HUVEC E-selectin expression by an antiserum against Lonomia.

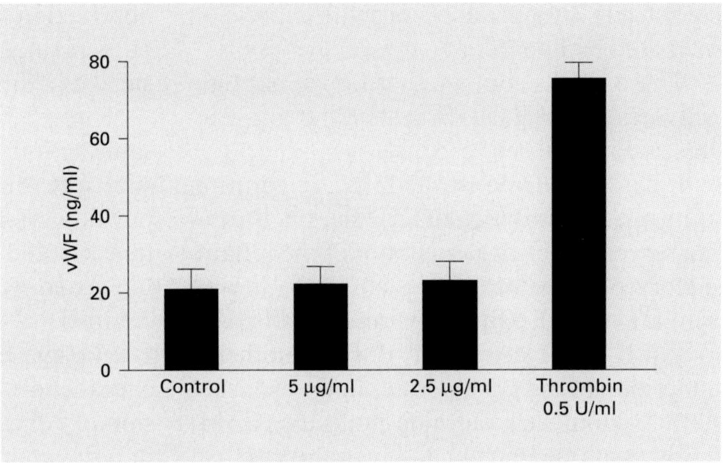

Fig. 5. vW factor levels in the supernatant of HUVEC incubated with different concentrations of Lopap for 1 h at 37 °C.

In the present study, Lopap showed no effect on platelet aggregation induced by collagen, ADP or thrombin, suggesting that it has no direct effect on platelet function.

Indeed, thrombin generated by the action of Lopap on prothrombin was able to induce a normal platelet aggregation response, showing that it can be recognized by platelet receptors.

It is possible that the platelet response to collagen was abolished in rats infused with Lopap because they had been previously activated by thrombin formation and possibly were already degranulated when used in ex vivo experiments.

On the other hand, fibrin degradation products can occupy the GPIIbIIIa fibrinogen receptors and contribute to platelet dysfunction [9].

Circulating activated platelets are not consumed, which can explain the modest fall in platelet count, both in animals and humans, despite a severe consumption coagulopathy.

Thrombin molecules generated by the action of Lopap on prothrombin are able to activate platelets normally and interact with AT as does the α-thrombin formed upon the action of factor Xa on prothrombin.

Why platelet count and AT levels were not reduced in animal models and in human envenomation is not explained. Analysis by mass spectrometry and sequencing of prothrombin fragments induced by Lopap could help to clarify the structural properties of the generated thrombin.

In our study, we used a purified system and we observed that AT was able to block the amidolytic action of the thrombin generated from prothrombin by Lopap, suggesting that it is quite similar to thrombin generated by factor Xa. However, this finding may not represent what happens within the prothrombinase complex on the platelet surface. The direct action of Lopap on endothelial cells might have an important role in the pathogenesis of the coagulation and inflammatory disorders observed during infusion of Lopap in rats or in human accidental contact with *L. obliqua*.

In fact, we observed that Lopap upregulates ICAM-1 and E-selectin expression on cultured endothelial cells, whereas it does not modify VCAM-1 expression.

This effect was specific for Lopap, since serum raised against whole venom almost completely suppressed E-selectin expression mediated by Lopap.

These results confirm that Lopap not only activates the coagulation system but also triggers endothelial cell responses.

It has been demonstrated that, in addition to its role in intravascular coagulation, thrombin serves as a critical mediator of the inflammatory process through its ability to induce activation of NF-κBp65, expression of ICAM-1- and ICAM-1-dependent PMN endothelial adhesion [10, 11]. Therefore, upon exposure to the venom, cell adhesion molecule (CAM) levels might be upregulated by a direct effect of Lopap on the endothelial surface and through intravascular thrombin formation mediated by Lopap.

ICAM-1 and E-selectin are essential adhesion molecules required for neutrophil recruitment and endothelial transmigration. The ability of Lopap to induce CAM expression correlates with our previous demonstration that in vivo Lopap infusion produced PMN sequestration in pulmonary microcirculation vessels [3].

On the other hand, Lopap showed no effect on vWF synthesis or release by cultured endothelial cells.

These data suggest that Lopap interacts only with some receptors on endothelial cells, causing specific effects, like the expression of few molecules such as ICAM-1 and E-selectin, but not of VCAM-1 or production of vWF.

In this regard, Lopap may behave differently from other mediators such as thrombin itself or TNF-α.

It would be interesting to evaluate the direct action of Lopap on the endothelial expression of other molecules, such as thrombomodulin, as well as its effects on endothelial cell fibrinolytic activity.

In summary, our results showed that Lopap does not affect washed platelet function in vitro, so that the impaired platelet function observed in animals infused with Lopap may be explained by the action of thrombin instead of a direct effect of Lopap.

Thrombin generation by Lopap shows the same action as α-thrombin on a chromogenic substrate, and can be inhibited by AT in a purified system. However, we do not rule out that it could behave differently from thrombin generated in vivo, by the prothrombinase complex.

Finally, Lopap exerts a direct effect on endothelial cells increasing the surface expression of adhesion molecules which participate in inflammatory responses, but it was not able to stimulate vWF synthesis nor its release from endothelial cells.

Further studies are necessary to elucidate other possible interactions of Lopap that could explain some important aspects of the envenoming and may clarify the hemorrhagic syndrome.

Acknowledgements

This work was supported by FAPESP and CNPq (Brazil) and CONICET and Fundación René Barón, (Argentina). The authors thank Dr. Beatriz L. Fernandes from Center for Applied Toxinology, CEPID-FAPESP for revision of the text.

References

1 Reis CV, Kelen EMA, Farsky SHP, Portaro FCV, Sampaio CAM, Fernandes BL, Camargo ACM, Chudzinski-Tavassi AM: A Ca^{++} activated serine protease (LOPAP) could be responsible for the haemorrhagic syndrome caused by the caterpillar *Lonomia obliqua. L. obliqua* prothrombin activator protease. Lancet 1999;353:1942.

2 Reis CV, Portaro FCV, Andrade SA, Fritzen M, Fernandes BL, Sampaio CAM, Camargo ACM, Chudzinski-Tavassi AM: A prothrombin activator serine protease from the *Lonomia obliqua* caterpillar venom (Lopap): Biochemical characterization. Thromb Res 2001;102:427–436.

3 Reis CV, Farsky SH, Fernandes BL, Santoro ML, Oliva ML, Mariano M, Chudzinski-Tavassi AM: In vivo characterization of Lopap, a prothrombin activator serine protease from the *Lonomia obliqua* caterpillar venom. Thromb Res 2001;102: 437–443.

4 Zannin M, Lourenço DM, Dalla-Costa LR, Grando M, Gamborge G, Noguti MA, Motta G, Chudzinski-Tavassi AM: Hemostatic disorder observed in patients, affected by accidental contact with *Lonomia obliqua* caterpillar in Santa Catarina, South Brazil (abstract). Thromb Haemost 2001 (suppl July):3352.

5 Kelen EMA, Picarelli ZP, Duarte AC: Hemorrhagic syndrome induced by contact with caterpillars of the genus *Lonomia* (Saturniidae, Hemoleucinae). J Toxicol Toxin Rev 1995;14:283–308.

6 Mustard JF, Kinlough-Rathbone RL, Packham MA: Isolation of human platelets from plasma by centrifugation and washing. Methods Enzymol 1989;169:3–11.

7 Born GVR: Aggregation of blood platelets by adenosine diphosphate and its reversal. Nature 1962;194: 927–929.

8 Jaffe EA, Nachman RL, Becker CG, Minick R: Culture of human endothelial cells derived from umbilical veins. Identification by morphologic and immunologic criteria. J Clin Invest 1973;52:2745–2756.

9 Levi M, Ten Cate H: Disseminated intravascular coagulation. N Engl J Med 1999;34:1586–1592.

10 Rahman A, Anwar KN, True AL, Malik AB: Thrombin-induced p65 homodimer binding to downstream NF-κB site of the promoter mediates endothelial ICAM-1 expression and neuthrophil adhesion. J Immunol 1999;162:5466–5476

11 Kaplanski G, Fabrigoule M, Boulay V, Dinarello CA, Bougrand P, Kaplanski S, Farnarier C: Thrombin induces endothelial type II activation in vitro: IL-1 and TNFα independent IL-8 secretion and E-selectin expression. J Immunol 1997; 158:5435–5446.

Anticoagulant Proteins and Diagnostic Agents

Hæmostasis

Haemostasis 2001;31:266–272

Protein C Activators from Snake Venoms and Their Diagnostic Use

P.M. Gempeler-Messina K. Volz B. Bühler C. Müller

Pentapharm Ltd., Basel, Switzerland

Key Words

Protein C activators · Snake venom enzymes · Diagnostic tools

Abstract

Proteinases converting the zymogen protein C (PC) of vertebrates into activated PC have been detected in several snake venoms. Most PC activators have been purified from venom of snake species belonging to the genera of the *Agkistrodon* complex. Unlike the physiological, thrombin-catalyzed PC activation reaction which requires thrombomodulin as a cofactor, most snake venom activators directly convert the zymogen PC into the catalytically active form which can easily be determined by means of coagulation or chromogenic substrate techniques. Due to this feature, the fast-acting PC activator Protac® from *Agkistrodon contortrix contortrix* (southern copperhead snake) venom has found a broad application in diagnostic practice for the determination of disorders in the PC pathway. Recently, screening assays for the PC pathway have been introduced, based on the observation that the PC pathway is probably the most important physiological barrier against thrombosis.

The Protein C Pathway

The protein C (PC) pathway belongs to the most important anticoagulant mechanisms of blood coagulation.

In the blood coagulation cascade, formation of thrombin – a serine proteinase with pronounced procoagulant properties – occurs in the final phase.

The plasma concentration of the enzymatically inactive prothrombin is approximately 100 times higher than would be necessary for

KARGER

Fax + 41 61 306 12 34
E-Mail karger@karger.ch
www.karger.com

0301–0147/01/0316–0266$17.50/0

Accessible online at:
www.karger.com/journals/hae

Dr. Patrizia M. Gempeler-Messina
Pentapharm Ltd., Engelgasse 109, P.O. Box
CH–4002 Basel (Switzerland)
Tel. +41 61 706 48 48, Fax +41 61 706 48 00
E-Mail patrizia.gempeler@pentapharm.com

complete blood coagulation – after total conversion into active thrombin. The healthy organism has the particular property of inhibiting the clotting system with an efficient negative regulation. Thereby, lesions to blood vessels are repaired by the controlled, locally limited formation of a blood clot. Several mechanisms that inhibit the propagation of the clotting process on the endothelium surface have been described. The anticoagulant PC pathway, one of the anticoagulation mechanisms, regulates blood coagulation by inactivating factors VIIIa and Va, and increasing the fibrinolytic activity [1, 2]. The mechanism can be described as follows: after binding of thrombin to thrombomodulin – a proteoglycan bound to the endothelium membrane – PC is activated [3]. Activated PC (APC) together with its cofactor protein S (PS) [4] inactivates factors VIIIa and Va by proteolytic degradation and therefore thrombin formation is reduced.

PC is a vitamin-K-dependent glycoprotein with a molecular weight of 62,000. In blood, it circulates as an inactive zymogen of a serine proteinase. PC is synthesized in the liver. Its half-life amounts to 8 h and its concentration in plasma is 4 μg/ml.

PS is also a vitamin-K-dependent blood clotting factor with a molecular weight of 70,000. Unlike other vitamin-K-dependent blood clotting factors, PS is not a zymogen of a serine proteinase; it is synthesized in the liver, in endothelial cells and in megakaryocytes. Its half-life amounts to 42 h and it circulates in blood at a concentration of 25 μg/ml. The cofactor activity of PS towards APC is regulated by the capacity of PS to bind to C4b-binding protein (C4bBP) [5]. C4bBP plays an important part in the complement system. The complex C4bBP PS has no APC cofactor activity. Approximately 60% of PS is bound to C4bBP.

PC Disorders

Two types of congenital heterozygous PC deficiencies can be distinguished: PC deficiency type I, where activity and antigen are equally reduced, and PC deficiency type II where clotting activity is lower than antigen concentration. In rare cases of homozygous PC deficiency, fulminant, fatal thromboembolism already occurs in the neonatal phase [6].

Acquired PC deficiency is relatively frequent, as PC belongs to the first factors that are reduced due to liver dysfunction. Disturbances of PC synthesis are mainly observed in case of hepatic diseases, vitamin K deficiency and coumarin therapy. Increased conversion of PC takes place above all in disseminated intravascular coagulation, postoperatively, in insulin-dependent diabetes mellitus and in renal failure.

Increased PC concentration may be due to drugs, e.g., ovulation inhibitors or anabolics. Increased PC levels are also found in pregnancy.

Concentrations between 65 and 150% correspond to the normal range. Concentrations between 20 and 70% imply an increased risk of thromboembolism.

PS Disorders

Three types of familiar, heterozygous PS deficiencies can be distinguished. In type I, total and free PS antigen level are reduced; in type II, PS activity is reduced, but the antigen level of free and total PS is normal; in type III, PS activity and free PS are reduced, but the total PS antigen remains normal [7, 8]. Homozygous PS deficiency is extremely rare.

Acquired PS deficiency is mainly observed in case of hepatic diseases, vitamin K deficiency, coumarin therapy, chronic inflammatory intestinal diseases, disseminated intravascular coagulation and nephrotic syndrome.

Concentrations between 60 and 160% correspond to the normal range, concentrations below 60% bring about an increased risk of thromboembolism.

APC Resistance

APC resistance is associated with the most common hereditary defect predisposing to venous thromboembolism [9]. More than 97% of APC resistance cases are caused by a point mutation in the factor V gene, replacing Arg 506 by Gln in the resulting factor V protein, also referred to as factor V Leiden [10]. Heterozygotes for this mutation have a 5- to 10-fold increased risk of venous thrombosis, while homozygotes are even 50–100 times more susceptible. Due to the high prevalence of factor V Leiden in the normal population [11], this factor should be the first parameter to be evaluated in any scheme assessing a hereditary risk.

PC Activators

The following snake venoms have been described to contain PC activators [12–17]: *Agkistrodon contortrix contortrix, Agkistrodon contortrix mokasen, Agkistrodon contortrix pictigaster, Agkistrodon piscivorus piscivorus, Agkistrodon piscivorus leucostoma, Agkistrodon bilineatus, Agkistrodon blomhoffii ussuriensis, Agkistrodon halys halys, Bothrops moojeni, Bothrops pradoi, Vipera lebetina, Daboia russelli, Cerastes cerastes, Trimeresurus mucrosquamatus.*

First results on the isolation of thrombomodulin-independent and fast-activating PC activator from *A. contortrix contortrix* (southern copperhead snake), and related PC-activating properties of enzymes from venom of *Agkistrodon* subspecies and in venom of *A. piscivorus* (eastern cottonmouth snake), *Agkistrodon bilineatus* (tropical moccasin snake), *B. moojeni*, *B. pradoi* (Prado's pit viper), *Vipera lebetina* (desert adder) and *Cerastes cerastes* (Egyptian sand viper) were presented in 1985 at the Symposium on Animal Venoms and Hemostasis in San Diego, Calif. [18].

The commercial preparation Protac® – the purified form of the fast-acting PC activator from *A. contortrix contortrix* – has considerably simplified the determination of PC [19–21]. Protac converts the zymogen PC very quickly, independently of thrombomodulin, into a serine proteinase which exerts the enzymatic and anticoagulant effects of PC. In contrast to thrombin, the activation product of Protac is not influenced by the inhibitory effect of PC inhibitor. Consequently, time-consuming adsorption steps to separate PC from its inhibitor for the determination of PC are no longer required. Protac also activates PC from cattle, sheep, goats, rabbits, horses and pigs.

Properties and Mode of Action of PC Activators

Protein C activators are mainly characterized from *Agkistrodon* snake venom. According to a comparative study, PC activators can be divided into two groups [16]:

PC Activators from New World Species of the Genus Agkistrodon

Protein C Activators from A. contortrix *Venom. A. contortrix contortrix* crude venom is composed of up to 80 different proteins [22, 23]. Characterizations made by different laboratories [24–28] revealed the following properties for the PC activator from *A. contortrix contortrix.* The activator is a serine proteinase with a molecular weight of 36,000–40,000 D and shows an isoelectric point in the pH range of 7.5–9. PC is activated by limited proteolysis and splits off the substrate Tos-Gly-Pro-Arg-pNA. The anticoagulant and amidolytic activity is inhibited by serine proteinase inhibitors such as NPGB, PMSF, DFP, H-D-Phe-Pro-Arg-chloromethylketone and the active center acylating inhibitor for serine proteinases, 2-ethoxy-6.7-dimethoxy-4H-3.1-benzoxazine-4-one [29]. The amino acid sequence of the PC activator from *A. contortrix* has been completely elucidated [30]. The protein is composed of 231 amino acids with a calculated molecular weight of 25,095 D. Similarly to mammalian serine proteinases, the PC activator has His-40, Asp-85 and Ser-177 in the catalytic active center. The primary structure of the PC activator from *A. contortrix contortrix* shows many homologies to other snake venom enzyme proteinases such as batroxobin from *Bothrops atrox*, flavoxobin from *Trimeresurus flavoviridis*, gyroxin from *Crotalus durissus terrificus*, crotalase from *C. adamanteus* and cerastobin from *Cerastes vipera*.

Protein C Activator from A. bilineatus. The PC activator from *A. bilineatus* [31] has also been characterized. This activator has been shown to have a molecular weight of 35,000–38,000 D by SDS-PAGE under reducing and nonreducing conditions. Analysis of the amino acid sequence revealed a strong homology to PC activator from *A. contortrix* venom.

Protein C Activators from Old World Species of the Genus Agkistrodon

Protein C Activator from A. blomhoffii. A PC activator with a molecular weight of 44,000 D has been determined by SDS-PAGE in *A. blomhoffii ussuriensis* (Ussuri mamushi) venom [15]. This PC activator was described to prolong the aPTT in human plasma, and after intravenous injection to rats the aPTT proved to be prolonged, plasminogen activity increased and FV procoagulant activity decreased.

Protein C Activator from Agkistrodon halys halys. A PC activator with a molecular weight of 36,000–40,000 D and an antithrombotic effect in rats [14, 32] has been isolated from *A. halys halys* (Asian pit viper) venom. The difference between PC activators from New and Old World snake venom consists in the fact that PC activators from *Agkistrodon* from the New World have a molecular weight of less than 40,000 D and a pH optimum of 7.0–8.5, and activate PC from rat plasma to a lesser extent than PC from human plasma. PC activators from *Agkistrodon* species from the Old World have molecular weights greater than 40,000 D and a pH optimum between 5.0 and 6.5, and activate PC from rat plasma to a higher extent than PC from human plasma.

PC Assays Using a PC Activator from Snake Venom

The presently applied methods of PC determination are either functional or immunochemical. Functional assays are either clotting assays or based on determinations using chromogenic assays [19, 33, 34]. They have been largely simplified by the introduction of Protac. The chromogenic assay is performed as follows: PC in plasma is activated with Protac and APC splits off a chromogenic substrate. This test is very easy and specific, can easily be automated for clinical use and shows an extraordinary accuracy.

In the PC clotting assay, patient plasma is prediluted with PC-deficient plasma to exclude any disturbing factors. PC is activated with Protac and the inhibitory effect of APC is determined by aPTT. Prolongation of the clotting time is proportional to the activity of PC in patient plasma. This assay too can be automated.

PS Assays Using a PC Activator from Snake Venom

Basically three types of assays can be recognized [35–40]: (1) Immunological assays for total PS; (2) Immunological assays for free PS; (3) PS functional assays. The APC cofactor activity can be measured as the prolongation of a prothrombin-time-based clotting assay [41], in which diluted test plasma, PS-depleted plasma containing Protac-activated PC, bovine brain tissue thromboplastin and calcium ions are mixed. The clotting time prolongation is proportional to the PS content in patient plasma. Chromogenic methods and aPTT methods have also been described.

Screening Assays for the PC Pathway

Recently, screening assays for the PC pathway have been introduced which are based on the observation that the PC pathway is probably the most important physiological barrier against thrombosis. For this reason, identification of a dysfunctional PC pathway should be the first step in order to assess an individual thrombosis risk. Several groups have described similar functional assays which are based on the following principle [42]: a screening test, e.g. aPTT, is performed on the sample with or without previous or simultaneous activation of PC by Protac.

The two clotting times are required in order to compensate any changes of coagulation factors which influence the individual aPTT of a sample. A functional PC pathway leads to a massive prolongation of the clotting time in the sample with the activator and thus to a PC ratio (the ratio of the clotting times with and without Protac) above the cut-off value for normals. The PC ratio is slightly instrument dependent and can be normalized with the help of a calibration plasma.

Pathological results are found in almost all patients with factor V Leiden or with PC deficiency, whereas the detection of PS dysfunction is less efficient. Other factors which influence the assay are lupus anticoagulants, the prothrombin mutation 20210 G/A or elevated factor VIII.

APC Resistance Assay

The following example of APC resistance assay (APC-R assay) was performed with the PC activator from *A. contortrix contortrix* [unpubl. data].

Sample plasma is prediluted with factor-VV-deficient plasma and incubated at 37°C with a specific factor V activator from *Daboia russelli* and phospholipids in the presence and absence of Protac. Coagulation is triggered by the addition of a factor-V-dependent prothrombin activator from *Notechis scutatus* in the absence of calcium. The time until clot formation is recorded. The ratio of clotting time with/without added Protac is calculated. Homozygous factor V Leiden gives a ratio below 1.0, heterozygous factor V, a ratio between 1.0 and 2.0, and wild-type, a ratio above 2.0.

Conclusions

The active component Protac is a protease which causes the activation of PC and is used for the determination of PC, PS, APC-R and in global assays for determination of PC pathway disorders. Properties of PC activators from snake venom have been determined in a relatively short time and the PC activator from *A. contortrix contortrix* is now widely used as a diagnostic tool in medical research and clinical use.

Acknowledgements

We thank Ms. A. Galmiche for assistance in preparing the manuscript.

References

1 Esmon CT: The protein C anticoagulant pathway. Arterioscler Thromb 1992;12:135–145.
2 Dahlbäck B: The protein C anticoagulant system: Inherited defects as basis for venous thrombosis. Thromb Res 1995;77:1–43.
3 Esmon CT, Owen WG: Identification of an endothelial cell cofactor for thrombin-catalyzed activation of protein C. Proc Natl Acad Sci USA 1981;78:2249–2252.
4 Walker FJ: Regulation of activated protein C by a new protein. A possible function of bovine protein S. J Biol Chem 1980;255:5521–5524.
5 Dahlbäck B: Inhibition of protein Ca cofactor function of human and bovine protein S by C4b-binding protein. J Biol Chem 1986;261: 12022–12027.
6 Reitsma PH, Bernardi F, Doig, RG, Gandrille S, Greengard JS, Ireland H, Krawczak M, Ling B, Long GL, et al: Protein C deficiency: A database of mutations, 1995 update. On behalf of the subcommittee on plasma coagulation inhibitors of the scientific and standardization committee of the ISTH. Thromb Haemost 1995;73:876–889.
7 Gandrille S, Borgel D, Sala N, Espinosa-Parrilla Y, Simmonds R, Rezende S, Lind B, Mannhalter C, Pabinger I, Reitsma PH, Formstone C, Cooper DN, Saito H, Suzuki K, Bernardi F, Aiach M: Protein S deficiency: A database of mutations – summary of the first update. Thromb Haemost 2000;84:918.
8 Gandrille S, Borgel D, Ireland H, Lane DA, Simmonds R, Reitsma PH, Mannhalter C, Pabinger I, Saito H, Suzuki K, Formstone C, Cooper DN, Espinosa Y, Sala N, Bernardi, F, Aiach M: Protein S deficiency. A database of mutations. For the plasma coagulation inhibitors subcommittee of the scientific and standardization committee of the international society on thrombosis and hemostasis. Thromb Haemost 1997; 77:1201–1214.
9 Dahlbäck B, Carlsson M, Svensson PJ: Familial thrombophilia due to a previously unrecognized mechanism characterized by poor anticoagulant response to activated protein C: Prediction of a cofactor to activated protein C. Proc Natl Acad Sci USA 1993;90:1004–1008.
10 Bertina RM, Koeleman BPC, Koster T, Rosendaal FR, Dirven RJ, de Ronde H, van der Veldon PA, Reitsma PH: Mutation in blood coagulation factor V associated with resistance to activated protein C. Nature 1994;369:64–66.
11 Rosen SB, Sturk A: Activated protein C resistance. A major risk factor for thrombosis. Eur J Clin Chem Clin Biochem 1997;35:501–516.
12 Stocker K, Fischer H, Meier J, Brogli M, Svendsen L: Protein C activators in snake venoms. Behring Inst Mitt 1986;79:7–47.
13 Meier J: Proteinases activating protein C; in Bailey GS (ed): The Enzymology of Snake Venoms. Fort Collins, Alaken, 1998, pp 253–285.
14 Bakker HM, Tans G, Yukelson LY, Janssen-Claessen TW, Bertina RM, Hemker HC, Rosing J: Protein C activation by an activator purified from the venom of *Agkistrodon halys halys.* Blood Coagul Fibrinolysis 1993;4:605–614.
15 Kogan AE, Bashkov GV, Bobruskin ID, Romanova EP, Makarov VA, Strukova SM: Protein C activator from the venom of *Agkistrodon blomhoffi ussuriensis* retards thrombus formation in the arterio-venous shunt in rats. Thromb Res 1993;70: 385–393.
16 Kogan AE, Makarov AN, Bobruskin ID, Strukova SM: Comparative study of protein C activators from the Agkistrodon snake venoms. Thromb Res 1991;2:75–80.
17 Sun L, Guang J, Huang S, Yu Q: Identification of protein C activator from nine species of Chinese snake venoms. Zhongguo Bingli Shengli Zazhi 2001;17:241–244.
18 Stocker KF, Meier J: Thrombin-like snake venom enzymes; in Pirkle H, Markland FS (eds): Hemostasis and Animal Venoms. New York, Dekker, 1988, pp 67–84.
19 Martinoli JL, Stocker K: Fast functional protein C assay using Protac®, a novel protein C activator. Thromb Res 1986;43:253–264.
20 Stocker K, Fischer H, Meier J, Brogli M, Svendsen L: Characterization of the protein C activator Protac® from the venom of the southern copperhead *(Agkistrodon contortrix)* snake. Toxicon 1987;25:239–252.
21 Stocker K, Fischer H, Meier J: Practical application of the protein C activator Protac® from *Agkistrodon contortrix* venom. Folia Haematol 1988;115:260–264.
22 Dyr JE, Suttnar J, Kornalik F: Assessment of a complex mixture of interacting proteins in the coagulant-active venom from *Agkistrodon contortrix contortrix.* Folia Haematol 1988;115:203–207.

23 Suttnar J, Dyr JE, Kornalik F: Evaluation of individual variability in the composition of *Agkistrodon contortrix contortrix* venom by means of HPLC and two-dimensional PAGE. Folia Haematol 1988;115:197–202.
24 Klein JD; Walker FJ: Purification of a protein C activator from the venom of the southern copperhead snake *(Agkistrodon contortrix contortrix)*. Biochemistry 1986;25:4175–4179.
25 Kisiel W, Kondo S, Smith KJ, McMullen BA, Smith LF: Characterization of a protein C activator from *Agkistrodon contortrix contortrix* venom. J Biol Chem 1987;262:12607–12613.
26 Kisiel W, Choi E, Kondo S: Isolation of a protein C activator from southern copperhead venom. Biochem Biophys Res Commun 1987;143:917–922.
27 Exner T, Vaasjoki R: Characterization and some properties of the protein C activator from *Agkistrodon contortrix contortrix* venom. Thromb Haemost 1988;59:40–44.
28 Orthner CL, Bhattacharya P, Strickland DK: Characterization of a protein C activator from the venom of *Agkistrodon contortrix contortrix.* Biochemistry 1988;27:2558–2564.
29 Stuerzebecher J, Neumann U, Meier, J: Inhibition of the protein C activator Protac®, a serine proteinase from the venom of the southern copperhead snake *Agkistrodon contortrix contortrix.* Toxicon 1991;29:151–155.
30 McMullen BA, Fujikawa K, Kisiel W: Primary structure of a protein C activator from *Agkistrodon contortrix contortrix* venom. Biochemistry 1989;28:674–679.
31 Nakagaki T, Kazim AL, Kisiel W: Isolation and characterization of a protein C activator from tropical moccasin venom. Thromb Res 1990;58:593–602.
32 Strukova SM, Kogan AE, Tara AA, Aaviksaar A: Antithrombotic effect of protein C activator from snake venom. Vopr Med Khim 1989;35:115–119.
33 Marlar RA, Adcock DM: Clinical evaluation of protein C: A comparative review of antigenic and functional assays. Progr Pathol 1989;20:1040–1047.
34 Francis RB, Seyfert U: Rapid amidolytic assay of protein C in whole plasma using an activator from the venom of *Agkistrodon contortrix.* Am J Clin Pathol 1987;87:619–625.
35 Bertina RM, Van Wijngaarden A, Reinalda-Poot J, Poort SR, Bom VJJ: Determination of plasma protein S – the protein cofactor of activated protein C. Thromb Haemost 1985;53:268–272.
36 Woodhams BJ: The simultaneous measurement of total and free protein S by ELISA. Thromb Res 1988;50:213–220.
37 Edson JR, Vogt JM, Huesman DA: Laboratory diagnosis of inherited protein S deficiency. Am J Clin Pathol 1990;94:176–186.
38 Suzuki K, Nishioka J: Plasma protein S activity measured using Protac®, a snake venom derived activator of protein C. Thromb Res 1988;49:241–251.
39 Wolf M, Boyer-Neumann C, Martinoli JL, Leroy-Matheron C, Amiral J, Meyer D, Larrieu MJ: A new functional assay for human protein S activity using activated factor V as substrate. Thromb Haemost 1989;62:1144–1145.
40 Kobayashi I, Amemiya N, Endo T, Okuyama K, Tamura K, Kume S: Functional activity of protein S determined with use of protein C activated by venom activator. Clin Chem 1989;35:1644–1648.
41 Preda L,Tripodi A, Valsecchi C, Lombardi A, Finotto E, Mannucci PM: A prothrombin time-based functional assay of protein S. Thromb Res 1990:60:19–32.
42 Dati F, Hafner G, Erbes H, Prellwitz W, Kraus M, Niemann F, Noah M, Wagner C: ProC Global: The first functional screening assay for the complete protein C pathway. Clin Chem 1997;43:1719–1723.

Anticoagulant Proteins and Diagnostic Agents

Hæmostasis

Haemostasis 2001;31:273–278

Interaction of Bothrojaracin with Prothrombin

Russolina B. Zingali M. Lucia Bianconi Robson Q. Monteiro

Departamento de Bioquímica Médica, ICB/CCS, Universidade Federal do Rio de Janeiro, Brazil

Key Words

Bothrojaracin · Prothrombin · Proexosite I · Factor Va · Prothrombinase

Abstract

Bothrojaracin (BJC) is a 27-kD protein from *Bothrops jararaca* venom that interacts with α-thrombin (K_D = 0.7 n*M*) through both anion-binding exosites I and II. Recently, it has been shown that BJC interacts with the exosite I precursor (proexosite I) on human prothrombin (K_D = 75 n*M*), forming a 1:1 Ca^{2+}-independent noncovalent complex with the zymogen. Complex formation was associated with inhibition of zymogen activation by *Oxyuranus scutellatus* venom. In addition, BJC strongly decreased the prothrombin activation by factor Xa only in the presence of factor Va. A similar effect was observed in the presence of phospholipids, suggesting that BJC specifically inhibits the interaction of prothrombin with factor Va. It is proposed that BJC has two independent mechanisms for anticoagulation: (1) inhibition of exosite-I-dependent activities on α-thrombin, and (2) inhibition of prothrombin activation through interaction with proexosite I.

Introduction

Thrombin is formed by the cleavage of two peptide bonds in prothrombin by blood coagulation factor Xa, with the release of prothrombin fragments 1 and 2. These reactions are modulated by calcium, membranes containing acidic phospholipids, and the protein cofactor, factor Va, which assemble with factor Xa and prothrombin to form the prothrombinase complex [1]. It is generally observed that phospholipids increase the substrate affinity while factor Va enhances the catalytic rate of the reaction [2].

During prothrombin activation, besides the exposure of the catalytic site, additional regulatory domains become available [3–4].

KARGER

Fax + 41 61 306 12 34
E-Mail karger@karger.ch
www.karger.com

0301–0147/01/0316–0273$17.50/0

Accessible online at:
www.karger.com/journals/hae

Dr. Russolina B. Zingali
Departamento de Bioquímica Médica/ICB/CCS
Bloco H – 2° Andar – Ilha do Fundão
Rio de Janeiro, RJ 21941-590 (Brazil), Tel. +55 21 2562 6782
Fax +55 21 2270 8647, E-Mail lzingali@bioqmed.ufrj.br

Anion-binding exosite I is a positively charged region responsible for the high specificity of the enzyme towards macromolecular substrates, cofactors and some inhibitors [5, 6]. Exosite II is referred to as heparin binding site [7]. Recently, it has been shown that exosite I is partially exposed on prothrombin in a precursor state (proexosite I) and is fully exposed on thrombin [8].

Binding of prothrombin to factor Va is believed to involve fragment 2 domain of prothrombin [9–11]. More recently, it was shown that C-terminal hirudin peptides inhibit prothrombin activation by factor Xa only in the presence of factor Va [12]. This observation suggested the involvement of proexosite I in substrate interactions with the cofactor.

Herein we show that the interaction of bothrojaracin (BJC) with prothrombin occurs through the proexosite 1, and also that the formation of this complex impairs prothrombin activation in the presence of factor Va.

Interaction of BJC with Thrombin

BJC was isolated from the venom of the Brazilian snake jararaca *(Bothrops jararaca)* [13]. It is a disulfide-linked heterodimer that belongs to the C-type (Ca^{2+}-dependent) lectin-like related protein family, which contains structurally homologous proteins displaying distinct biological functions. BJC has been characterized as a potent and selective thrombin inhibitor (K_D = 0.7 n*M*) since it binds to both exosites I and II [14], forming a 1:1 noncovalent complex. Binding of BJC to thrombin modulates, but does not impair the cleavage of tripeptide chromogenic substrates [16]. On the other hand, complex formation results following inhibition of exosite-I-related functions, such as fibrinogen clotting, platelet aggregation, thrombomodulin-dependent protein C activation and factor V activation [13, 15]. In addition, exosite II-related functions are also impaired, as demonstrated by blockage of heparin-dependent antithrombin inactivation of the thrombin-BJC complex [14].

Interaction of BJC with Prothrombin

BJC forms a calcium-independent, 1:1 complex with human prothrombin [14, 17]. Nevertheless, no interaction was observed with prothrombin fragment 1, which contains the Gla and first kringle domains, or fragment 2, that is constituted by the second kringle domain. Thus these data suggest that the site of interaction with BJC is located in the catalytic domain of prothrombin. Isothermal titration calorimetry (ITC) showed that the binding of prothrombin to BJC is endothermic (ΔH = 13.5 ± 1.8 kcal/mol), indicating that the binding is entropically driven, with free energy ΔG = –9.7 kcal/mol, and entropy term – $T\Delta S$ = –3.8 kcal/mol [17].

Anderson et al. [8] recently characterized the partially exposed exosite I (proexosite I) on prothrombin. Binding experiments showed that BJC and sulfated C-terminal hirudin compete for the same site. In fact, BJC presents the highest affinity for proexosite I described to date (table 1). Interestingly, BJC and C-terminal hirudin peptides display a similar ~100-fold increase in affinity for exosite I when prothrombin is activated to thrombin (table 1). These observations strongly suggest that proexosite I is a low-affinity, precursor state of exosite I.

Effect of BJC on Prothrombin Activation

It has previously been shown that BJC inhibits factor V activation by thrombin [15]. Therefore the thrombin formation by the pro-

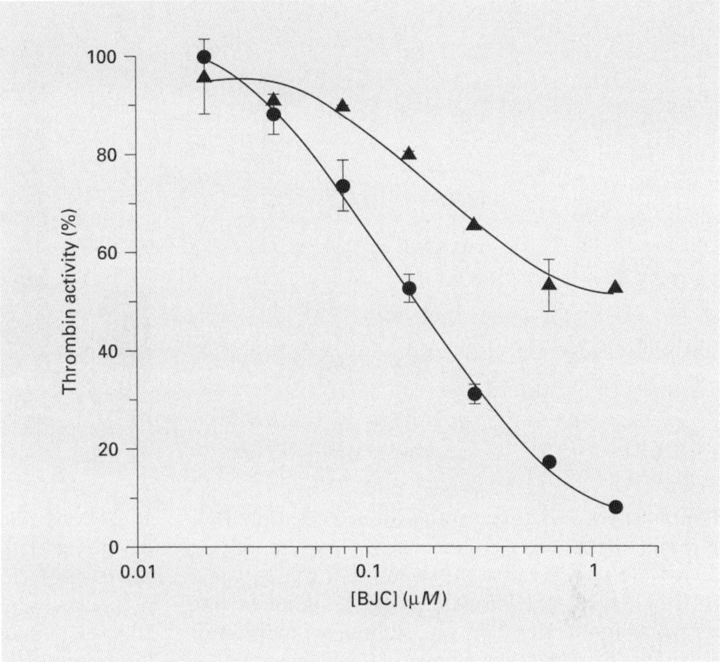

Fig. 1. Anticoagulant X anti-prothrombin-converting activity of BJC. Prothrombin (100 n*M*) activation was carried out in 50 m*M* Tris-HCl, 150 m*M* NaCl, 5 m*M* $CaCl_2$, 1 mg/ml PEG 8000, pH 7.5 containing varying amounts of BJC and 0.2 m*M* S-2238 (▲) or 2 mg/ml human fibrinogen (●). The reaction was started by addition of 10 μg/ml *O. scutellatus* venom. The experiments performed in the absence of BJC were taken as 100%. Data represent mean ± SD of two independent determinations.

Table 1. Proexosite I ligands and their affinities for prothrombin and thrombin binding

Ligand	K_D, μ*M* prothrombin/thrombin	Ratio	Reference
$Hirudin^{54-65}$	45/0.65	69	8
$Hirudin^{54-65}(SO^{3-})$	3.2/0.025	128	8
BJC	0.076/0.00062	126	17, 14

thrombinase complex in the presence of factor V is decreased [15]. Since BJC also interacts with prothrombin it was reasonable to speculate that this protein could interfere with prothrombin activation. Several prothrombin activators isolated from snake venoms are currently known [18]. Scuterin is a 300-kD enzyme from *Oxyuranus scutellatus* venom that is structurally related to a factor Xa-factor Va complex [19]. Formation of the BJC-prothrombin complex was associated with partial inhibition of zymogen activation by *O. scutellatus* venom (fig. 1 and 2) [20], thus indicating that BJC can also block prothrombin activation. In addition, thrombin formed in the assay showed a progressive decrease in the fibrinogen-clotting activity. Thus, at ~1 μ*M* BJC, prothrombin activation by *O. scutellatus* was decreased by ~50% while the thrombin formed was completely inactive towards fibrinogen (fig. 1). In a purified system, BJC decreased prothrombin activation by factor

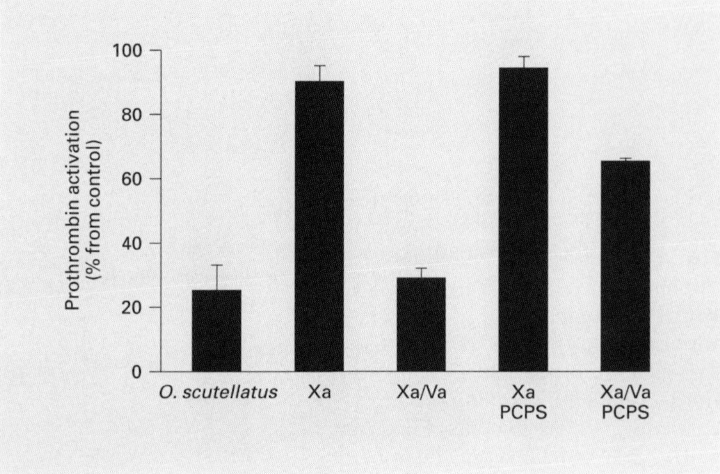

Fig. 2. Effect of BJC on prothrombin activation. Prothrombin (500 n*M*) activation was carried out in 50 m*M* Tris-HCl, 150 m*M* NaCl, 5 m*M* $CaCl_2$, 1 mg/ml PEG 8000, pH 7.5 containing 1 mg/ml bovine serum albumin. Reaction was started by addition of 10 µg/ml *O. scutellatus* venom, or 10 n*M* factor Xa, or 10 n*M* factor Xa/1 n*M* factor Va, or 10 n*M* factor Xa/10 µ*M* phospholipid vesicles (PCPS; 75% phosphatidylcholine, 25% phosphatidylserine), or 100 p*M* factor Xa/10 p*M* factor Va/10 µ*M* phospholipid vesicles. Experiments were performed in the absence (taken as 100% activation) or in the presence of 1.5 µ*M* BJC. Data represent mean ± SD of three independent determinations.

Xa in the presence of factor Va, with 70% inhibition. However, no effect was observed in the absence of the cofactor (fig. 2). This result agrees with previous studies in which C-terminal hirudin peptides also inhibited factor-Va-accelerated prothrombin activation [12]. We suggested that the effect of BJC is specific towards prothrombin interaction with factor Va. Experiments performed in the presence of model membranes (75% phosphatidylcholine, 25% phosphatidylserine) showed a similar profile although the inhibitory effect of BJC was significantly lower than that observed in the absence of phospholipids (fig. 2). Thus, BJC caused a 35% inhibition of prothrombin activation by factor Xa in the presence, but not in the absence of factor Va. Altogether these data reinforce the hypothesis that proexosite I participates in the interaction of prothrombin with factor Va.

BJC as Anticoagulant

We have previously shown that BJC specifically interacts with prothrombin in human plasma [20]. As expected, BJC is an effective inhibitor of plasma clotting [15, 20], which is also able to reduce thrombin formation after activation of the intrinsic pathway of blood coagulation [20]. Studies of BJC interaction with either thrombin or prothrombin suggest that, in vivo, BJC produces an anticoagulant effect. Two distinct mechanisms may thus be involved (fig. 3): (1) BJC binds to the generated thrombin, and inhibits the exosite-1-dependent activities such as fibrinogen clotting, factor V activation, platelet activation and possibly factor VIII activation. It is unclear whether BJC inhibits factor XI and/or factor XIII activation, and (2) whether BJC interacts with prothrombin and interferes

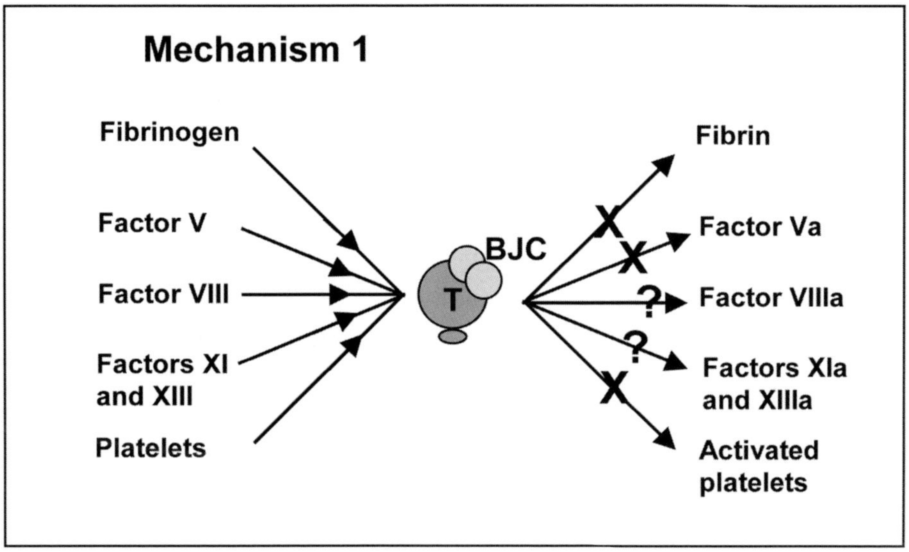

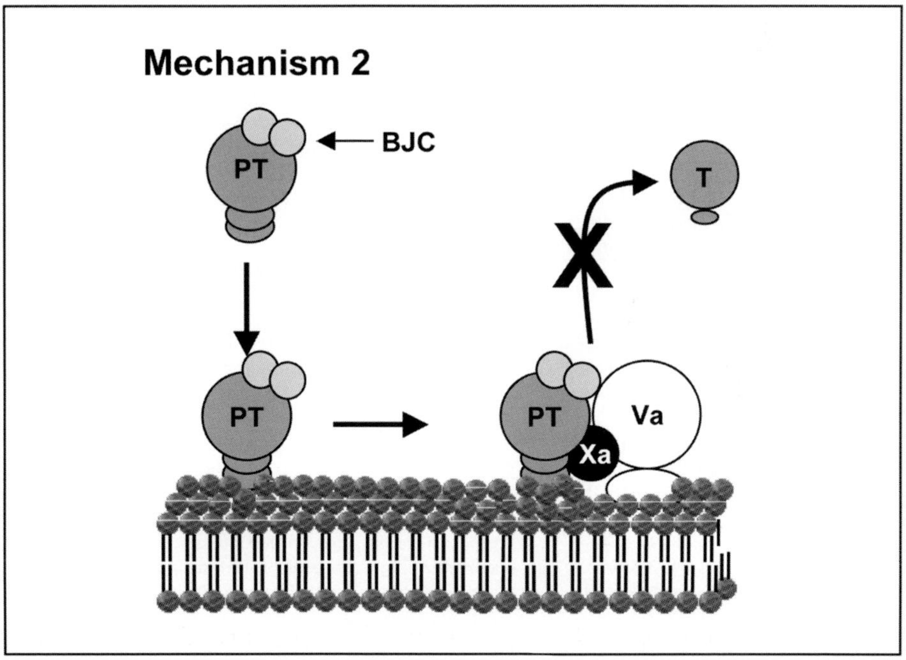

Fig. 3. Anticoagulant mechanism of BJC. Mechanism 1 is based on the inhibition of exosite-I-related activities. Mechanism 2 is based on the inhibition of prothrombin activation through interaction with proexosite I.

with proexosite-I-mediated interactions with factor Va, thus inhibiting the zymogen activation.

Future Perspectives

The physiological role of proexosite I in prothrombin function/activation remains unclear. The specific interference of BJC in factor-Va-accelerated prothrombin activation suggests that this region might modulate the interaction of the cofactor with the zymogen in vivo during its activation. At this point, proexosite I may be a potential target for the development of new antithrombotic drugs. Furthermore, the higher binding affinity of BJC when compared to $Hir^{54-65}(SO^{3-})$ point to this molecule as a tool for future studies on that newly characterized prothrombin site.

References

1 Kalafatis M, Swords NA, Rand MD, Mann KG: Membrane-dependent reactions in blood coagulation: Role of vitamin-K-dependent enzyme complexes. Biochim Biophys Acta 1994;1227:113–129.

2 Rosing J, Tans G, Govers-Riemslag JW, Zwaal RF, Hemker HC: The role of phospholipids and factor Va in the prothrombinase complex. J Biol Chem 1980;255:274–283.

3 Liu LW, Ye J, Johnson AE, Esmon CT: Proteolytic formation of either of the two prothrombin activation intermediates results in formation of a hirugen-binding site. J Biol Chem 1991;266:23632–23636.

4 Wu Q, Picard V, Aiach M, Sadler JE: Activation-induced exposure of the thrombin anion-binding exosite. J Biol Chem 1994;269:3725–3730.

5 Stubbs MT, Bode W: A player of many parts: The spotlight falls on thrombin's structure. Thromb Res 1993;69:1–58.

6 Guillin MC, Bezeaud A, Bouton MC, Jandrot-Perrus M: Thrombin specificity. Thromb Haemost 1995; 74:129–133.

7 Sheehan JP, Sadler JE: Molecular mapping of the heparin-binding exosite of thrombin. Proc Natl Acad Sci USA 1994;91:5518–5522.

8 Anderson PJ, Nesset A, Dharmawardana KR, Bock PE: Characterization of proexosite I on prothrombin. J Biol Chem 2000;275:16428–16434.

9 Esmon CT, Jackson CM: The conversion of prothrombin to thrombin. IV. The function of the fragment 2 region during activation in the presence of factor V. J Biol Chem 1974;249:7791–777.

10 Kotkow KJ, Deitcher SR, Furie B, Furie BC: The second kringle domain of prothrombin promotes factor Va-mediated prothrombin activation by prothrombinase. J Biol Chem 1995;270:4551–4557.

11 Krishnaswamy S, Walker RK: Contribution of the prothrombin fragment 2 domain to the function of factor Va in the prothrombinase complex. Biochemistry 1997;36: 3319–3330.

12 Anderson PJ, Nesset A, Dharmawardana KR, Bock PE: Role of proexosite I in factor Va-dependent substrate interactions of prothrombin activation. J Biol Chem 2000; 275:16435–16442.

13 Zingali RB, Jandrot-Perrus M, Guillin MC, Bon C: Bothrojaracin, a new thrombin inhibitor isolated from *Bothrops jararaca* venom: Characterization and mechanism of thrombin inhibition. Biochemistry 1993; 32:10794–107802.

14 Arocas V, Zingali RB, Guillin MC, Bon C, Jandrot-Perrus M: Bothrojaracin: A potent two-site-directed thrombin inhibitor. Biochemistry 1996;35:9083–9089.

15 Arocas V, Lemaire C, Bouton MC, Bezeaud A, Bon C, Guillin MC, Jandrot-Perrus M: Inhibition of thrombin-catalyzed factor V activation by bothrojaracin. Thromb Haemost 1998;79:1157–1161.

16 Monteiro RQ, Rapôso JG, Wisner A, Guimarães JA, Bon C, Zingali RB: Allosteric changes of thrombin catalytic site induced by interaction of bothrojaracin with anion-binding exosites I and II. Biochem Biophys Res Commun 1999;262:819–822.

17 Monteiro RQ, Bock PE, Bianconi ML, Zingali RB: Characterization of bothrojaracin interaction with human prothrombin. Protein Sci 2001; 10:1897–1904.

18 Rosing J, Tans G: Structural and functional properties of snake venom prothrombin activators. Toxicon 1992;30:1515–1527.

19 Speijer H, Govers-Riemslag JW, Zwaal RF, Rosing J: Prothrombin activation by an activator from the venom of *Oxyuranus scutellatus* (Taipan snake). J Biol Chem 1986; 261:13258–13267.

20 Monteiro RQ, Zingali, RB: Inhibition of prothrombin activation by bothrojaracin, a C-type lectin from *Bothrops jararaca* venom. Arch Biochem Biophys 2000;382:123–128.

Anticoagulant Proteins and Diagnostic Agents

Hæmostasis

Haemostasis 2001;31:279–287

Anticoagulant Venom and Mammalian Secreted Phospholipases A_2: Protein- versus Phospholipid-Dependent Mechanism of Action

Carine M. Mounier[a] Cassian Bon[b] R. Manjunatha Kini[c,d]

[a]Département de Biologie, Université de Cergy-Pontoise, Cergy-Pontoise, [b]Unité des Venins, Institut Pasteur, Paris, France; [c]National University of Singapore, Singapore; [d]Department of Biochemistry and Molecular Biophysics, Medical College of Virginia, Virginia Commonwealth University, Richmond, Va., USA

Key Words

Anticoagulant · Secreted phospholipase A_2 · Snake venom · Mammalian · Factor Xa · Factor Va

Abstract

Some venom and mammalian-secreted phospholipases A_2 ($sPLA_2$) have been described to exert an anticoagulant effect. This review will discuss and compare phospholipid-dependent versus protein-dependent mechanisms of action of these $sPLA_2$ on the coagulation cascade. The importance of venom proteins, and of the study of their pharmacological effects, to explore the physiological functions of homologous mammalian proteins is also pointed out.

Introduction

Snake venoms contain numerous components provoking various pharmacological effects [1–3]. This review will focus on venom and mammalian secreted phospholipases A_2 (sPLA2) that exert an anticoagulant effect. sPLA2 most likely affect blood coagulation through hydrolysis of or binding to procoagulant phospholipids (PL) [4]. However, some data previously suggested that certain venom sPLA2 might exhibit a PL-independent mechanism [4]. The evidence of the existence of a protein-dependent mechanism of action came from studies on the basic sPLA2 isoform (CM-IV) of *Naja nigricollis* venom [5], while the first demonstration of direct binding to a coagulation factor was obtained with the human group IIA sPLA2 (hGIIA). This sPLA2 binds to factor Xa leading to a deficiency in the formation of a fully active pro-

KARGER

Fax + 41 61 306 12 34
E-Mail karger@karger.ch
www.karger.com

0301–0147/01/0316–0279$17.50/0

Accessible online at:
www.karger.com/journals/hae

Dr. Carine Mounier
Unité des Venins, Institut Pasteur, 25, rue du Docteur-Roux
F–75724 Paris Cedex (France)
Tel. +33 1 45 68 86 85, Fax +33 1 40 61 30 57
E-Mail cmounier@pasteur.fr

thrombinase complex [6]. Such a mechanism was also described to be involved for the CM-IV isoform [7]. The importance of a protein-dependent versus PL-dependent mechanism of action will be compared for different anticoagulant venom sPLA2. Anticoagulant effects of other recently cloned mammalian sPLA2 will also be discussed. Finally, these data highlight the interest of studying venom components in order to propose biological functions for homologous mammalian proteins.

Classification of Venom and Mammalian sPLA2

sPLA2 are enzymes found in mammals and animal venoms; they hydrolyse the fatty acids esterified at the *sn*-2 position of PL. They share common characteristics including low molecular mass (14–17 kD), calcium-dependent catalytic activity, the presence of 5–8 disulphide bridges and a number of conserved amino acids located in the calcium-binding loop and the catalytic site [8, 9]. Venom sPLA2 from Elapidae and Hydrophiidae have been classified into group IA with the exception of a few enzymes that belong to group IB [10]. Venom sPLA2 from Crotalidae and Viperidae are found in group IIA and group IIB contains some sPLA2 from *Bitis* species. Finally, sPLA2 from bee and lizard venoms belong to group III. Despite common catalytic properties, venom sPLA2 differ greatly in their pharmacological effects such as neurotoxic, myotoxic, cardiotoxic or anticoagulant properties.

In mammals, different sPLA2 have been reported and classified into groups IB, IIA, IIC, IID, IIE, IIF, V, X and XII [9]. Their distribution is both tissue-specific, indicating different functions in different organs, and species-dependent, suggesting that a particular sPLA2 may not necessarily have the same function in all the species [9]. Although a function in the generation of eicosanoids has been proposed for some mammalian sPLA2, the precise biological function of each sPLA2 has still to be investigated.

Blood Coagulation Cascade

Platelet activation plays a central role in haemostasis and thrombosis, leading to the formation of the primary plug. In addition, the appearance of anionic PL at the membrane surface of activated platelets increases the efficiency of the coagulation process [11–13]. Since the plug formed by activated platelets is unstable, the initiation of the coagulation process is needed for the definitive blockade of bleeding. The coagulation process can be initiated by the extrinsic or intrinsic pathway, both leading to the activation of factor X to factor Xa (FXa) (fig. 1). Then, the prothrombinase complex, which is composed of FXa, factor Va (FVa), PL and calcium, constitutes an important step of the blood coagulation cascade leading to the generation of thrombin that converts fibrinogen into fibrin. The fibrin generated will therefore consolidate the primary plug. Three steps of the coagulation cascade require the formation of a complex between an active protease, a zymogen substrate and another protein cofactor on cell surfaces: the intrinsic tenase, the extrinsic tenase and the prothrombinase reactions. Because the assembly of these complexes occurs on a PL surface, these complexes may constitute potential targets for the action of anticoagulant sPLA2. However, the dependency on PL of the extrinsic tenase is more strict than for the prothrombinase and the intrinsic tenase since the two latter ones remain active when PL are replaced by lysoPL and fatty acids [14].

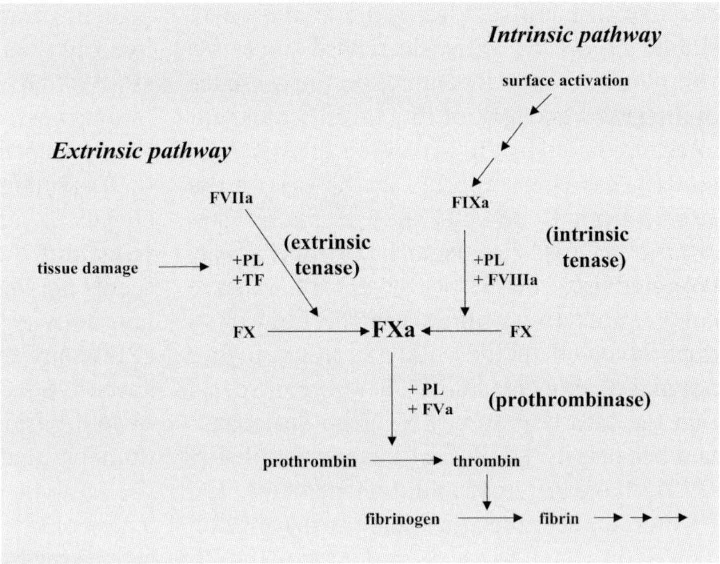

Fig. 1. The coagulation cascade. The initiation of the coagulation cascade can occur after vascular injury leading to the exposure of tissue factor (extrinsic pathway), or when small amounts of thrombin are generated which activate factor IX (intrinsic pathway). The two pathways lead to the generation of FXa, which in turn will allow thrombin to be activated from prothrombin. TF = Tissue factor; FVIIa = activated factor VII; FX = factor X; FVIIIa = activated factor VIII; FIXa = activated factor IX.

Anticoagulant Effect of Venom sPLA2

Some reviews discuss anticoagulant activities of venom sPLA2 and can be consulted for more details [2, 4]. Briefly, anticoagulant sPLA2 have been described in all major snake groups, i.e. Viperids, Crotalids, Elapids and Hydrophids. In Viperids, three sPLA2 from *Vipera berus, Deboia russelli* and *Bitis caudalis* have been described to be strongly anticoagulant. In Crotalids, *Trimeresurus mucrosquamatus* venom contains one strongly anticoagulant sPLA2. Venoms from *Crotalus durissus terrificus, Agkistrodon halys blomhoffii* and *Agkistrodon halys pallas* contain several sPLA2 isoforms, the most basic one being the most efficient in inhibiting blood coagulation. Other sPLA2 with anticoagulant activity have been isolated from the venoms of *Bothrops asper, Bothrops atrox, Bothrops godmani* and *Bothrops jararacussu.* In Elapids, the venom of *N. nigricollis* contains several sPLA2 isoforms, the basic one, CM-IV, accounting for the major anticoagulant effect. The basic sPLA2 from *N. m. mossambica* venom, designated as CM-III, is also a strong anticoagulant. Finally, the only sPLA2 from hydrophid venom described as weakly anticoagulant is found in *Enhydrina schistosa.*

While the anticoagulant activity of venom sPLA2 is well established, some controversies exist in the literature concerning the mechanism of their anticoagulant action. Some observations suggest that the sPLA2 catalytic activity is important, while others support the notion that the catalytic activity is not required. In fact, the mechanism of inhibition depends on each particular sPLA2 and therefore both hypotheses are correct.

PL-Dependent Mechanism of Anticoagulant Venom sPLA2

Since sPLA2 hydrolyze PL, it has first been suggested that venom sPLA2 inhibited blood coagulation due to their ability to destroy procoagulant PL. In such case, the higher effect should be on the extrinsic tenase since this complex is strictly dependent on the presence of integral procoagulant PL. An anticoagulant effect mediated by the hydrolysis of procoagulant PL has been reported in the case of the sPLA2 from *V. berus* [15], *A. h. pallas* [16], *B. jararacussu* [17], *N. n. atra* [18], and *T. mucrosquamatus* [19]. However, all sPLA2 have enzymatic activity and share the same residues involved in the catalytic reaction, but not all are anticoagulants. Therefore, it seems that the idea that an sPLA2 is an anticoagulant because it hydrolyses PL is too simplistic. sPLA2 have different ability to penetrate the PL membrane, called the 'penetrating ability'. This allows them to access and destroy the PL even in tightly packed membranes. It was proposed that the penetrating ability of sPLA2 was critical for exerting anticoagulant effects since this increases their efficiency to hydrolyse the procoagulant PL present on the surface of activated platelets or platelet-derived microvesicles [15].

Some sPLA2 have higher affinity for anionic PL (like phosphatidylserine), which mainly support the procoagulant activity. Therefore, the anticoagulant effect of these sPLA2 may be due to their competition with coagulation factors for binding to such PL and is not necessarily related to the hydrolysis [14, 20]. This scenario was proposed for the anticoagulant effects of sPLA2 from *T. mucrosquamatus* [19], *N. m. mossambica* [20] and *D. russelli* [21], *V. berus* [14] venoms. The difference between a PL hydrolysis and PL-binding mechanism is difficult to establish experimentally since both mechanisms may be used in combination by sPLA2. The use of chemically modified sPLA2 on the histidine48 of the active site in order to eliminate the catalytic activity is questionable, because the chemical group may modify the binding properties and/or the local conformation of the protein in addition to the suppression of its hydrolytic activity. The use of recombinant sPLA2 that will be modified by site-directed mutagenesis may be useful to better elucidate the PL-dependent mechanism.

It is interesting to note that all strong anticoagulant venom sPLA2 are usually basic proteins and the overall basicity of the sPLA2 seems to be necessary for their anticoagulant activity [15, 22]. However, not all basic sPLA2 are anticoagulants, supporting that a specific effect is involved in addition to the overall basicity. It was proposed that sPLA2 do have pharmacological sites in addition to the catalytic site to explain specific biological actions [23]. By sequence comparison analysis, the region 55–80, which is a basic exposed loop, appears important to explain the anticoagulant activity of venom sPLA2 and may constitute their 'anticoagulant region' [24]. This hypothesis has also been supported by the analysis of three-dimensional structures of few venom sPLA2 [22, 25], but no direct experimental evidence is available to support this hypothesis. The presence of this basic loop that seems to be related to strong anticoagulant action may either increase the penetrating ability or the binding affinity for anionic PL. The chemical modification of lysines (by carbamylation, acetylation or guanidination) only slightly affected the catalytic activity, but strongly decreased in the anticoagulant effect of sPLA2 from *D. russelli, B. asper, B. gadmani* and *N. nigricollis* venoms [21, 26, 27]. This indicates that the positive charges of lysines are important for the anticoagulant effect of sPLA2 and suggests that dissociation between hydrolytic activity and pharmacological properties of anticoagulant venom sPLA2 can be achieved.

Protein-Dependent Mechanism for the Anticoagulant Effect of the Basic CM-IV Isoform

The venom of *N. nigricollis* contains several sPLA2 isoforms, CM-I, CM-II and CM-IV. The CM-I and CM-II isoforms are weak anticoagulants and are likely to act mainly on the tenase complex due to PL hydrolysis [28, 29]. In contrast, the basic CM-IV isoform, which accounts for the major anticoagulant effect, inhibits both the tenase and prothrombinase reactions, and its anticoagulant effect was observed when only low PL hydrolysis was achieved [18]. Interestingly, when a chemically modified CM-IV was prepared in order to abolish the catalytic activity, it partially failed to inhibit the tenase reaction [29], but still remained a potent inhibitor for the prothrombinase complex [5]. Moreover, when the prothrombinase complex was reconstituted in the absence of PL, a strong inhibitory effect of CM-IV was observed. These observations offered the first evidence that the anticoagulant effect of CM-IV was due, at least partly, to a PL-independent mechanism on the prothrombinase. Recently, it was shown that CM-IV binds to bovine FXa but not prothrombin or FVa [7]. It forms a 1:1 complex with FXa with a K_d of 500 n*M*. Moreover, increased amounts of added FVa to the mixture reverse the inhibitory effect on the prothrombinase complex, suggesting that CM-IV may compete with FVa for binding to FXa [7].

Protein-Dependent Mechanism for the Anticoagulant Effect of the hGIIA

hGIIA has been detected in α-granules of platelets and is secreted within a few minutes upon activation by physiological agonists. Once secreted at the platelet plug, hGIIA is expected to adsorb to the plasma membrane rather than being diluted into the systemic circulation [30]. Moreover, it has been reported that other cell types (such as inflammatory cells) also secrete high amounts of hGIIA. Indeed, high concentrations of hGIIA have been detected at local sites, such as synovial and ascitic fluids and in tears, and localization of hGIIA appears to occur during local thrombolytic events [31].

hGIIA has been reported to exert an anticoagulant effect and might play a negative feedback role during the initiation of thrombosis upon secretion from activated platelets [32, 33]. hGIIA specifically inhibits the prothrombinase complex and this effect is still observed in the absence of phospholipids [6]. hGIIA forms a 1:1 complex with FXa (K_d of 230 n*M*) and prevents the formation of the FXa/FVa complex [6]. Wild-type enzyme, as well as the catalytically inactive H48Q hGIIA obtained by site-directed mutagenesis, efficiently delays the formation of thrombin measured in human platelet-rich plasma. These observations indicate that the anticoagulant effect of hGIIA occurs under conditions close to the physiological situation and is independent of its catalytic activity. This supports an inhibitory mechanism based on the interaction of hGIIA with FXa even when natural phospholipids from plasma and platelet membrane are present [34].

hGIIA contains 13 lysine and 10 arginine residues scattered over its entire surface that form cationic clusters. Several of these basic clusters lie on the surface of the molecule that contacts the lipid membrane, the so-called 'interfacial binding surface' (IBS) [35]. Mutations of these basic residues lead to a significant reduction in the ability of hGIIA to inhibit prothrombinase activity and to bind to FXa. Further, increased salt concentrations completely abolished the binding of hGIIA to FXa, supporting an important role for elec-

Table 1. Effect of various mammalian and venom sPLA2 on prothrombinase activity

	pI	Inhibition of prothrombinase activity (IC_{50}, nM)	
		absence of PL	presence of PL
hGIB	8.7	>1,000	>1,000
hGIIA	10.6	50±6	80±10
rGIIA	10.3	80±15	130±30
mGIIA	10.3	80±10	150±10
mGIID	10	105±12	160±25
hGV	9.8	110±20	100±5
rGV	9.7	150±10	350±40
hGX	4.8	>1,000	35±5
Ba IV	basic[a]	105±10	140±20
CB	9.9	80±10	20±3
Vbb	9.8	4±0.5	3±0.5
CM-III	9.9	>1,000	0.5±0.1
Apis melli.	7.8	>1,000	2.5±0.7

[a] Ba IV sPLA2 from *B. asper* has not been entirely sequenced but has been shown to be basic. CB is the basic subunit of crotoxin from *C. d. terrificus*, Vbb the sPLA2 from *V. berus*, CM-III the basic isoform from *N. mossambica* venom, and *A. melli.* refers to the sPLA2 of the bee venom *A. mellifera*. The data are reproduced from reference 34 with some additions.

trostatic interactions [34]. High concentrations of FVa reversed the inhibitory effect of hGIIA on the prothrombinase complex, suggesting that hGIIA may compete with FVa for binding to FXa. Therefore, hGIIA is expected to act at early stages of the coagulation process by delaying the formation of a fully active prothrombinase complex, until enough FVa is generated. hGIIA has several basic residues on the loop proposed as the anticoagulant site for venom sPLA2. Substitution by mutagenesis showed no significant involvement of these residues in the interaction with FXa. This loop of hGIIA is indeed exposed on the opposite face to the IBS, which is involved in the interaction with FXa and therefore is not likely to belong to the binding area.

Prothrombinase Inhibition by Other Venom and Mammalian sPLA2

It was observed that the naturally catalytically inactive basic sPLA2, Ba IV from *B. asper* venom, inhibits prothrombinase activity in the absence of PL, suggesting it may act through direct binding to FXa [34]. A similar inhibition of prothrombinase was also observed with the basic subunit, CB, of the crotoxin (table 1). Recent observations in our laboratory show that CB binds to FXa like hGIIA and CM-IV [Grazyna Faure, pers. commun.]. It is thus likely that some other basic venom sPLA2 inhibit blood coagulation through direct binding to FXa. The overall basicity of the molecule is not sufficient to explain the inhibitory effect through binding to FXa since the basic CM-III sPLA2 from *N. m. mossambica*, the sPLA2 from *B. caudalis* and *Notechis scutatus scutatus* are devoid of inhibitory effect or weakly inhibitory in the absence of PL (table 1). Similarly, the bee venom sPLA2 inhibits prothrombinase activity through a PL-dependent mechanism (table 1).

The recently cloned mammalian sPLA2 may also exert anticoagulant effects. Indeed, mammalian rGIIA, mGIIA, hGV, rGV and mGIID sPLA2, were found to inhibit prothrombinase activity in the absence of PL with efficiency similar to that of hGIIA (table 1). This suggests that they may also act through direct binding to FXa. These sPLA2 are all basic proteins as noticed above for anticoagulant venom sPLA2. Amino acid alignment of the hGIIA with other mammalian sPLA2 shows that several of the basic residues shown to be involved the hGIIA/FXa complex are conserved between basic sPLA2 ac-

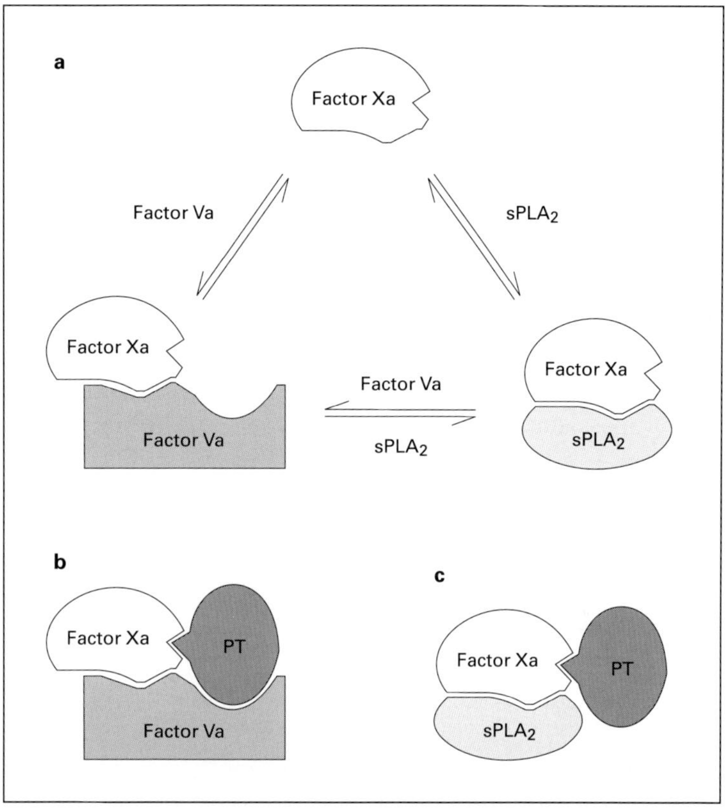

Fig. 2. Model showing the mechanism of inhibition of the prothrombinase complex by the hGIIA and the venom CM-IV isoform. These sPLA2 bind to FXa with a K_d value of 200–500 nM (**a**). They can replace FVa in the normal prothrombinase complex (bottom left), resulting in an inactive sPLA2/FXa complex (bottom right). Interaction of the substrate prothrombin (PT) with normal prothrombinase complex (**b**) or with sPLA2/FXa complex (**c**) is shown. sPLA2/FXa complex binds to prothrombin with the same affinity as FXa and sPLA2 do not interact with prothrombin. This explains the inability of these two sPLA2 to interfere in the prothrombinase activity of FXa alone. Reproduced with permission from Kerns et al. [7].

tive on prothrombinase via a PL-independent mechanism [34]. It is however expected that the precise interaction site of other mammalian and venom sPLA2 is not identical to that of hGIIA. The basic residues of hGIIA which are involved in the interaction with FXa, although all located on the IBS, are distributed throughout the primary sequence of the protein. It is thus possible that basic residues located at different positions on the primary sequence of other sPLA2 may be involved in similar basic cluster patterns. Studies based on the molecular modeling of these sPLA2 will be useful to test this hypothesis. On the other hand, neutral or acidic mammalian sPLA2 like hGIB and hGX were unable to inhibit prothrombinase activity in the absence of PL (table 1). This further supports the critical role of basic residues for the PL-independent antiprothrombinase activity of sPLA2. In agreement with this observation, these two sPLA2 do not share with hGIIA most of the basic residues that are important for binding to FXa. Furthermore, the hGX inhibits the prothrombinase complex in the presence of PL only, indicating that its inhibitory effect is likely to be related to binding or hydrolysis of procoagulant PL (table 1). This raises the question of the circulating levels of these sPLA2 in blood and their possible roles in thrombolytic events. Thus, the data on the circulating levels of GIID, GV, or GX sPLA2 and their presence in blood platelets will be of interest.

Perspectives

A microcalorimetry and molecular docking approaches are currently under investigation in order to better characterize the thermodynamic parameters of the hGIIA/FXa interaction and to define the part(s) of the FXa involved in formation of the complex. Similar approaches, as well as the production of a cocrystal with FXa, could be used with CB and CM-IV to define the nature of the residues involved in the interaction with FXa. The proposal antiprothrombinase mechanism for hGIIA and CM-IV is presented on figure 2. The elucidation of the parts of FXa involved in such an interaction may help to better define the interaction between FVa and FXa, since these sPLA2 compete with FVa for binding to FXa. These sPLA2 are small proteins compared to FVa and may offer an alternative approach. Finally, the study of venom sPLA2 has led to the investigation of anticoagulant effect of some mammalian sPLA2. However, whether such inhibitory effects may occur under physiological/pathological conditions remains to be determined.

References

1 Meier JM, Stocker K: Effects of snake venoms on hemostasis. Crit Rev Toxicol 1991;21:171–182.
2 Ouyang C, Teng CM, Huang TF: Characterization of snake venom components acting on blood coagulation and platelet function. Toxicon 1992;30:945–966.
3 Markland FS: Snake venoms and the hemostatic system. Toxicon 1998;36:1749–1800.
4 Evans HJ, Kini RM: The anticoagulant effects of snake venom phospholipase A_2; in Kini RM (ed): Venom Phospholipase A2 Enzymes: Structure, Function and Mechanism. New York, Wiley, 1997, pp 353–367.
5 Stefansson S, Kini RM, Evans HJ: The basic phospholipase A_2 from *Naja nigricollis* venom inhibits the prothrombinase complex by a novel nonenzymatic mechanism. Biochemistry 1990;29:7742–7746.
6 Mounier C, Hackeng TM, Schaeffer C, Faure G, Bon C, Griffin JH: Inhibition of prothrombinase by human secretory phospholipase A2 involves binding to factor Xa. J Biol Chem 1998;273:23764–23772.
7 Kerns RT, Kini RM, Stefansson S, Evans H: Targeting of venom phospholipases: The strongly anticoagulant phospholipase A_2 from *Naja nigricollis* venom binds to coagulation factor Xa to inhibit the prothrombinase complex. Arch Biochem Biophys 1999;36:107–113.
8 Valentin E, Lambeau G: Increasing molecular diversity of secreted phospholipases A_2 and their receptors and binding proteins. Biochim Biophys Acta 2000;1488:59–70.
9 Gelb MH, Valentin E, Ghomashchi F, Lazdunski M, Lambeau G: Cloning and recombinant expression of a structurally novel human secreted phospholipase A2. J Biol Chem 2000;275:39823–39829.
10 Huang MZ, Gopalakrishnakone P, Chung MCM, Kini RM: Complete amino acid sequence of an acidic, cardiotoxic phospholipase A_2 from the venom of *Ophiophagus hannah* (king cobra): a novel cobra venom enzyme with 'pancreatic loop'. Arch Biochem Biophys 1997;338:150–156.
11 Mann KG, Nesheim ME, Church WR, Haley P, Krishnaswamy S: Surface-dependent reactions of the vitamin K-dependent enzyme complexes. Blood 1990;76:1–16.
12 Zwaal RFA, Comfurius P, Bevers EM: Platelet procoagulant activity and microvesicle formation – its putative role in hemostasis and thrombosis. Biochim Biophys Acta 1992;1180:1–8.
13 Davie EW: Biochemical and molecular aspects of the coagulation cascade. Thromb Haemost 1995;74:1–6.
14 Boffa MC, Boffa GA: A phospholipase A_2 with anticoagulant activity. II. Inhibition of the phospholipid activity in coagulation. Biochim Biophys Acta 1976;429:839–852.
15 Verheij HM, Boffa MC, Rothen C, Bryckaert MC, Verger R, de Haas GH: Correlation of enzymatic activity and anticoagulant properties of phospholipase A2. Eur J Biochem 1980;112:25–32.
16 Zhong X, Liu J, Wu X, Zhou Y: Expression, purification and biochemical characterization of a recombinant phospholipase A_2 with anticoagulant activity from *Agkistrodon Halys Pallas*. J Nat Toxins 2001;10:17–25.
17 Andriao-Escarso SH, Soares AM, Rodrigues VM, Angulo Y, Diaz C, Lomonte B, Gutierrez JM, Giglio JR: Myotoxic phospholipase A_2 in *Bothrops* snake venoms: Effect of chemical modifications on the enzymatic and pharmacological properties of bothropstoxins from *Bothrops jararacussu*. Biochimie 2000;82:755–763.
18 Condrea E, Yang C-C, Rosenberg P: Lack of correlation between anticoagulant activity and phospholipid hydrolysis by snake venom phospholipases A_2. Thromb Haemost 1981;45:82–85.

19 Ouyang C, Jy W, Zan Y-P, Teng C-M: Mechanism of the anticoagulant action of phospholipase A purified from *Trimeresurus mucrosquamatus* (formosan habu) snake venom. Toxicon 1981;19:113–120.

20 Prigent-Dachary J, Boffa MC, Boisseau MR, Dufourcq J: Snake venom phospholipases A_2. A fluorescence study of their binding to phospholipid vesicles correlation with their anticoagulant activities. J Biol Chem 1980;255:7734–7739.

21 Babu AS, Gowda TV: Dissociation of enzymatic activity from toxic properties of the most basic phospholipase A_2 from *Vipera russelli* snake venom by guanidination of lysine residues. Toxicon 1994;32:749–752.

22 Carredano E, Westerlund B, Persson B, Saarinen M, Ramaswamy S, Eaker D, Eklund H: The three-dimensional structures of two toxins from snake venom throw light on the anticoagulant and neurotoxic sites of phospholipase A_2. Toxicon 1998;36:75–92.

23 Kini RM, Evans HJ: A model to explain the pharmacological effects of snake venom phospholipases A2. Toxicon 1989;27:613–635.

24 Kini RM, Evans HJ: Structure-function relationships of phospholipases. The anticoagulant region of phospholipases A_2. J Biol Chem 1987;262:14402–14407.

25 Singh G, Gourinath S, Sharma S, Paramasivam M, Srinivasan A, Singh TP: Sequence and crystal structure determination of a basic phospholipase A_2 from common krait at 2.4 Å resolution: Identification and characterization of its pharmacological sites. J Mol Biol 2001;307:1049–1059.

26 Diaz-Oreiro C, Gutierrez JM: Chemical modification of histidine and lysine residues of myotosic phospholipases A_2 isolated from *Bothrops asper* and *Bothrops godmani* snake venoms: effects on enzymatic and pharmacological properties. Toxicon 1997;35:241–252.

27 Condrea E, Fletcher JE, Rapuano BE, Yang C-C, Rosenberg P: Dissociation of enzymatic activity from lethality and pharmacological properties by carbamylation of lysines in *Naja nigricollis* and *Naja naja atra* snake venom phospholipases A_2. Toxicon 1981;19:705–720.

28 Stefansson S, Kini RM, Evans HJ: The inhibition of clotting complexes of the extrinsic coagulation cascade by the phospholipase A_2 isoenzymes from *Naja nigricollis* venom. Thromb Res 1989;55:481–491.

29 Kini RM, Evans HJ: The role of enzymatic activity in inhibtion of the extrinsic tenase complex by phospholipase A_2 isoenzymes from *Naja nigricollis* venom. Toxicon 1995;33:1585–1590.

30 Emadi S, Mirshahi M, Elalamy I, Nicolas C, Vargaftig BB, Hatmi M: Cellular source of human platelet secretory phospholipase A_2. Br J Haematol 1998;100:365–373.

31 Gelb MH, Mounier CM, Hefner Y, Watson SP: Platelet phospholipases A_2; in Gresele P, Page CP, Fuster V, Vermylen J (eds): Platelets in Thrombotic and Non-Thrombotic Disorders: Pathophysiology, Pharmacology and Therapeutics. Cambridge, Cambridge University Press, in press.

32 Cirino G, Cicala C, Sorrentino L, Browning JL: Human recombinant non pancreatic secreted platelet phospholipase A_2 has anticoagulant activity in vitro on human plasma. Thromb Res 1993;70:337–342.

33 Mounier C, Franken PA, Verheij HM, Bon C: The anticoagulant effect of the human secretory phospholipase A2 on blood plasma and on a cell-free system is due to a phospholipid-independent mechanism of action involving the inhibition of factor Va. Eur J Biochem 1996;237:778–785.

34 Mounier CM, Luchetta P, Lecut C, et al: Basic residues of human group IIA phospholipase A_2 are important for binding to factor Xa and prothrombinase inhibition. Comparison with other mammalian secreted phospholipases A_2. Eur J Biochem 2000;267:4960–4969.

35 Snitko Y, Koduri RS, Han SK, Othman R, Baker SF, Molini BJ, Wilton DC, Gelb MH, Cho W: Mapping the interfacial binding surface of human secretory group IIa phospholipase A2. Biochemistry 1997;36:14325–14333.

Anticoagulant Proteins and Diagnostic Agents

Hæmostasis

Haemostasis 2001;31:288–293

Lonomia Genus Caterpillar Envenomation: Clinical and Biological Aspects

Carmen Luisa Arocha-Piñango Belsy Guerrero

Centro de Medicina Experimental, Laboratorio de Fisiopatologia, Instituto Venezolano de Investigaciones Científicas (IVIC), Caracas, Venezuela

Key Words

Lonomia caterpillars · Hemostasis · Bleeding syndrome · Animal venoms

Abstract

Persons who have been in contact with *Lonomia achelous* or *Lonomia obliqua* caterpillars present external and internal bleeding and opening of recently healed wounds. Hematological tests show normal platelet count, prolonged prothrombin time, activated partial thromboplastin time and thrombin time, totally corrected by normal plasma. Decreased fibrinogen (Fg), factor (F) V, FXIII, plasminogen and α_2-antiplasmin with increased FVIII: C, von Willebrand factor, Fg degradation products and D dimers. Tissue plasminogen activator, plasminogen activator inhibitor and protein C varied. In *L. achelous* biological fluids, compounds with anticoagulant or procoagulant properties have been identified. In *L. obliqua* bristle extracts, mainly procoagulant activities have been identified. Subcutaneous injections of *L. achelous* crude extracts and a semipurified fraction reduce Fg, plasminogen and FXIII in rabbits. Intravenous injections of a very purified fraction of *L. achelous* in rabbits produce lysis of preformed thrombi, a decrease of Fg, plasminogen, α_2-antiplasmin, FXIII and inhibition of post-thrombolytic thrombus growth. Subcutaneous injections of *L. obliqua* bristle extracts prolong prothrombin time and activated partial thromboplastin time and reduce FXIII. Intravenous injections of crude bristle extract and a purified fraction of *L. obliqua* induce disseminated intravascular coagulation.

Introduction

In 1967, we reported 5 cases of a severe bleeding syndrome occurring in humans who have been in contact with spiny-hair caterpillars in Venezuela [1]. Since 1972, new patients have been reported from different areas

KARGER

Fax + 41 61 306 12 34
E-Mail karger@karger.ch
www.karger.com

0301–0147/01/0316–0288$17.50/0

Accessible online at:
www.karger.com/journals/hae

C.L. Arocha-Piñango
IVIC, Apartado 21827
Caracas 1020A (Venezuela)
Tel. +58 212 504 1140/1180/1087, Fax +58 212 504 1086
E-Mail carocha@ivic.ve and bguerrer@medicina.ivic.ve

of Venezuela and other South American countries: Brazil, French Guyana [2–9] and also cases from Peru, Paraguay, Argentina and Colombia (congress reports and personal communications).

Lemaire [10] identified the caterpillars from Venezuela as the larvae of the Saturnidae moth *Lonomia achelous* (Cramer), those from the north of Brazil as *Lonomia diabolous* (brown variety of *L. achelous*), and those from the south of Brazil as *Lonomia obliqua* (Walker).

Clinical Aspects

The patients experienced pain and a burning sensation at the site of contact, headache, malaise, and fever in children, followed by obnubilation or coma. External and internal bleeding started within a period of 1 h to 10 days after contact [1–3, 6–8]. In South Brazil, about 12% of the patients developed renal insufficiency, whereas in Venezuela, it has been detected only in 1 case [4, 5, 9]. Laboratory tests show anemia, leukocytosis, normal glycemia and liver function, and high values of creatinine and urea in cases of renal insufficiency, as well as prolonged prothrombin time, activated partial thromboplastin time, and thrombin time, which are totally corrected by the addition of normal plasma. The levels of platelets, factors (F) II, FVII, FIX, FX, FXI, FXII, and antithrombin III were normal. However, the tests reflected a marked decrease in fibrinogen (Fg), FV, FXIII, plasminogen (Pg) and α_2-antiplasmin with a moderate increase in FVIII:C and von Willebrand factor, short euglobulin lysis time, and high titers of Fg degradation products and D dimers. Tissue plasminogen activator, plasminogen activator inhibitor-1 and protein C varied [1–3, 6–8]. Some patients presented thrombin-antithrombin complexes [11].

The treatment has changed in Venezuela according to clinical response and because new drugs have been discovered. The first cases were treated with whole blood or fresh frozen plasma. The patients clinically deteriorated with a decrease in platelets, FII, FV, FVII, FVIII:C, FXIII, Pg and α_2-antiplasmin, and a slight increase in Fg. Some patients died of cerebral hemorrhage, and those who recovered did so very slowly. In the following years, ε-amino caproic acid (Capramol, Choay, France) and aprotinin (Trasylol, Bayer, Germany) began to be used together with replacement therapy. A slight decrease in platelet counts without alteration of the other clotting factors and slow clinical recovery were observed. Years later, only purified human Fg (Immune AG, Austria) together with aprotinin have been used. In some patients, ε-amino caproic acid was administered after aprotinin. A very fast clinical recovery and normalization of hemostatic parameters were observed; with the exception of Fg which improved slowly [7, 8]. In Brazil, ε-amino caproic acid has been the drug of choice, but recently, the Instituto Butantan in Sao Paulo has developed an anti-*lonomia* antiserum from *L. obliqua* bristle extract. The antiserum has been tested in animals as well as in humans and seems to be effective [12, 13].

Biochemical Studies

Several compounds with anticoagulant or procoagulant activities have been detected in the venom of both *L. achelous* and *L. obliqua* caterpillars (table 1).

Anticoagulant Compounds

The venom of *L. achelous* contains a urokinase-like serine protease (lonomin I) [14, 15] and two serine proteases, named achelase I and II (lonomin II), showing direct fibrinolyt-

Table 1. Compounds identified in *L. achelous* and *L. obliqua* venoms

Anticoagulants	*L. achelous*	plasmin-like (lonomin II) urokinase-like/F XIII protease (lonomin I/lonomin V) FV inhibitor (lonomin VI:i)
	L. obliqua	fibrino(geno)lytic protease
Procoagulants	*L. achelous*	direct prothrombin activator (lonomin III) FXa-like (lonomin IV) FV activator (lonomin VI:a)
	L. obliqua	FX activator prothrombin activator (Lopap)
Miscellaneous	*L. achelous*	kallikrein-like (lonomin VII) extracelular matrix protease
	L. obliqua	phospholipase A_2-like (lonomia toxin)

ic activity. Their amino acid sequences exhibit high similarity to other lepidopteran larva serine proteinases [16]. Lonomin II degrades Fg, forming fragments different from those produced by plasmin and trypsin with a very rapid degradation of the α-chain [17]. These findings may explain the noninhibitory effect of high levels of Fg degradation products found in the patients as demonstrated by total correction of prothrombin time, activated partial thromboplastin time and thrombin time by normal plasma [1, 7, 8]. The venom of *L. achelous* also contains a serine protease (lonomin V) which produces a dose-dependent decrease in FXIII activity from plasma and platelets and shares many characteristics with lonomin I. It is likely that the same molecule (lonomin I and lonomin V) has both activities. Therefore, we decided to name this protease lonomin I/V [18–21]. Its N-terminal sequence shows 75% homology to the achelases (lonomin II) [unpubl. results]. Finally, the venom of *L. achelous* contains a FV inhibitor (lonomin Vl:i) which also seems to be a serine protease [22]. On the other hand, the venom of *L. obliqua* contains a fibrino(geno)lytic activity which produces fragments different from plasmin and similar to those reported for *L. achelous* [6, 23].

Procoagulant Compounds

Two types of prothrombin activators are present in *L. achelous:* a direct activation (lonomin III) and a FXa-like (lonomin IV) which have not yet been completely characterized [24]. In addition, there is a FV activator (lonomin VI:a) which seems to be a metalloproteinase [22]. In *L. obliqua,* a calcium-independent FX activator [25] and a calcium-dependent prothrombin activator named *L. obliqua* prothrombin activator protease (Lopap) have been identified (table 1). The N-terminal sequence of Lopap showed no similarity to any other prothrombin activator [26, 27]. This prothrombin activator differs from that of *L. achelous* in that the later is calcium independent [24–27].

Miscellaneous

L. achelous hemolymph, hair secretions or purified fractions and *L. obliqua* crude bristle extracts and purified fractions do not clot purified Fg nor hydrolyze chromogenic substrates specific for thrombin, and do not induce platelet aggregation in vitro [6, 7, 13, 14]. A kallikrein-like compound (lonomin VII) has been identified in *L. achelous* [15] and a protease which hydrolyzes laminin and fibronectin [unpubl. results]. Zymographic studies of *L. achelous* hemolymph and its semipurified fractions show areas of lysis of different molecular weights, some of which are plasminogen dependent [28]. These results correspond to the fibrinolytic activities of lonomin I and lonomin II [14, 16]. *L. achelous* crude hemolymph and a semipurified fraction lyse whole blood clots. This lysis is not inhibited by human plasma, thus indicating that the fibrinolytic physiological inhibitors do not act on them [29]. In *L. obliqua,* a phospholipase A_2-like activity (lonomiatoxin*)* has also been identified [30]. An antiserum has been prepared at the Instituto Butantan, Sao Paulo, Brazil, in horses immunized with *L. obliqua* bristle extracts [31].

Animal Experiments

L. achelous

Subcutaneous injections of crude venom and lonomin I in rabbits induce a dose-dependent decrease in Fg, plasminogen and FXIII without bleeding manifestations, followed by a rapid return to normal basal values [32]. Intravenous injections of lonomin I/V induce a dose-dependent thrombolysis of preformed thrombi in combination with a decrease in Fg, plasminogen, α_2-antiplasmin and FXIII, without an increase in Fg degradation products nor any bleeding manifestations [33]. The thrombolytic effect is followed by an inhibition of ^{125}I-fibrinogen accretion [34].

L. obliqua

In studies performed in Brazil, subcutaneous injections of *L. obliqua* bristle extracts (Loche) in rabbits cause a prolongation of prothrombin time and activated partial thromboplastin time without any detectable Fg. Similar to the effect of *L. achelous,* the tests returned to normal values within 24–48 h [32]. Intravenous injections of Loche deplete Fg and FXIII, but do not induce lysis of preformed thrombi [35]. Intravenous injections of purified Lopap induce disseminated intravascular coagulation in rats [36].

Conclusion

The presence of fibrinolytic compounds in *L. achelous* venom (plasmin-like and plasminogen activators), a FV inhibitor and a FXIII protease, along with activators of the prothrombinase complex (FII and FV activators and FXa-like), seems to indicate that in the pathogenesis of the clinical picture, two processes coexist: a very severe fibrinolytic syndrome and mild disseminated intravascular coagulation due to the procoagulant agents, which, in most cases, is masked by the former and is manifested when all coagulation factors are infused. *L. obliqua* venom seems to contain mainly activators of the prothrombin complex; therefore, disseminated intravascular coagulation might present as a primary pathological state.

Acknowledgments

We thank Z. Carvajal, A. Ojeda and A. Gil for their technical support, Dr. U. Lundberg for his useful comments on the manuscript, and Mrs. L. Rybak for her secretarial work.

References

1 Arocha-Piñango CL: Fibrinolisis producida por contacto con orugas. Acta Cient Ven 1967;18:136–1391.

2 Fraiha-Neto H, Costa D, Quiroz de Leao RN, Ballarini AJ: Acidentes hemorrágicos por larvas de *Lonomia*; in Schvartsman S (ed): Plantas Venenosas e Animais Peçohentos. Sao Paulo, Sarvier, 1992, pp 241–244.

3 Hommel D, Bouchareine L, Hulin A: Envenimation par une chenille de lépidoptère: *Lonomia achelous*. Revue de la littérature. A propos de deux cas en Guyane française. Sem Hôp Paris 1995;71:9–12.

4 Duarte CA, Caovilla J, Lorini I, Lorini D, Mantorani G, Sumida J, Manfre PC, Silveira R, Moura SP: Insuficiencia renal aguda por acidentes com lagartas. J Bras Nefrol 1990;12:184–187.

5 Burdmann EA, Antunes I, Saldanha LB, Abdulkader RC: Severe acute renal failure induced by the venom of *Lonomia* caterpillars. Clin Nephrol 1996;46:337–339.

6 Kelen EMA, Picarelli ZP, Duarte AC: Hemorrhagic syndrome induced by contact with caterpillars of the genus *Lonomia* (Saturnidae, Hemileucinae). J Toxicol Rev 1995; 14:283–308.

7 Arocha-Piñango CL, Blumenfeld-Bosh N, Nouel AL, Torres A, Perales J, Alonso ME, Rodriguez S, Carvajal Z, Ojeda A, Tasayco ML, Chitty W: Fibrinolytic and procoagulant agents from a Saturnidae moth caterpillar; in Pirkle H, Markland FS (eds): Hemostasis and Animal Venoms. New York, Dekker, 1988, pp 223–240.

8 Arocha-Piñango CL, de Bosch NB, Torres A, Goldstein C, Nouel A, Arguello A, Carvajal Z, Guerrero B, Ojeda A, Rodriguez A, et al: Six new cases of a caterpillar-induced bleeding syndrome. Thromb Haemost 1992;67:402–407.

9 García J, Rendon C, Brewster F, Guipe S, Gonzalez A, Nazzoure J, Flores L, Frontado C, Guillen A, Tilac C, Pernalete N: Acute renal insufficiency due to *Lonomia achelous* envenomations. 2nd Lat Am Congr on Acute Renal Failure. Rio de Janeiro, 2000.

10 Lemaire C: Révision du genre *Lonomia* Walker (Lep attacidae). Ann Soc Ent Fr 1972;8:767–861.

11 Valença Reis C, Kelen EMA, Chudzinski-Tavassi AM: Isolamento da proteina procoagulante presente das cerdas da lagarta *Lonomia obliqua*. Abstr 12° ReunAn da Federacao de Soc de Biol Exp, Caxaribu, 1995, p 18023.

12 Duarte AC, Coavilla JJ, Kelen EMA, Wen FH, Machado SMS, Herrman F, Renner LO, Santos SR, Tefilli AP, Rodriguez C, Butzke LMW: Estudo comparativo entre o soro anti-lonomico e o acido aminocaproico. Abstr 2° Encont Nac de acidentes com animais peçonhentos, Porto Alegre, 1997, p 65.

13 Prezoto BC, Maffei FHA, Mattar L, Kelen EMA, Picarelli ZP, Curi PR: Comparaçao das vias de administraçao e determinaçao da dose anticoagulante do extracto de cerdas de *Lonomia obliqua* (Walker) em Coelhos. Abstr 11° Reun An da Federaçao de Soc de Biol Exp, Caxambu, 1996, p 361

14 Arocha-Piñango CL, Marsh NA, Robinson D: A fibrinolytic agent from a Saturnid caterpillar. Partial purification and characterization. Thromb Diath Haemorrh 1973;29: 135–142.

15 Arocha-Piñango CL, Pepper DS: Studies of a fibrinolytic enzyme from the larvae of *Lonomia achelous* (Cramer) using chromogenic peptide substrates. Thromb Haemost 1981;46:710–713.

16 Amarant T, Burkhart W, LeVine H 3rd, Arocha-Piñango CL, Parikh I: Isolation and complete amino acid sequence of two fibrinolytic proteinases from the toxic Saturnid caterpillar *Lonomia achelous*. Biochim Biophys Acta 1991;1079:214–221.

17 Arocha-Piñango CL, Perales J, Carvajal, Z: Studies on the degradation of fibrinogen by proteolytic enzymes from the larvae of *Lonomia achelous* (Cramer). Thromb Haemost 1981;45:233–236.

18 Guerrero B, Arocha-Piñango CL, Gil San Juan A: *Lonomia achelous* caterpillar venom (LACV) selectively inactivate blood clotting factor XIII. Thromb Res 1997;87:83–93.

19 Guerrero BA, Arocha-Piñango CL, Gil San Juan A: Degradation of human factor XIII by lonomin V, a purified fraction of *Lonomia achelous* caterpillar venom. Thromb Res 1997;87:171–181.

20 Guerrero B, Perales J, Gil A, Arocha-Piñango CL: Effect on platelet FXIII and partial characterization of Lonomin V, a proteolytic enzyme from *Lonomia achelous* caterpillars. Thromb Res 1999;93:243–252.

21 Arocha-Piñango CL, Marval E, Guerrero B: *Lonomia* genus caterpillars toxins. Biochemical aspects. Biochimie 2000;82:937–942.

22 López M, Gil A, Arocha-Piñango, CL: The action of *Lonomia achelous* caterpillars venom on human factor V. Thromb Res 2000;98:103–110.

23 Pinto AFM, Veiga ABG, Blochtein B, Guimaraes JA: Protease from different secretion are involved on the paradoxical (pro-coagulant and hemorrhagic) envenomation caused by *Lonomia obliqua* caterpillars. Abstr 13th World Congr of the Int Soc on Toxinology, Paris, 2000, p 178.

24 Guerrero B, Arocha-Piñango CL: Activation of human prothrombin by the venom of *Lonomia achelous* (Cramer) caterpillars. Thromb Res 1992;66:169–177.

25 Donato JL, Moreno RA, Hyslop S, Duarte A, Antunes E, Le Bonniec BF, Rendu F, de Nucci G: *Lonomia obliqua* caterpillar spicules trigger human blood coagulation via activation of factor X and prothrombin. Thromb Haemost 1998;79:539–542.

26 Reis CV, Kelen EMA, Parasky SHP, Portaro FCV, Sampaio CAM, Fernandez BL, Camargo ACM, Chudzinski-Tavassi AM: A Ca^{2+} activated serin protease (LOPAP) could be responsible for the haemorrhagic syndrome caused by the caterpillar *Lonomia obliqua*. Lancet 1995;353: 1942.

27 Reis CV, Portaro FC, Andrade SA, Fritzen M, Fernándes BL, Sampaio CA, Camargo AC, Chudzinski-Tavassi AM: A prothrombin activator serine protease from the *Lonomia obliqua* caterpillar venom (Lopap). Thromb Res 2001;102:427–436.
28 Carvajal Z, Arocha-Piñango CL: Estudio zimográfico de las fracciones purificadas del veneno de la *Lonomia achelous.* Rev Iberoamer Tromb Hemostas 1991;4:182–184.
29 Coll-Sangrona E, Arocha-Piñango CL: Fibrinolytic action on fresh human clots of whole body extracts and two semipurified fractions from *Lonomia achelous* caterpillars. Braz J Med Biol Res 1998;31:779–784.
30 Rocha-Campos ACM, Alves EW, Melo PD, Machado OLT, Diaz da Silva W: Lonomiatoxin: A 20 Kd protein present in bristle extracts of *Lonomia obliqua* extracts. Abstr 12th World Congr of the Intern Soc Toxin, Cuernavaca, 1997.
31 Da Silva WD, Campos CM, Gonçalves LR, Sousa-e-Silva MC, Higashi HG, Yamagushi IK, Kelen EM: Development of an antivenom against toxins of *Lonomia obliqua* caterpillars. Toxicon 1996;34:1045–1049.
32 Marval E, Guerrero B, Arocha-Piñango CL: The action of *Lonomia achelous* caterpillar venom on some blood coagulation and fibrinolytic parameters of the rabbit. Toxicon 1999;37:1491–1504.
33 Guerrero B, Alves-Pinto M, Müller CA, Perales J, Gil A, Arocha-Piñango CL: Evaluation of dose and administration scheme of Lonomia V as thrombolytic agent in a rabbit jugular vein thrombosis model. Abstr 20th Blankenese Conf Anim venoms, Hamburg-Blankenese, 2000, p 37.
34 Guerrero B, Perales J, Alves Pinto M, Müller CA, Gil San-Juan A, Amorin S, Arocha-Piñango CL: Enhancement of thrombolysis and inhibition of thrombus growth by Lonomin V. 18th Congr of the Int Soc on Thromb and Haemost, Paris, July 2001.
35 Presoto BC, Maffei FHA, Mattar L, Chudzinski-Tavassi AM, Kelen EMA, Curi PR: Effects of *Lonomia obliqua* hair extracts (LOCHE) in experimental venous thrombosis. 13th Congr of the Int Soc on Thromb and Haemost, Washington, 1999, Abstract Search CD Rom.
36 Reis CV, Farsky SH, Fernandes BL, Santoro ML, Oliva ML, Mariano M, Chudzinski-Tavassi AM: In vivo characterization of lopap, a Prothrombin Activator Serine Portease from *Lonomia obliqua* caterpillar venom. Thromb Res 2001;102:437–443.

Anticoagulant Proteins and Diagnostic Agents

Hæmostasis

Haemostasis 2001;31:294–305

Identification of Anticoagulant Activities in Salivary Gland Extracts of Four Horsefly Species (Diptera, Tabanidae)

M. Kazimírová[a] M. Šulanová[b] M. Kozánek[a] P. Takáč[a] M. Labuda[a] P.A. Nuttall[c]

[a]Institute of Zoology, Slovak Academy of Sciences and [b]Department of Animal Physiology and Etology, Faculty of Natural Sciences, Comenius University, Bratislava, Slovakia; [c]Centre for Ecology and Hydrology, NERC, Oxford, UK

Key Words

Tabanidae · Arthropods · Anticoagulant · Thrombin inhibition

Abstract

Anticoagulant activities against the extrinsic and intrinsic coagulation pathways were identified in salivary gland extracts (SGE) prepared from four tabanids *(Hybomitra muehlfeldi, Tabanus autumnalis, Haematopota pluvialis, Heptatoma pellucens).* All extracts prolonged human plasma clotting time in a dose-dependent manner and inhibited thrombin activity in the chromogenic substrate assay. Horsefly SGE did not inhibit factor Xa. Partial purification of SGE proteins using reversed-phase high-performance liquid chromatography revealed species-specific differences in the elution profiles and range of fractions with anticoagulant activities.

Introduction

Haematophagous arthropods have major health and economic impacts on humans and animals. To facilitate blood feeding, such arthropods produce a wide array of antihaemostatic compounds in their salivary glands which they secrete into the skin feeding site via their saliva. These substances developed during the evolution of the ectoparasite-host relationship and serve to overcome haemostatic mechanisms and immune reactions of the host [1–3]. Studies on the activity, mode of action and characteristics of these compounds have been carried out more extensively during the last decade and were driven by their potential therapeutic use in treatment of defects in haemostasis and diseases of the cardiovascular and immune systems.

Antihaemostatic compounds of blood-sucking arthropods have been distinguished into several groups such as inhibitors of fac-

KARGER

Fax + 41 61 306 12 34
E-Mail karger@karger.ch
www.karger.com

0301–0147/01/0316–0294$17.50/0

Accessible online at:
www.karger.com/journals/hae

Dr. Mária Kazimírová
Institute of Zoology, Slovak Academy of Sciences
SK–84206 Bratislava (Slovakia)
Tel. +421 2 5930 2645, Fax +421 2 5930 2646
E-Mail uzaemkaz@savba.sk

tors VII and V, thrombin (antithrombins), the prothrombinase complex (mainly inhibitors of factor Xa), platelet functions and fibrinolytic enzymes [1, 3, 4]. Most haematophagous arthropods have been found to produce several different anticoagulants each inhibiting different factors of the coagulation pathways. The majority of anticoagulants isolated from arthropods have been peptides and small proteins.

There have been comparatively few studies on antihaemostatic factors produced by tabanids (horseflies, clegs and deerflies). Prior to feeding, tabanids make an incision in the skin of the host and feed on the blood that wells up [5]. Females of the larger species are able to ingest up to 200 mg of blood within only 1–3 min [6], suggesting that they must possess very potent anticlotting mechanisms. Potent anticoagulant activity was discovered in the salivary glands of some representatives of the Tabaninae subfamily, which was attributed to antithrombin [7]. A thrombin inhibitor, isolated from *Tabanus bovinus,* was named tabanin [8]. Tabanin is thought to be an approximately 7-kD peptide [4]; however, its amino acid sequence is unknown. In addition to tabanin, a platelet aggregation inhibitor and a potent glycoprotein IIb/IIIa fibrinogen receptor antagonist have been isolated from deerflies, *Chrysops* sp. of the Chrysopsinae subfamily [9, 10].

We report identification of anticoagulation activities in crude and partially purified salivary gland extracts from four horsefly species from the Tabaninae subfamily.

Material and Methods

Salivary Gland Extracts

Female tabanids were collected using Manitoba-type traps with carbon dioxide as the attractant at selected sites in SW Slovakia during May-September 1999–2000. The flies were transported to the laboratory and immediately processed. Females were immobilized on ice and salivary glands (SG) were dissected under a stereomicroscope in PBS pH 7.2 (0.01 *M* phosphate buffer, 0.15 *M* NaCl). Dissected SGs were transferred to Eppendorf tubes and stored in PBS at –70 °C. Prior to coagulation assays, SGs were homogenized and centrifuged in a bench-top centrifuge at 10,000 rpm for 15 min. This procedure was repeated twice. Supernatants were pooled and referred to as salivary gland extracts (SGE). Serial dilutions of SGE were tested in coagulation assays.

RP-HPLC

For partial purification and identification of anticoagulant compounds, SGE of the species *Hybomitra muehlfeldi* (173 SG pairs), *Tabanus autumnalis* (63 SG pairs), *Heptatoma pellucens* (50 SG pairs) and *Haematopota pluvialis* (355 SG pairs) were used.

SG samples stored at –70 °C in PBS were incubated in a water bath at approximately 90 °C for 5 min, homogenized and centrifuged in a bench-top microcentrifuge at 10,000 rpm for 15 min. The supernatants (crude SGE) were filtered through a Millex-LG syringe driven filter unit (0.20 µm, 4 mm). Filtered SGE was further processed by HPLC.

The SGE samples were diluted in 500 µl of 10% acetonitrile (ACN) with 0.1% trifluoroacetic acid (TFA) (buffer A) and loaded onto a Beckman Instruments 126/168 DAD HPLC system. SGE were applied to a Vydac C-4, 250 × 4.6 mm ID, 5-µm particle size column. Proteins were eluted with a gradient of ACN (10–100%) in 0.1% TFA at a flow rate of 1 ml/min and 1% increments of ACN/min. Peaks were detected at UV 210 and 220 nm. Fractions were collected and dried in a Savant Instruments Speed-Vac to a volume of approximately 50 µl.

Coagulation Assays

Pooled human citrated plasma showing normal coagulation times was provided by the Department of Haematology and Transfusion of the Slovak Institute of Cardiovascular Diseases in Bratislava, Slovakia.

Thrombin time (TT, thrombin activity assay), prothrombin time (PT, extrinsic pathway assay) and activated partial thromboplastin time (APTT, intrinsic pathway assay) were used to determine the anticoagulant activities in tabanid SGE and SGE fractions. The assays were carried out with 50 µl (PT, APTT) or 100 µl (50 µl for HPLC fractions) (TT) plasma. The plasma was pre-incubated with 1–5 µl SGE, 1 µl HPLC fraction or the same volume of PBS, pH 7.2 (control), in a water bath at 37 °C for 1 min. The reaction was started with the addition of the appropriate pre-warmed reagent: 100 µl (50 µl for HPLC fractions) Dade Thromboclotin reagent (lyophilized bovine thrombin, 2.5 NIH U/ml, Dade AG, Düdingen, Switzerland) in the TT assay, 100 µl Dade Thromboplastin IS reagent (lyophilized thromboplastin from rabbit brain, 1.0 × 10^{-2} mol/l Ca^{2+}, Dade International Inc., Miami, Fla., USA) in the PT assay, and 50 µl Dade Actin FS Activated PTT reagent (purified soy phosphatides in 1.0 × 10^{-4} *M* ellagic acid, Dade) +50 µl of pre-warmed 0.02 *M* $CaCl_2$ after 3 min of incubation in the APTT assay. The time required for the formation of the fibrin clot was determined in triplicate samples visually using a stopwatch. In all assays, fibrin clot formation was monitored for a maximum of 2 min. If no clot formation occurred by that time, SGE or fraction was considered to inhibit coagulation by 100%. Prolongation of coagulation in comparison with the control was expressed as percentage of inhibition in the range 0–100%.

Chromogenic Substrate Assays for Thrombin and Factor Xa

The assays were carried out using 96-well microplates. For screening of antithrombin activity, 190 µl Tris buffer (50 m*M* Tris, 227 m*M* NaCl, pH 8.3, containing 0.1% BSA and 0.1% sodium azide), 10 µl Tris buffer containing 0.029 NIH units of thrombin from human plasma (Sigma, Steinheim, Germany), 10 µl PBS (control), 10 µl of SGE (6 µg protein), or 2 µl (<0.1 µg protein) of HPLC fractions +8 µl PBS were pipetted into a well and incubated at room temperature (RT) for 10 min. Twenty microliters of substrate, 1.9 m*M* N-*p*-tosyl-gly-pro-arg-*p*-nitroanilide (Sigma), dissolved in redistilled water were added. For screening of anti-factor Xa activity, 190 µl Tris buffer (50 m*M* Tris, 150 m*M* NaCl, pH 8.3, containing 0.1% BSA and 0.1% sodium azide), 20 µl Tris buffer containing 0.05 NIH units of factor Xa from bovine plasma (Sigma) and 10 µl PBS (control) or 10 µl of SGE (6 µg protein) were pipetted into a well and incubated at RT for 10 min. Twenty microliters of substrate, 2 m*M* Chromozym X (Boehringer, Mannheim, Germany), dissolved in redistilled water were then added. After addition of the substrates, changes in absorbance were monitored in two parallel samples at 405 nm for 30 min using an ELISA reader (Multiskan RC, Labsystems, Helsinki, Finland). The results are expressed as progress curves showing substrate hydrolysis by thrombin or factor Xa (i.e. changes in absorbance at 405 nm dependent on time, means of two replicates).

Protein Assays

Protein concentrations were determined for SGE and SGE fractions by the method of Bradford [11], adapted to 96-well microplates. Bovine serum albumin was used as standard. Absorbance was read by an ELISA reader (Multiskan RC, Labsystems) at 595 nm.

Statistics

Statistical analyses were performed using Statgraphics, version V software (Statistical Graphics Corporation, Rockville, Md., USA). For statistical analysis, percentages were transformed to arcsine. Two-factor analysis of variance (ANOVA), followed by Tukey's test was used.

Results

Anticoagulant Activities in Horsefly SGE

Anticoagulant activities against thrombin and the extrinsic and intrinsic coagulation pathways were found in horsefly SGE in the TT, PT and APTT assays, respectively (fig. 1). All extracts prolonged human plasma clotting time in a dose-dependent manner. Comparison of the four species showed significant differences in the anticlotting activities of their SGE. The strongest anticoagulant activity in all assays was exhibited by *H. pellucens,* and the weakest by *H. pluvialis.* Salivary anticlotting activity of *H. pluvialis* was significantly less in both the TT and PT assay, while in the APTT assay differences were significant between all examined species ($p < 0.001$, Tukey's test).

Based on comparisons of the salivary anticlotting activity of the four tabanids investigated in this study by two-way ANOVA, followed by Tukey's test, considering species and dilution as factors, the species were ranked *H. pellucens* > *T. autumnalis* > *H. muehlfeldi* > *H. pluvialis* ($F_{species}$ = 22.71***, $F_{dilution}$ = 26.92***, $p < 0.001$). The differences were significant between the first three species and *H. pluvialis* and between *H. pellucens* and *H. muehlfeldi.*

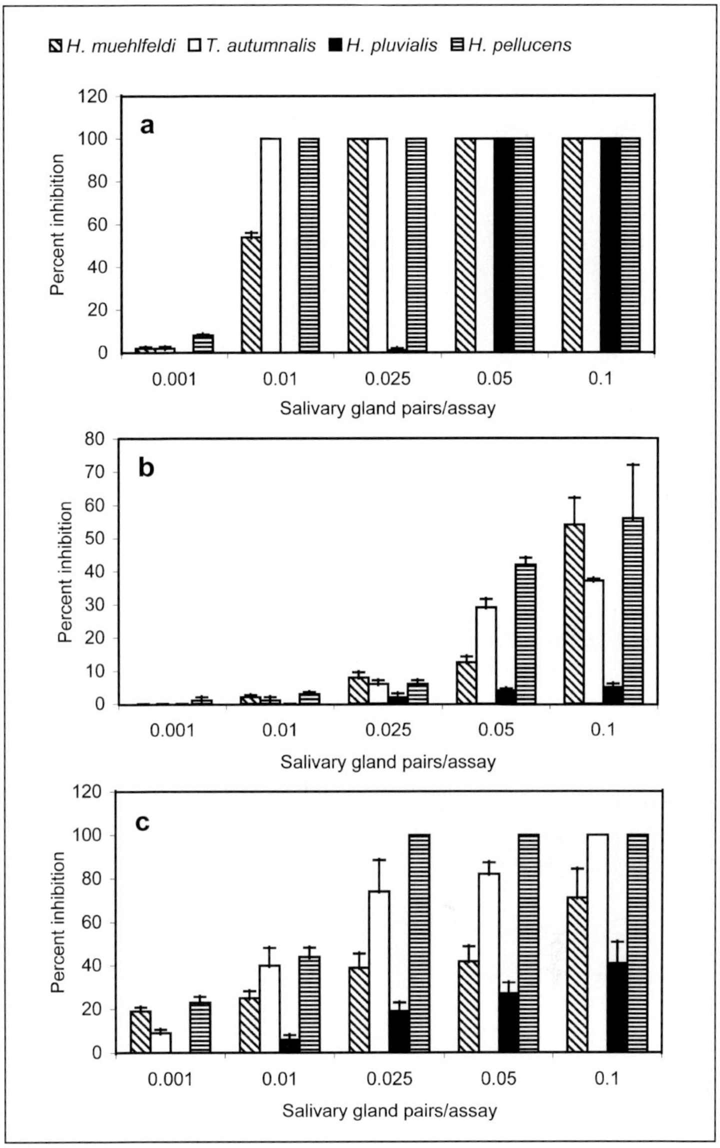

Fig. 1. Effect of salivary gland extracts of *H. muehlfeldi, T. autumnalis, H. pluvialis* and *H. pellucens* on inhibition of fibrin clot formation of human plasma. **a** TT assay ($F_{species}$ = 60.79*, $F_{dilution}$ = 66.84***). **b** PT assay ($F_{species}$ = 15.13***, $F_{dilution}$ = 52.97***). **c** APTT assay ($F_{species}$ = 62.54***, $F_{dilution}$ = 55.16***). $F_{species}$, $F_{dilution}$ – F ratio for the factor of species and SGE dilution, respectively; *** $p < 0.001$ (two-way ANOVA). Values represent means ± SD of three assays. During analysis, percentages were transformed to arcsine. Actual percentages are presented.

RP-HPLC of Horsefly SGE

Anticoagulant activities were identified in RP-HPLC fractions for all examined species. For individual species, the chromatograms and anticoagulant activities of fractions in the TT, PT and APTT assays are shown in figures 2–5. The elution profiles and activity spectrum of fractions were species-specific, although the most potent fractions were those with retention times between 17 and 25 min. The exception was *T. autumnalis,* for which strong anticoagulation activity was spread across fractions with retention times between 27 and 44 min. Protein concentrations in the fractions were <0.1 µg/µl.

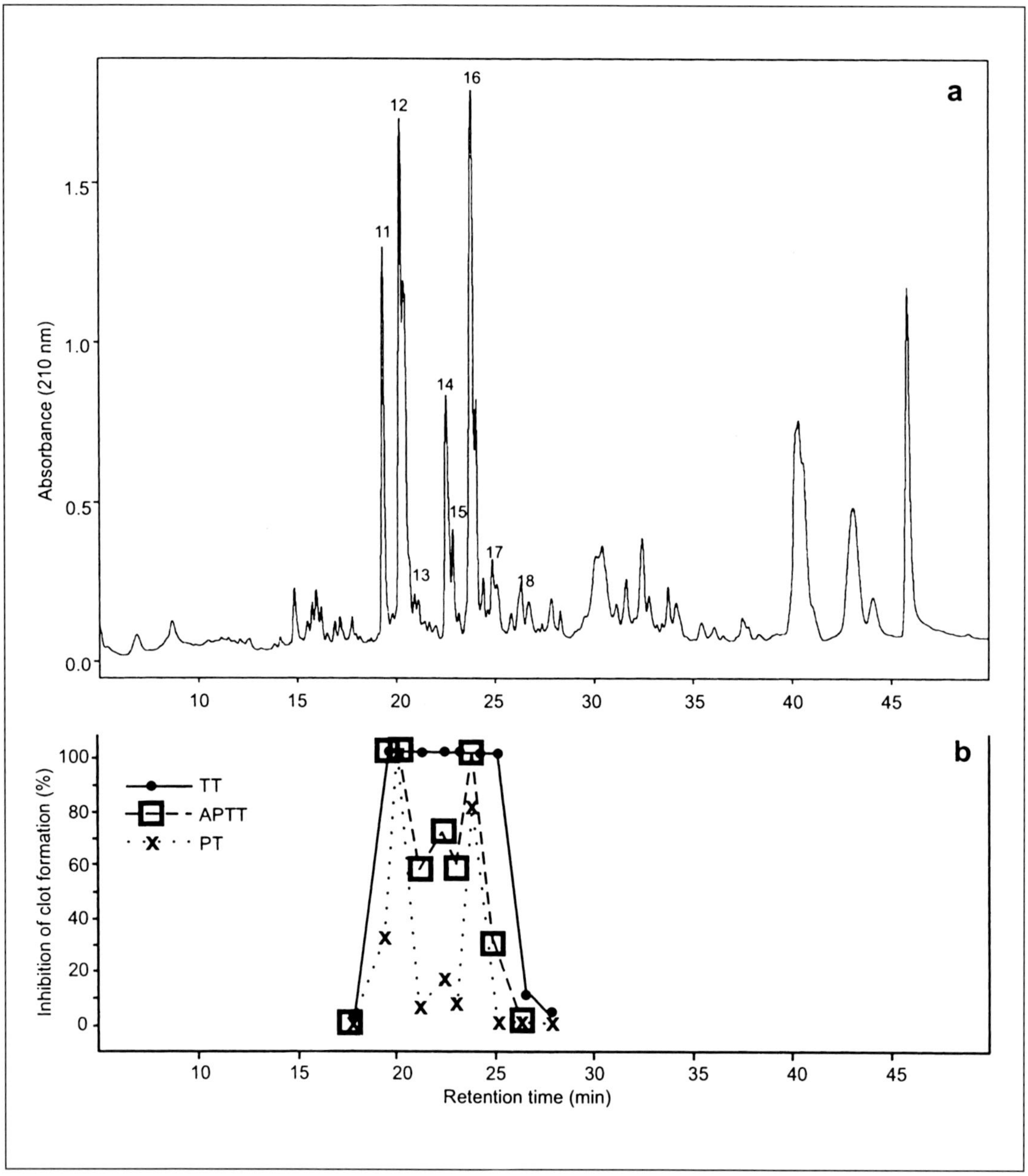

Fig. 2. RP-HPLC of anticoagulant activities in salivary gland extracts from *H. muehlfeldi*. **a** 173 SG pairs were processed as described in Material and Methods and loaded onto a Vydac C-4 column. The sample was eluted with a linear 10–100% ACN gradient with 0.1% TFA at a flow rate of 1 ml/min. Eluted fractions were concentrated to ~50 µl in a Speed-Vac. **b** One microliter of each fraction was used for the coagulation assays.

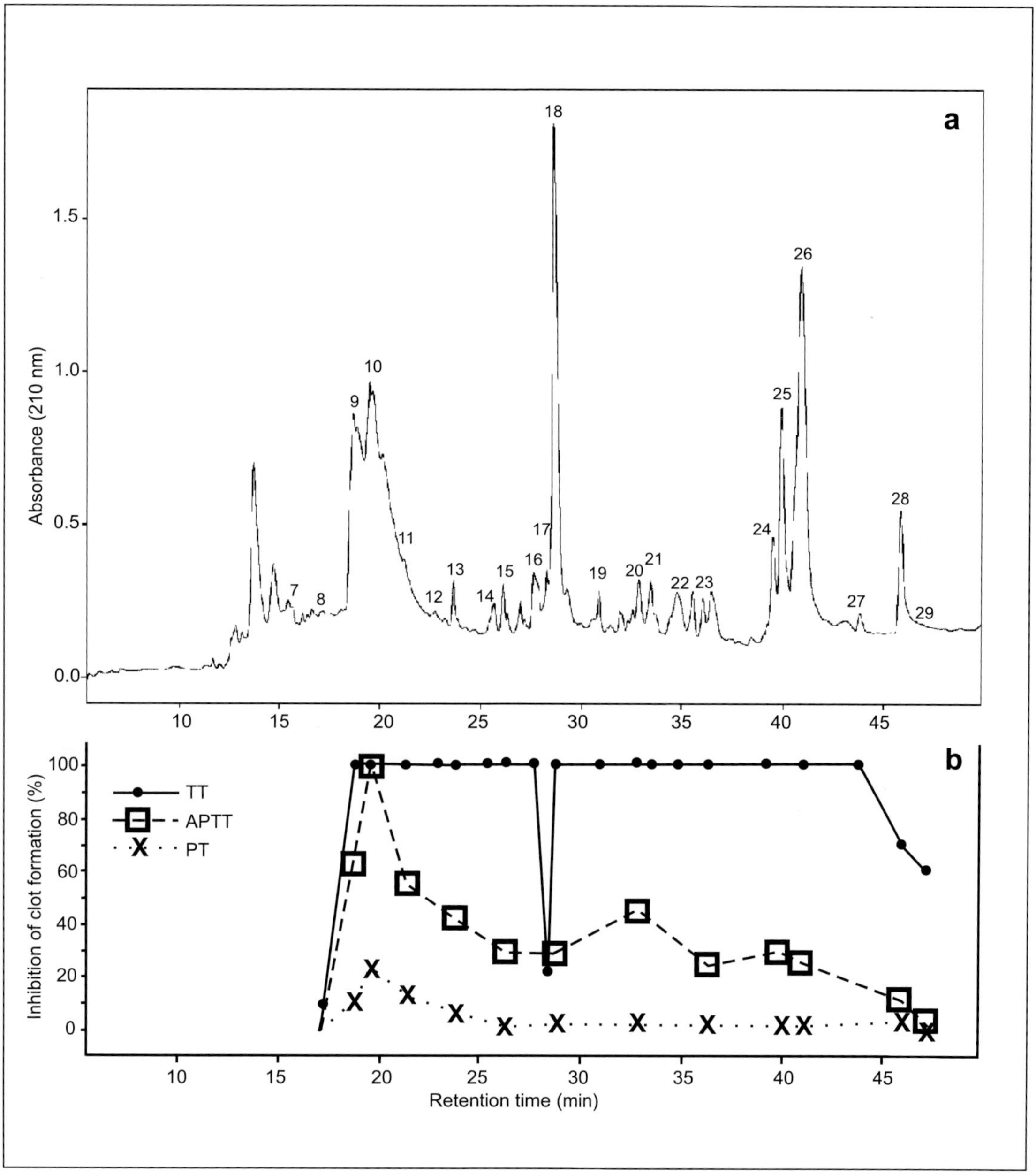

Fig. 3. RP-HPLC of anticoagulant activities in salivary gland extracts from *T. autumnalis.* **a** 63 SG pairs were processed as described in Material and Methods and loaded onto a Vydac C-4 column. The sample was eluted with a linear 10–100% ACN gradient with 0.1% TFA at a flow rate of 1 ml/min. Eluted fractions were concentrated to ~50 μl in a Speed-Vac. **b** One microliter of each fraction was used for the coagulation assays.

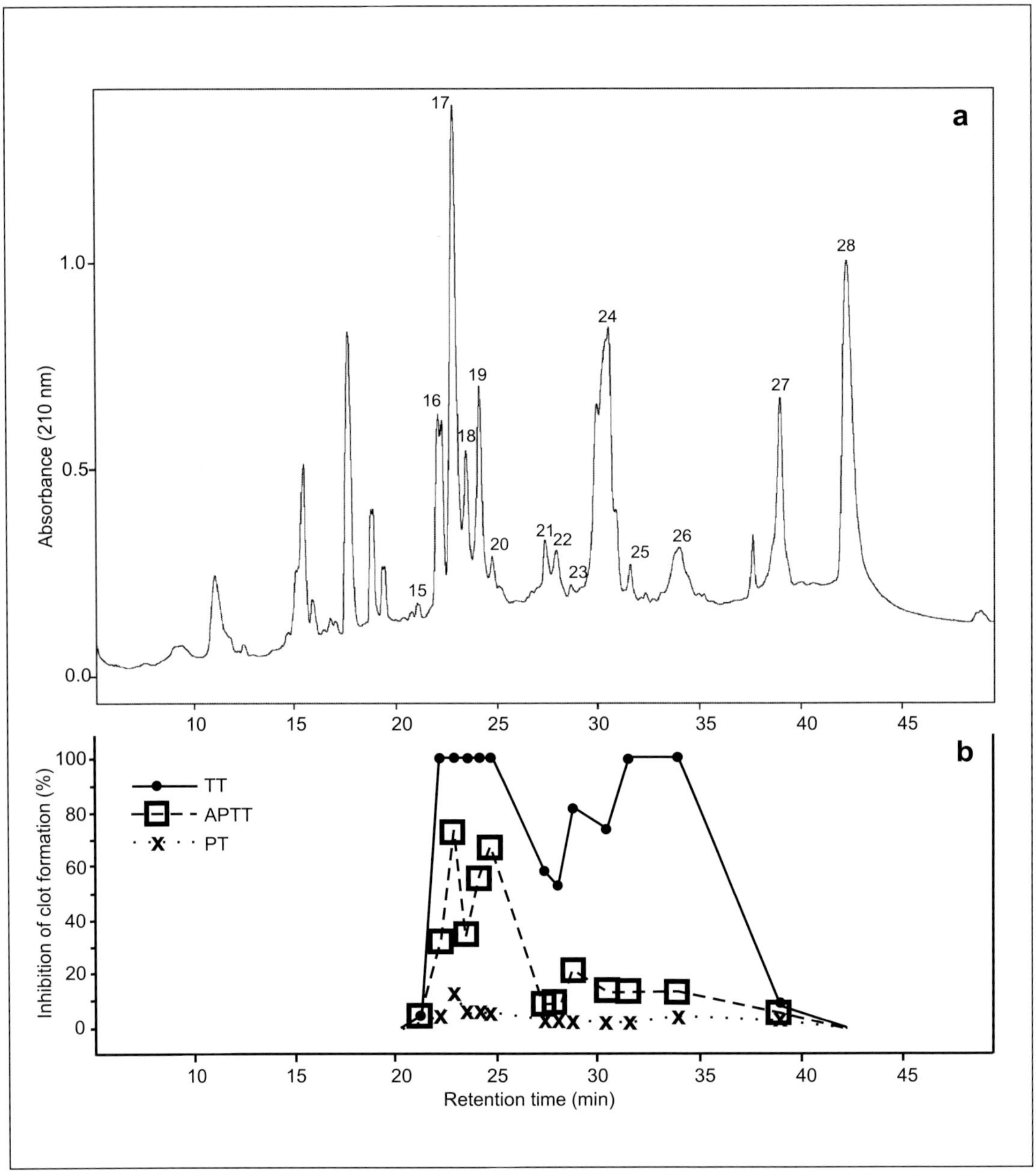

Fig. 4. RP-HPLC of anticoagulant activities in salivary gland extracts from *H. pluvialis.* **a** 355 SG pairs were processed as described in Material and Methods and loaded onto a Vydac C-4 column. The sample was eluted with a linear 10–100% ACN gradient with 0.1% TFA at a flow rate of 1 ml/min. Eluted fractions were concentrated to ~50 μl in a Speed-Vac. **b** One microliter of each fraction was used for the coagulation assays.

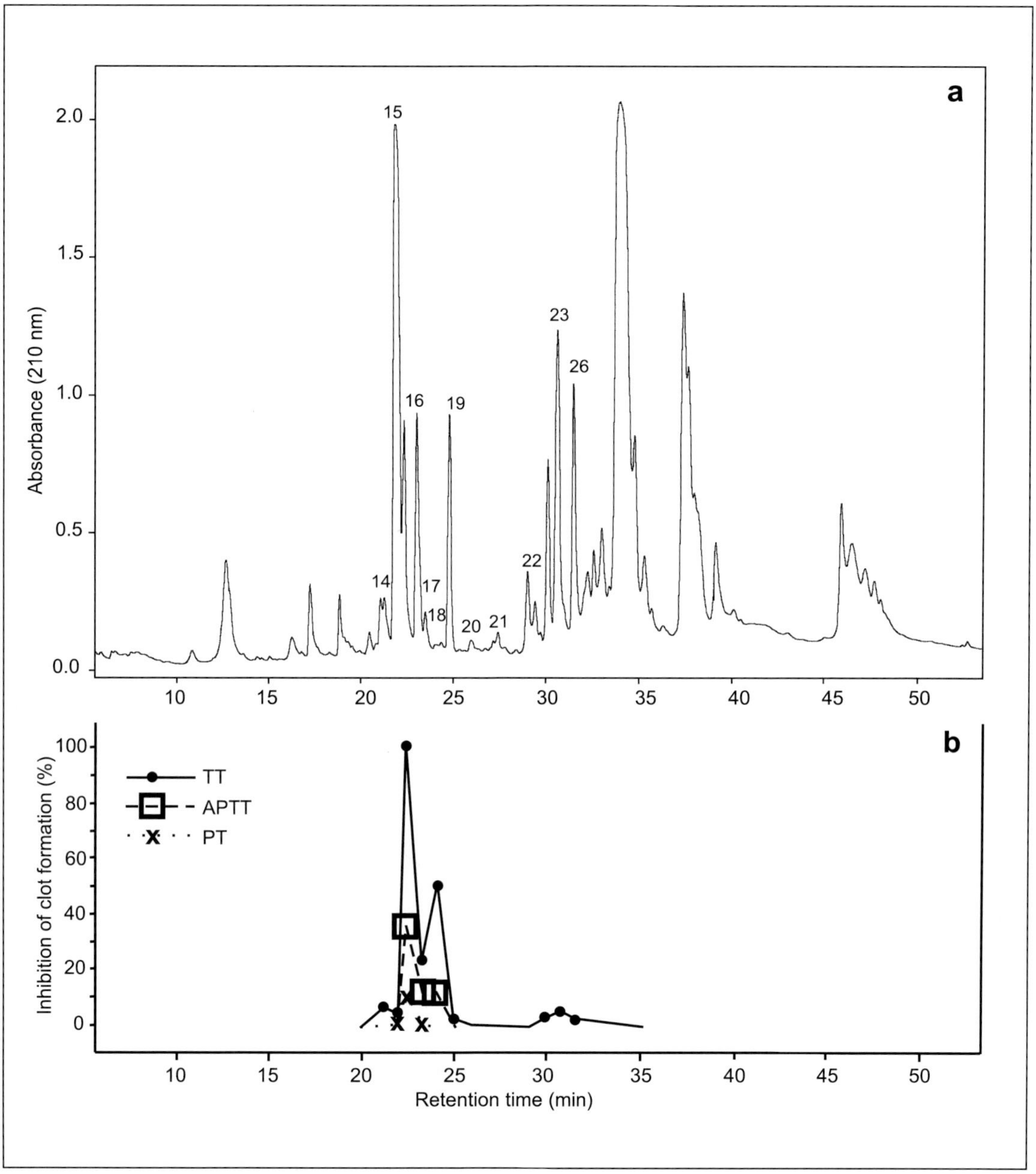

Fig. 5. RP-HPLC of anticoagulant activities in salivary gland extracts from *H. pellucens.* **a** 50 SG pairs were processed as described in Material and Methods and loaded onto a Vydac C-4 column. The sample was eluted with a linear 10–100% ACN gradient with 0.1% TFA at a flow rate of 1 ml/min. Eluted fractions were concentrated to ~50 μl in a Speed-Vac. **b** One microliter of each fraction was used for the coagulation assays.

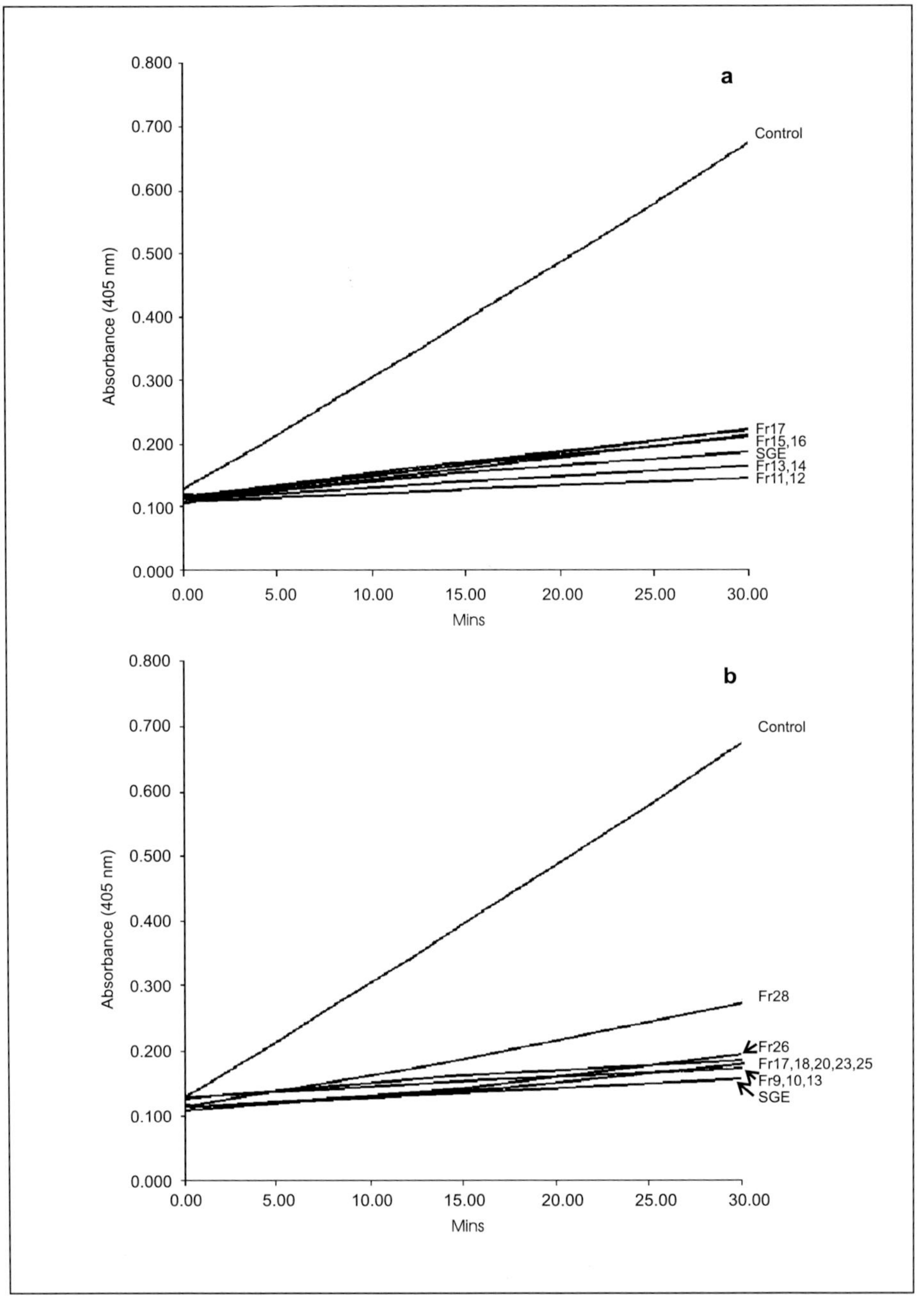

Fig. 6. Inhibitory effect of horsefly salivary gland extracts and fractions on substrate hydrolysis by thrombin (human thrombin –0.029 U/well, substrate –20 m*M* Tosyl-glycyl-prolyl-arginine-4-nitroanilide-acetate –20 µl/wel). Control –10 µl PBS, SGE (~6 µg protein in 10 µl PBS/well), Fr – HPLC fractions (<0.1 µg protein in 2 µl fraction diluted to 10 µl in PBS/well). For the numbering of fractions, see the corresponding chromatograms in figures 2–5. **a** *H. muehlfeldi.* **b** *T. autumnalis.* **c** *H. pluvialis.* **d** *H. pellucens.*

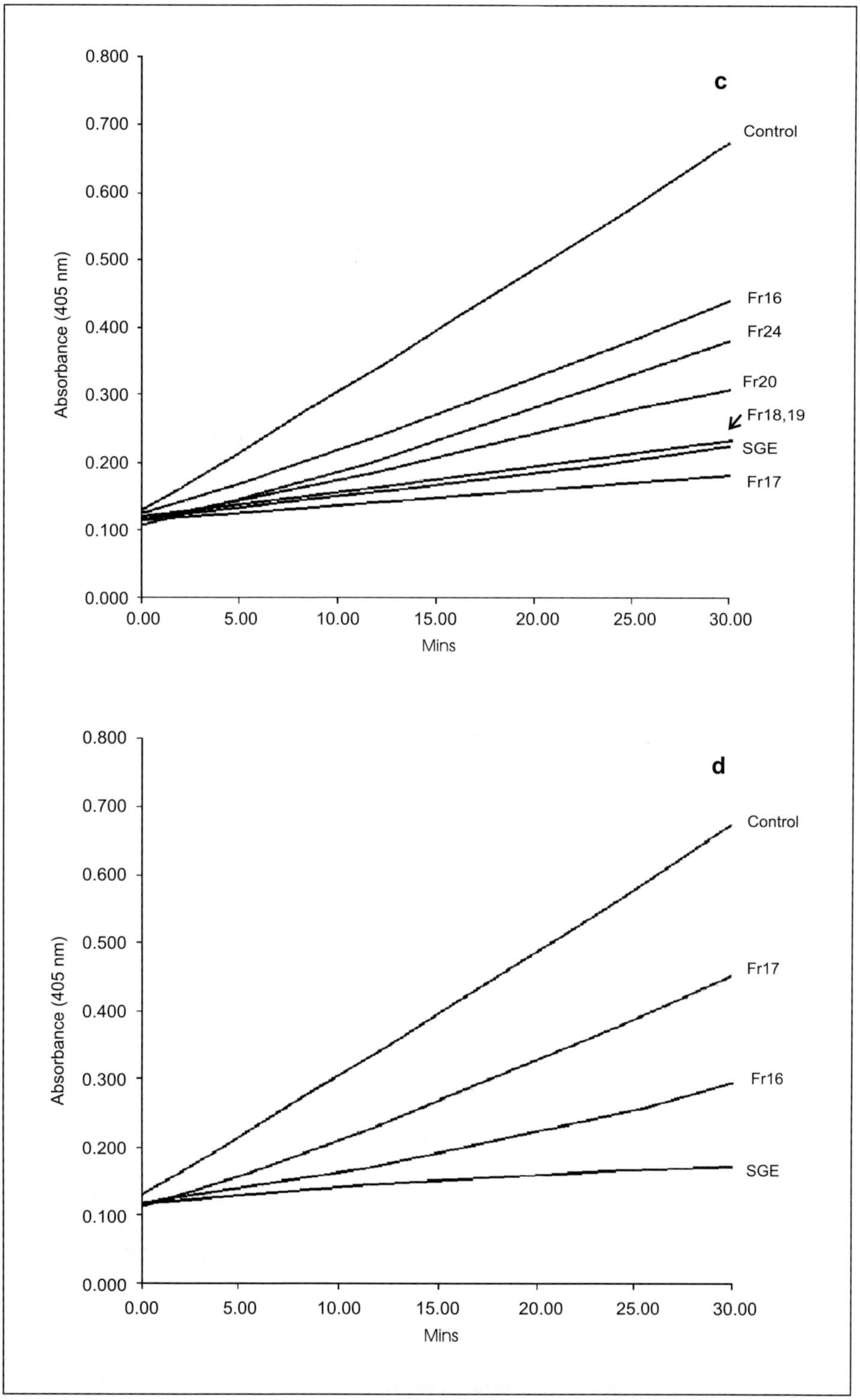

c
Absorbance (405 nm)
0.800
0.700
0.600
0.500
0.400
0.300
0.200
0.100
0.000
0.00
5.00
10.00
15.00
20.00
25.00
30.00
Mins
Control
Fr16
Fr24
Fr20
Fr18,19
SGE
Fr17
d
Absorbance (405 nm)
0.800
0.700
0.600
0.500
0.400
0.300
0.200
0.100
0.000
0.00
5.00
10.00
15.00
20.00
25.00
30.00
Mins
Control
Fr17
Fr16
SGE

Chromogenic Substrate Assay

Incubation of bovine factor Xa with its chromogenic substrate Chromozym X resulted in an increase in absorbance at 405 nm. The reaction was not inhibited by any of the tabanid SGE and progress curves of substrate hydrolysis were similar to the control (progress curves not shown).

Chromogenic substrate assay for human thrombin activity showed inhibition of substrate hydrolysis by SGE of all investigated species (fig. 6). The presence of compounds with antithrombin activity was also confirmed in HPLC fractions (fig. 6).

Discussion

Haematophagous arthropods possess various antihaemostatics in their saliva to overcome host haemostatic systems. Most of the compounds identified so far are direct inhibitors of coagulation factors (mainly antithrombins and inhibitors of factor Xa), or they inhibit platelet aggregation or act as vasodilators [2, 12]. In Diptera, inhibition of thrombin and factor Xa seems to be the most common mode of action of anticoagulants [3, 12]. The strategy of tabanids to counteract host haemostasis might be similar to other pool feeders (e.g. black flies, ticks, leeches) and probably differs from other higher Diptera.

Black flies show activities against factor Xa [13], thrombin [14, 15], factor V [16], platelet aggregation mediated by apyrase [17] and vasodilation [18]. In tabanids at least two anticoagulation strategies seem to be present, based on: (1) potent antithrombins in species of the genera *Tabanus, Hybomitra, Haematopota* and *Heptatoma* [8, this paper]; (2) inhibition of platelet aggregation by deerflies (*Chrysops* spp.) [9, 10]. Apart from the discovery of tabanin, a potent inhibitor of thrombin [4, 7, 8] in one of the largest species *(Tabanus bovinus)* which shows similarities to the mode of action of hirudin, horsefly anticoagulants have not been studied. Our results confirmed the existing knowledge on horsefly anticoagulants in routine coagulation assays (TT, PT, APTT). The results of the chromogenic substrate assay suggest that horseflies prevent blood coagulation mainly through the presence of a potent antithrombin, but not through inhibition of factor Xa. *Hybomitra* spp. and *Tabanus* spp. SGE also inhibit platelet aggregation induced by collagen and ADP, which is suggested to be a consequence of its strong antithrombin activity [P. Martin, unpubl. results]. Nevertheless, further extensive studies are needed to clarify the mode of action of horsefly anticoagulants.

Partial purification of the anticoagulants from the horsefly SGE by RP-HPLC showed species-specific differences in their activity spectrum. The results suggest that more than one anticoagulant or several isoforms are present. Moreover, horsefly anticoagulants are highly potent at extremely low concentrations. Molecular weights of the peptides in the active fractions could not be detected by electrophoresis, but preliminary studies using mass spectrometry indicate that they are low molecular weight peptides [P. Takáč et al., unpubl. results]. Further purification and amino acid sequence analyses will reveal whether homology exists between horsefly anticoagulants and other naturally occurring thrombin inhibitors.

In conclusion, our study demonstrated interspecific differences in the anticoagulant activities in the SGE of four horsefly species. The mode of action of the anticoagulant(s) is probably based mainly on the inhibition of thrombin; however, tabanids may target other factors of the coagulation cascade. Horsefly saliva may also offer a source of new compounds for potential pharmaceutical use.

Acknowledgments

The authors wish to thank Ľ. Vidlička, F. Čiampor and L. Roller for their help in fieldwork and species identification and Dr. E. Silvanová and Dr. I. Balanová from the Slovak Institute of Cardiovascular Diseases in Bratislava for providing plasma samples. The research was financially supported by Evolutec Ltd. and partly by grant VEGA, No. 2/1129/21, from the Slovak Academy of Sciences.

References

1 Ribeiro JMC: Role of saliva in blood-feeding by arthropods. Annu Rev Entomol 1987;32:463–478.
2 Wikel SK: The immunology of host-ectoparasitic arthropod relationships. Wallingford, CAB International, 1996.
3 Arocha-Piñango CL, Marchi R, Carvajal Z, Guerrero B: Invertebrate compounds acting on the hemostatic mechanism. Blood Coagul Fibrinolysis 1999;10:43–68.
4 Markwardt F: Coagulation inhibitors from animals feeding on blood. Rev Iberoamer Tromb Hemost 1994;7:225–231.
5 Chvála M, Lyneborg L, Moucha J: The horse flies of Europe (Diptera, Tabanidae). Ent Soc Copenhagen, 1972.
6 Chvála M, Hůrka K, Chalupský J, Knoz J, Minář J, Országh I: Fauna ČSSR, Svazek 22, Krevsající mouchy a střečci – Diptera. Academia Nakl ČSAV, Praha, 1980.
7 Markwardt F, Leberecht E: Untersuchungen über den blutgerinnungshemmenden Wirkstoff der Tabaniden. Naturwissenschaften 1959;46:17–18.
8 Markwardt F, Schulz E: Über den Mechanismus der blutgerinnungshemmenden Wirkung des Tabanins. Naunyn-Schmiedebergs Arch Exp Pathol Pharmakol 1960;238:320–328.
9 Grevelink SA, Youssef DE, Loscalzo J, Lerner EA: Salivary gland extracts from the deerfly contain a potent inhibitor of platelet aggregation. Proc Natl Acad Sci USA 1993; 90:9155–9158.
10 Reddy VB, Kounga K, Mariano F, Lerner EA: Chrysoptin is a potent glycoprotein IIb/IIIa fibrinogen receptor antagonist present in salivary gland extracts of the deerfly. J Biol Chem 2000;275:15861–15867.
11 Bradford MM: A rapid and sensitive method for the quantitation of microgram quantities of protein utilizing the principle of protein-dye binding. Anal Biochem 1976;72: 248–254.
12 Beaty BJ, Marquardt WC: The Biology of Disease Vectors. Niwot, University Press of Colorado, 1996.
13 Jacobs JW, Cupp EW, Sardana M, Friedman PA: Isolation and characterization of a coagulation factor Xa inhibitor from black fly salivary glands. Thromb Haemost 1990;64: 235–238.
14 Abebe M, Cupp MS, Ramberg FB, Cupp EW: Anticoagulant activity in salivary gland extracts of black flies (Diptera: Simuliidae). J Med Entomol 1994;31:908–911.
15 Abebe M, Cupp MS, Champagne D, Cupp EW: Simulidin: A black fly *(Simulium vittatum)* salivary gland protein with anti-thrombin activity. J Insect Physiol 1995;41:1001–1006.
16 Abebe M, Ribeiro JMC, Cupp MS, Cupp EW: Novel anticoagulant from salivary glands of *Simulium vittatum* (Diptera: Simuliidae) inhibits activity of coagulation factor V. J Med Entomol 1996;33:173–176.
17 Cupp MS, Cupp EW, Ramberg FB: Salivary gland apyrase in black flies *(Simulium vittatum)*. J Insect Physiol 1993;39:817–821.
18 Cupp MS, Ribeiro JMC, Cupp EW: Vasodilative activity in black fly salivary glands. Am J Trop Med Hyg 1994;50:241–246.

Author Index Vol. 31, No. 3–6, 2001

Hæmostasis

Aaspõllu, A. 123
Andrews, R.K. 155
Arocha-Piñango, C.L. 288

Berlanga, O. 155
Bianconi, M.L. 273
Bon, C. 133, 279
Brando, C. 177
Braud, S. 133
Budzynski, A.Z. 117
Bühler, B. 266

Chudzinski-Tavassi, A.M. 257
Clemetson, J.M. 148
Clemetson, K.J. 148

Dörmann, D. 148
Du, X.-Y. 148

El Ayeb, M. 173, 207

Fathallah, M. 207
Fritzen, M. 257

Gempeler-Messina, P.M. 266
Golubkov, V. 183
Govers-Riemslag, J.W.P. 241
Guerrero, B. 288

Huang, T.F. 192

Joseph, J.S. 218, 234

Kamiguti, A.S. 155
Kazimírová, M. 294
Kini, R.M. 218, 234, 279
Kozánek, M. 294
Kuchar, M.A. 177

Labuda, M. 294
Lazzari, M.A. 257
Leduc, M. 155
Lourenço, D. 257
Luis, J. 173

Mabrouk, K. 207
McLane, M.A. 177
Markland, F.S. 183
Maroun, R.C. 247
Marrakchi, N. 173, 207
Marsh, N.A. 211
Marvaldi, J. 173
Miele, M.E. 177
Monteiro, R.Q. 273
Mounier, C.M. 279
Müller, C. 266

Navdaev, A. 148
Nuttall, P.A. 294

Paquette-Straub, C.A. 177
Pozner, R. 257

Rao, V.S. 218
Regaya, I. 207
Reis, C.V. 257
Richters, V. 183
Rochat, H. 207
Rosing, J. 225, 241

Samel, M. 123
Santoli, D. 177
Sarray, S. 173, 207
Schattner, M. 257
Schleuning, W.-D. 118
Sherwin, R.P. 183
Shieh, K. 183
Siigur, E. 123
Siigur, J. 123
Sposto, R. 183
Srairi, N. 173
Subbi, J. 123
Šulanová, M. 294

Takáč, P. 294
Tans, G. 225, 241
Theakston, R.D.G. 155
Tõnismägi, K. 123
Toombs, C.F. 141
Trummal, K. 123

Vija, H. 123
Volz, K. 266

Watson, S.P. 155
Wisner, A. 133
Wu, W.B. 192

Yeh, C.H. 192
Yukelson, L. 241

Zhou, Q. 183
Zingali, R.B. 273

KARGER

Fax + 41 61 306 12 34
E-Mail karger@karger.ch
www.karger.com

Accessible online at:
www.karger.com/journals/hae

Hæmostasis

Subject Index Vol. 31, No. 3–6, 2001

Adhesion 183
Alfimeprase 141
Angiogenesis 183, 192
Animal model 183
– venoms 288
Anti-agglutination protein 173
Anticoagulant 279, 294
Antiplatelet aggregation 173, 207
Arthropods 294

Binding sites 148
Bleeding syndrome 288
Blood coagulation 218
– – factor V 241
– – – X 225
Bothrojaracin 273
Botrocetin 155

C-lectin protein 173
C-type lectin 148, 155
Caterpillar 241
Clotting cascade 218
Coagulation assays 211
Collagen 155
Convulxin 155
Cytokine 192

Desmoteplase 118
Diagnostic tools 266
Disintegrin(s) 177, 183, 192, 211
DSPA 118

E-selectin 257
Endothelial cell(s) 192, 257
Eristostatin 177

Factor V activation 241
– – activator 123
– – inactivation 241
– Va 273, 279
– X activator 123, 225
– Xa 234, 279

Fibrin specificity 118
Fibrino(geno)lytic enzymes 123
Fibrinolysis 141
Fibrinolytic therapy 118
Fibrolase 141

Glycoprotein 123
– Ib 192
GPIb-IX-V 155
GPVI 155

H(a)emostasis 133, 211, 288

ICAM-1 257
Integrin 177, 183, 192
Invasion 183

Lebetase 123
Lebetin peptides 207
Lonomia caterpillars 288
– *obliqua* 257
Lopap 257

MALDI-TOF MS 123
Mammalian 279
Mechanism 148
Melanoma 177
Metalloprotease 141, 225
Metalloproteinase 155, 192
Metastasis 177
Molecular modeling 247
– recognition 247

NK cells 177

Ovarian cancer 183

Plasminogen activator 118, 133
Platelet(s) 257
– function tests 211
– receptor 148
Procoagulants 234

Proexosite I 273
Proteases 123
Protein C activators 266
Proteinase 133
Prothrombin 273
– activator(s) 234, 257
Prothombinase 273

Secreted phospholipase A_2 279
Serine protease 225
Snake toxin 155
– venom(s) 123, 133, 148, 173, 183, 207, 225, 234, 241, 247, 279
– – enzymes 266
– – thrombin-like enzymes 211
– – toxins 211
Stroke 118
Structure-function 133
Substrate specificity 123

Tabanidae 294
Thrombin 247
– inhibition 294
Thrombin-like serine proteinases 247
Thrombocytopenia 207
Thrombolysis 141
Thrombolytic therapy 118
Thrombosis 141

Venom 241
– proteinase 218
Vipera lebetina 123

von Willebrand factor 155

KARGER
Fax + 41 61 306 12 34
E-Mail karger@karger.ch
www.karger.com

Accessible online at:
www.karger.com/journals/hae

Hæmostasis

Author Index Vol. 31, 2001

Aaspõllu, A. 123
Albert, F.W. 12
Alvarez, C. 18
Anderson, D.R. 42
Andrews, R.K. 155
Arocha-Piñango, C.L. 288
Audebert, A. 32

Barbera, N. 49
Berlanga, O. 155
Bianconi, M.L. 273
Bielska Falda, H. 69
Bombeli, T. 90
Bon, C. 133, 279
Boneu, B. 61
Brando, C. 177
Braud, S. 133
Budzynski, A.Z. 117
Bühler, B. 266
Burton, E. 42

Chudzinski-Tavassi, A.M. 257
Clemetson, J.M. 148
Clemetson, K.J. 148
Cvirn, G. 1

Dörmann, D. 148
Du, X.-Y. 148
Dubecq, F. 32
Dubreuil, M. 32

El Ayeb, M. 173, 207

Fathallah, M. 207
Fehr, J. 90
Fernandez, M.A. 18
Fransson, S.G. 55
Freyburger, G. 32
Frisina, N. 49
Fritzen, M. 257

Gallistl, S. 1
Gemmati, D. 99
Gempeler-Messina, P.M. 266
Giddings, J. 106
Giddings, J.C. 80
Golubkov, V. 183
Govers-Riemslag, J.W.P. 241
Groupe d'Etude sur l'Hémostase et la Thrombose 61
Guerrero, B. 288

Hajduk, B. 69
Hashimoto, M. 80
Huang, T.F. 192

Iriarte, J.A. 18

Järemo, P. 55
Joseph, J.S. 218, 234

Kamiguti, A.S. 155
Kanakura, Y. 26
Kawasaki, T. 26
Kazimírová, M. 294
Kikuchi, H. 71
Kini, R.M. 218, 234, 279
Kishi, T. 80
Kitajima, I. 71
Kozánek, M. 294
Kuchar, M.A. 177

Labrouche, S. 32
Labuda, M. 294
Lafuente, P.J. 18
Laville, C. 32
Lazzari, M.A. 257
Leduc, M. 155
Lindahl, T.L. 55
Logander, E. 55
Loreth, R.M. 12
Lourenço, D. 257
Luis, J. 173

Mabrouk, K. 207
McLane, M.A. 177
Markland, F.S. 183
Maroun, R.C. 247
Marrakchi, N. 173, 207
Marsh, N.A. 211
Martinez-Brotons, F. 18
Maruyama, I. 71
Marvaldi, J. 173
Matsui, H. 106
Matsunaga, T. 106
Meyer, M. 12
Miele, M.E. 177
Mii, Y. 106
Milovanovic, M. 55
Minowa, H. 106
Molinari, I. 32
Monteiro, R.Q. 273
Moratelli, S. 99
Mounier, C.M. 279
Müller, C. 266
Muntean, W. 1

Nasu, M. 71
Navdaev, A. 148
Nuttall, P.A. 294

Ohta, Y. 80
Oiwa, K. 80
Okamoto, T. 80
Ongaro, A. 99

Paquette-Straub, C.A. 177
Pistre, J.C. 32
Pizzoleo, M.A. 49
Polack, B. 61
Pozner, R. 257
Przedlacki, J. 69

Raddatz Mueller, P. 90
Rao, V.S. 218
Regaya, I. 207
Reis, C.V. 257
Richter, A. 55
Richters, V. 183
Ripoll, F. 18
Rochat, H. 207
Rosing, J. 225, 241

Samel, M. 123
Santoli, D. 177

KARGER

Fax + 41 61 306 12 34
E-Mail karger@karger.ch
www.karger.com

Accessible online at:
www.karger.com/journals/hae

Sarker, K.P. 71
Sarray, S. 173, 207
Scamardi, R. 49
Scapoli, G.L. 99
Schattner, M. 257
Schleuning, W.-D. 118
Schved, J.-F. 61
Serino, M.L. 99
Sherwin, R.P. 183
Shieh, K. 183
Siigur, E. 123
Siigur, J. 123
Sottilotta, G. 49
Sposto, R. 183
Srairi, N. 173
Subbi, J. 123
Suehisa, E. 26
Sugimoto, M. 106
Šulanová, M. 294

Takáč, P. 294
Takahashi, Y. 106
Tanaka-Horie, T. 106
Tans, G. 225, 241
Tezono, K. 71
Theakston, R.D.G. 155
Tognazzo, S. 99
Toku, M. 26
Tomkowski, W. 69
Tõnismägi, K. 123
Toombs, C.F. 141
Torbicki, A. 69
Trifiletti, A. 49
Trummal, K. 123
Tsukino, R. 106

Vacas, M. 18
Vija, H. 123
Villanove, X. 32
Volz, K. 266

Watanabe, S. 80
Watson, S.P. 155
Wawrzynska, L. 69
Wilbur, K. 42
Wilson, S.J.-A. 42
Wisner, A. 133
Wu, W.B. 192

Yamamoto, J. 80
Yeh, C.H. 192
Yoshioka, A. 106
Yukelson, L. 241

Zhou, Q. 183
Zingali, R.B. 273

Hæmostasis

Subject Index Vol. 31, 2001

Activated protein C 1, 80
Adhesion 183
Alfimeprase 141
Angiogenesis 183, 192
Animal model 183
– venoms 288
Anti-agglutination protein 173
Anticoagulant 279, 294
– treatment, oral 18
Antiplatelet aggregation 173, 207
Anti-Xa levels 42
Arthropods 294

Binding sites 148
Bleeding syndrome 288
– tendency 12
Blood coagulation 218
– – factor V 241
– – – X 225
– pressure 49
– sampling 61
Bothrojaracin 273
Botrocetin 155

C-lectin protein 173
C-type lectin 148, 155
Caterpillar 241
Clotting cascade 218
Coagulation 49, 61
– activation markers 90
– assays 211
Collagen 155
Combined defects 99
Congenital deficiency of plasminogen activator inhibitor-1 106
Convulxin 155
Coronary angiography 55
Cytokine(s) 71, 192

D-dimers 90
Desmoteplase 118
Diagnostic tools 266
Disintegrin(s) 177, 183, 192, 211
DSPA 118
Dysfibrinogenemia 12

E-selectin 257
Endogenous thrombolysis 80
Endothelial cell(s) 192, 257
– monocytes 71
Eristostatin 177

Factor V activation 241
– – activator 123
– – G1691A 99
– – inactivation 241
– Va 273, 279
– X activator 123, 225
– Xa 234, 279
Fibrin clot 71
– specificity 118
Fibrinogen 12
– variant 12
Fibrino(geno)lytic enzymes 123
Fibrinolysis 32, 141
Fibrinolytic therapy 118
Fibrinopeptide A 12
Fibrolase 141
Flow cytometer 55

Gene-environment interaction 99
Glycoprotein 123
– Ib 192
GPIb-IX-V 155
GPVI 155
Gynaecology 32

H(a)emostasis 32, 133, 211, 288
–, laboratory 61
Hypertension 49

ICAM-1 257
Inherited thrombophilia 99
Integrin 177, 183, 192
Invasion 183

Laparoscopic surgery 32
Lebetase 123
Lebetin peptides 207
Lonomia caterpillars 288
– *obliqua* 257

Lopap 257
Low-molecular-weight heparin 42, 90

α_2-Macroglobulin 1
MALDI-TOF MS 123
Mammalian 279
Mechanism 148
Melanoma 177
Metalloprotease 141, 225
Metalloproteinase 155, 192
Metastasis 177
Molecular modeling 247
– recognition 247
Mutation 12

NK cells 177

Ovarian cancer 183

Pigmented villonodular synovitis 106
Plasminogen activator 118, 133
Platelet(s) 55, 257
– density 55
– function tests 211
– receptor 148
Portable monitor 18
Preanalytical testing 61
Pregnancy 90
Procoagulants 234
Proexosite I 273
Proteases 123
Protein C 26
– – activators 266
– S 26
Proteinase 133
Prothrombin 273
– activation 1
– activator(s) 234, 257
– G20210A 99
– time 18
Prothombinase 273

KARGER

Fax +41 61 306 12 34
E-Mail karger@karger.ch
www.karger.com

Accessible online at:
www.karger.com/journals/hae

Secreted phospholipase A_2 279
Serine protease 225
Snake toxin 155
– venom(s) 123, 133, 148, 173, 183, 207, 225, 234, 241, 247, 279
– – enzymes 266
– – thrombin-like enzymes 211
– – toxins 211
Stroke 118
Structure-function 133
Substrate specificity 123

Tabanidae 294
Thrombin 247
– generation test 32
– inhibition 294
– inhibitors 80
– potential 1
Thrombin-antithrombin 90
Thrombin-like serine proteinases 247
Thrombocytopenia 207
Thrombolysis 141
Thrombolytic therapy 118
Thrombomodulin 26
Thromboprophylaxis 90
Thrombosis 141
Tranexamic acid 106

VEGF 71
Venom 241
– proteinase 218
Venous thromboembolism 42
Vipera lebetina 123

von Willebrand factor 155
Wound healing 71

Vol. 31, 2001

Official Journal of the 'Mediterranean League against Thromboembolic Diseases'

Hæmostasis

International Journal on Haemostasis and Thrombosis Research

KARGER

Basel · Freiburg · Paris · London · New York · New Delhi · Bangkok · Singapore · Tokyo · Sydney

Contents Vol. 31, 2001

Hæmostasis

No. 1

Original Papers

1 **Alpha-2-Macroglobulin Inhibits the Anticoagulant Action of Activated Protein C in Cord and Adult Plasma**
Cvirn, G.; Gallistl, S.; Muntean, W. (Graz)

12 **Fibrinogen Kaiserslautern III: A New Case of Congenital Dysfibrinogenemia with Aa 16 Arg → Cys Substitution**
Loreth, R.M. (Kaiserslautern); Meyer, M. (Jena); Albert, F.W. (Kaiserslautern)

18 **Comparative Study of a Portable Prothrombin Time Monitor Employing Three Different Systems in Oral Anticoagulant Units**
Vacas, M. (Bilbao); Fernández, M.A. (Valencia); Martínez-Brotons, F. (Barcelona); Lafuente, P.J. (Bilbao); Ripoll, F. (Valencia); Alvarez, C. (Barcelona); Iriarte, J.A. (Bilbao)

26 **Measurement of a Newly Developed Thrombomodulin Addition Activated Partial Thromboplastin Time Assay in Patients with Deep Venous Thrombosis**
Suehisa, E.; Toku, M.; Kawasaki, T.; Kanakura, Y. (Osaka)

32 **Changes in Haemostasis after Laparoscopic Surgery in Gynaecology: Contribution of the Thrombin Generation Test**
Freyburger, G.; Dubreuil, M.; Audebert, A.; Labrouche, S.; Pistre, J.C.; Molinari, I.; Dubecq, F.; Laville, C.; Villanove, X. (Bordeaux)

42 **Effect of Patient Weight on the Anticoagulant Response to Adjusted Therapeutic Dosage of Low-Molecular-Weight Heparin for the Treatment of Venous Thromboembolism**
Wilson, S.J-A. (Halifax, N.S.); Wilbur, K. (Vancouver, B.C.); Burton, E.; Anderson, D.R. (Halifax, N.S.)

49 **Increased Indexes of Thrombin Activation in Advanced Stages of Hypertension**
Trifiletti, A.; Scamardi, R.; Pizzoleo, M.A.; Sottilotta, G.; Barbera, N.; Frisina, N. (Messina)

55 **Inverse Relationship between Platelet Density and Reactivity Alterations at Coronary Angiography**
Järemo, P.; Lindahl, T.L.; Fransson, S.G.; Milovanovic, M.; Logander, E.; Richter, A. (Linköping)

Special Communication

61 **Preanalytical Recommendations of the 'Groupe d'Etude sur l'Hémostase et la Thrombose' (GEHT) for Venous Blood Testing in Hemostasis Laboratories**
Polack, B.; Schved, J.-F.; Boneu, B., on behalf of the 'Groupe d'Etude sur l'Hémostase et la Thrombose' (GEHT) (Grenoble)

Letter to the Editor

69 **Low-Molecular-Weight Heparins, Acenocoumarol and Bone Density**
Wawrzy'nska, L.; Przedlacki, J.; Hajduk, B.; Bielska Falda, H.; Tomkowski, W.; Torbicki, A. (Warsaw)

No. 2

Original Papers

71 **Bioactivity of the Vascular Endothelial Growth Factor Trapped in Fibrin Clots: Production of IL-6 and IL-8 in Monocytes by Fibrin Clots**
Tezono, K. (Oita); Sarker, K.P. (Kagoshima); Kikuchi, H.; Nasu, M. (Oita); Kitajima, I.; Maruyama, I. (Kagoshima)

80 **Enhanced Thrombolysis Induced by Argatroban or Activated Protein C in the Presence or Absence of Staphylokinase, Measured in an in vivo Animal Model Using Mesenteric Arterioles**
Hashimoto, M.; Watanabe, S.; Oiwa, K. (Kobe); Ohta, Y. (Osaka); Kishi, T.; Okamoto, T. (Kobe); Giddings, J.C. (Cardiff); Yamamoto, J. (Kobe)

90 **Evaluation of an Optimal Dose of Low-Molecular-Weight Heparin for Thromboprophylaxis in Pregnant Women at Risk of Thrombosis Using Coagulation Activation Markers**
Bombeli, T.; Raddatz Mueller, P.; Fehr, J. (Zurich)

Case Reports

99 **Coexistence of Factor V G1691A and Factor II G20210A Gene Mutations in a Thrombotic Family Is Associated with Recurrence and Early Onset of Venous Thrombosis**
Gemmati, D.; Serino, M.L.; Moratelli, S.; Tognazzo, S.; Ongaro, A.; Scapoli, G.L. (Ferrara)

KARGER

Fax + 41 61 306 12 34
E-Mail karger@karger.ch
www.karger.com

Access to full text and tables of contents, including tentative ones for forthcoming issues:
www.karger.com/journals/hae/hae_bk.htm

106 **Successful Arthroscopic Treatment of Pigmented Villonodular Synovitis of the Knee in a Patient with Congenital Deficiency of Plasminogen Activator Inhibitor-1 and Recurrent Haemarthrosis**
Matsui, H.; Takahashi, Y.; Matsunaga, T.; Tanaka-Horie, T.; Minowa, H.; Sugimoto, M. (Kashihara); Tsukino, R. (Arita); Mii, Y. (Kashihara); Giddings, J. (Cardiff); Yoshioka, A. (Kashihara)

No. 3–6

International Conference on Exogenous Factors Affecting Thrombosis and Haemostasis
Paris, July 12–13, 2001
In memoriam Stefan Niewiarowski, 1926–2001
Guest Editors: Bon, C. (Paris); Kini, M. (Singapore); Markland, F.S., Jr. (Los Angeles, Calif.); Marsh, N.A. (Brisbane); Rosing, J. (Maastricht)

In memoriam

117 **Stefan Niewiarowski**
1926–2001
Budzynski, A.Z. (Philadelphia, Pa.)

Fibrinolysis

118 **Vampire Bat Plasminogen Activator DSPA-Alpha-1 (Desmoteplase): A Thrombolytic Drug Optimized by Natural Selection**
Schleuning, W.-D. (Berlin)

123 **Proteases from *Vipera lebetina* Venom Affecting Coagulation and Fibrinolysis**
Siigur, J.; Aaspõllu, A.; Tõnismägi, K.; Trummal, K.; Samel, M.; Vija, H.; Subbi, J.; Siigur, E. (Tallinn)

133 **Snake Venom Proteinases as Tools in Hemostasis Studies: Structure-Function Relationship of a Plasminogen Activator Purified from *Trimeresurus stejnegeri* Venom**
Wisner, A.; Braud, S.; Bon, C. (Paris)

141 **Alfimeprase: Pharmacology of a Novel Fibrinolytic Metalloproteinase for Thrombolysis**
Toombs, C.F. (Thousand Oaks, Calif.)

Substances Affecting Platelet Function

148 **Multifunctional Snake C-Type Lectins Affecting Platelets**
Clemetson, K.J.; Navdaev, A.; Dörmann, D.; Du, X.-Y.; Clemetson, J.M. (Berne)

155 **The Use of Snake Venom Toxins as Tools to Study Platelet Receptors for Collagen and von Willebrand Factor**
Andrews, R.K. (Melbourne); Kamiguti, A.S. (Liverpool); Berlanga, O. (Oxford); Leduc, M. (Paris); Theakston, R.D.G. (Liverpool); Watson, S.P. (Oxford)

173 **Lebecetin, a C-Lectin Protein from the Venom of *Macrovipera lebetina* That Inhibits Platelet Aggregation and Adhesion of Cancerous Cells**
Sarray, S.; Srairi, N. (Tunis); Luis, J.; Marvaldi, J. (Marseille); El Ayeb, M.; Marrakchi, N. (Tunis)

177 **New Insights on Disintegrin-Receptor Interactions: Eristostatin and Melanoma Cells**
McLane, M.A.; Kuchar, M.A.; Brando, C.; Santoli, D.; Paquette-Straub, C.A.; Miele, M.E. (Newark, Del./Philadelphia, Pa.)

183 **A Novel Snake Venom Disintegrin That Inhibits Human Ovarian Cancer Dissemination and Angiogenesis in an Orthotopic Nude Mouse Model**
Markland, F.S.; Shieh, K.; Zhou, Q.; Golubkov, V.; Sherwin, R.P.; Richters, V.; Sposto, R. (Los Angeles, Calif.)

192 **Viper Venom Components Affecting Angiogenesis**
Huang, T.F.; Yeh, C.H.; Wu, W.B. (Taipei)

207 **Lebetin Peptides: Potent Platelet Aggregation Inhibitors**
Marrakchi, N. (Tunis); Mabrouk, K. (Marseille); Regaya, I.; Sarray, S. (Tunis); Fathallah, M.; Rochat, H. (Marseille); El Ayeb, M. (Tunis)

Procoagulant Proteins and Diagnostic Agents

211 **Diagnostic Uses of Snake Venom**
Marsh, N.A. (Brisbane)

218 **Procoagulant Proteins from Snake Venoms**
Kini, R.M. (Singapore/Richmond, Va.); Rao, V.S.; Joseph, J.S. (Singapore)

225 **Snake Venom Activators of Factor X: An Overview**
Tans, G.; Rosing, J. (Maastricht)

234 **Snake Venom Prothrombin Activators Homologous to Blood Coagulation Factor Xa**
Joseph, J.S. (Singapore); Kini, R.M. (Singapore/Richmond, Va.)

241 **Factor V Activation and Inactivation by Venom Proteases**
Rosing, J.; Govers-Riemslag, J.W.P.; Yukelson, L.; Tans, G. (Maastricht)

247 **Molecular Basis for the Partition of the Essential Functions of Thrombin among Snake Venom Serine Proteinases.** The Case of Thrombin-Like Enzymes
Maroun, R.C. (Paris)

257 **Effects of Lopap on Human Endothelial Cells and Platelets**
Chudzinski-Tavassi, A.M. (São Paulo); Schattner, M. (Buenos Aires); Fritzen, M. (São Paulo); Pozner, R.G. (Buenos Aires); Reis, C.V.; Lourenço, D. (São Paulo); Lazzari, M.A. (Buenos Aires)

Anticoagulant Proteins and Diagnostic Agents

266 **Protein C Activators from Snake Venoms and Their Diagnostic Use**
Gempeler-Messina, P.M.; Volz, K.; Bühler, B.; Müller, C. (Basel)

273 **Interaction of Bothrojaracin with Prothrombin**
Zingali, R.B.; Bianconi, M.L.; Monteiro, R.Q. (Rio de Janeiro)

279 **Anticoagulant Venom and Mammalian Secreted Phospholipases A_2: Protein- versus Phospholipid-Dependent Mechanism of Action**
Mounier, C.M. (Cergy-Pontoise); Bon, C. (Paris); Kini, R.M. (Singapore/Richmond, Va.)

288 ***Lonomia* Genus Caterpillar Envenomation: Clinical and Biological Aspects**
Arocha-Piñango, C.L.; Guerrero, B. (Caracas)

294 **Identification of Anticoagulant Activities in Salivary Gland Extracts of Four Horsefly Species (Diptera, Tabanidae)**
Kazimírová, M.; Šulanová, M.; Kozánek, M.; Takáč, P.; Labuda, M. (Bratislava); Nuttall, P.A. (Oxford)

306 **Author Index Vol. 31, No. 3–6, 2001**
307 **Subject Index Vol. 31, No. 3–6, 2001**

308 **Author Index Vol. 31, 2001**
310 **Subject Index Vol. 31, 2001**

Suppl. 1

Thrombosis and Haemostasis Issues in Cancer
International Conference, Bergamo
November 2–4, 2001
Lectures and Abstracts
Guest Editors: Falanga, A. (Bergamo); Brenner, B. (Haifa); Rickles, F.R. (Washington, D.C.)

S. Karger
Medical and Scientific Publishers
Basel · Freiburg · Paris · London
New York · New Delhi · Bangkok
Singapore · Tokyo · Sydney

Drug Dosage
The authors and the publisher have exerted every effort to ensure that drug selection and dosage set forth in this text are in accord with current recommendations and practice at the time of publication. However, in view of ongoing research, changes in government regulations, and the constant flow of information relating to drug therapy and drug reactions, the reader is urged to check the package insert for each drug for any change in indications and dosage and for added warnings and precautions. This is particularly important when the recommended agent is a new and/or infrequently employed drug.

Printed in Switzerland on acid-free paper by Reinhardt Druck, Basel

From epidemiology to pathogenesis and treatment

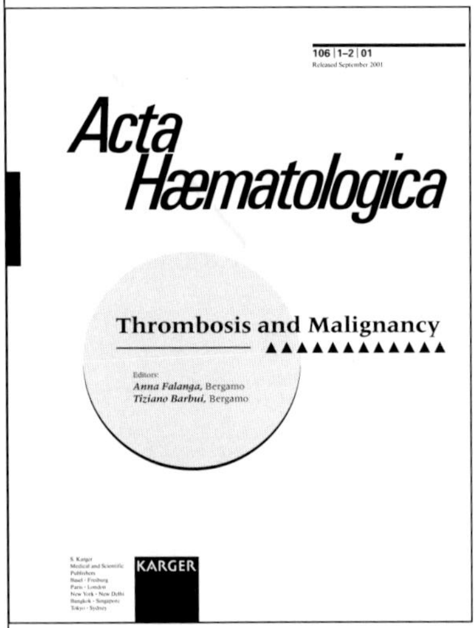

Thrombosis and Malignancy

Editors
Anna Falanga
Tiziano Barbui

Hematology; Oncology

Contents

Preface

Epidemiology of Thrombosis in Cancer: ***Rickles, F.R.; Levine, M.N.***
Venous Thromboembolism as First Manifestation of Cancer: ***Piccioli, A.; Prandoni, P.***
Pathogenetic Mechanisms of Thrombosis in Malignancy: ***Donati, M.B.; Falanga, A.***
Update on Tumor Cell Procoagulant Factors: ***Gale, A.J.; Gordon, S.G.***
The Thrombophilic State in Cancer Patients: ***Gouin-Thibault, I.; Achkar, A.; Samama, M.M.***
Coagulopathy of Acute Promyelocytic Leukemia: ***Falanga, A.; Barbui, T.***
Thrombotic Microangiopathy in the Cancer Patient: ***Kwaan, H.C.; Gordon, L.I.***
Hepatic Veno-Occlusive Disease following Hematopoietic Stem Cell Transplantation: ***Richardson, P.; Guinan, E.***
Thrombotic Complications of Central Venous Catheters in Cancer Patients: ***Monreal, M.; Davant, E.***
Thromboprophylaxis in the Cancer Patient: ***Thodiyil, P.A.; Walsh, D.C.; Kakkar, A.K.***
Treatment of Venous Thrombosis in the Cancer Patient: ***Levine, M.N.; Lee, A.Y.Y.***

Author Index
Subject Index

For over a hundred years, thrombosis has been known to contribute significantly to the morbidity and mortality of malignant disease. In the last decades, major advances in the investigation of the relationship between tumor growth and blood coagulation have been achieved, showing this complex phenomenon to be the second most important cause of death in cancer patients.
This special issue of Acta Haematologica covers the most relevant aspects of thromboembolism in malignancy, from epidemiology to pathogenesis and treatment. All important epidemiological studies are critically reviewed, and it is demonstrated that thrombosis often represents the earliest clinical manifestation of cancer or can be triggered by surgery or other predisposing conditions such as chemotherapy. Further, pathogenetic mechanisms are described, and the plausibility of the association of thrombosis with malignancy on the basis of various known tumor tissue-specific clot-promoting properties is shown. The issue revisits all varieties of clinical pictures associated with malignancy: subclinical hypercoagulable conditions as well as systemic syndromes like disseminated intravascular coagulation, thrombotic microangiopathy, and hepatic veno-occlusive disease. Finally, new concepts for the prophylaxis in surgical or medical conditions are discussed, and suggestions for treating these complications are provided.
'Thrombosis and Malignancy' is a valuable source of information for vascular biologists, physiologists and molecular biologists and other scientists working in the fields of thrombogenesis and tumor progression. But also oncologists, hematologists, internists or surgeons interested in the identification, prevention and treatment of thrombohemorrhagic complications in cancer patients might greatly benefit from this publication.

www.karger.com/journals/aha

KI 01265

Acta Haematologica
Editor: Ben-Bassat, I. (Tel Hashomer)
ISSN 0001–5792
Special issue: Vol. 106, No. 1–2 (2001)
Included in subscription

Thrombosis and Malignancy
Editors: Falanga, A.; Barbui, T. (Bergamo)
88 p., 11 fig., 13 tab., soft cover, 2001
CHF 39.– / EUR 28.– / USD 34.00
Prices subject to change
EUR price for Germany, USD price for USA only
ISBN 3–8055–7305–7

KARGER

Order Form

Please send: ___ copy/ies

Postage and handling free with prepayment

Payment:
☐ Check enclosed ☐ Please bill me

Please charge to my credit card
☐ American Express ☐ Diners ☐ Eurocard
☐ MasterCard ☐ Visa

Card No.: ______

Exp. date: ______

☐ Please send a free sample copy of the journal **Acta Haematologica**

Orders may be placed with any bookshop, subscription agency, directly with the publisher or through a Karger distributor.

Fax: +41 61 306 12 34
S. Karger AG, P.O. Box, CH–4009 Basel (Switzerland)
E-Mail karger@karger.ch, **www.karger.com**

Name/Address:

Date:

Signature:

Hæmostasis

Haemostasis
ISSN 0301–0147
31(3–6) 113–312 (2001)

International Conference on Exogenous Factors Affecting Thrombosis and Haemostasis

This publication highlights recent research on the biochemistry, molecular biology, structure, function and mechanism of action of exogenous factors affecting thrombosis and haemostasis. These factors include numerous components of snake venoms, which can either activate or inhibit platelets or coagulation factors, and anti-haemostatic compounds from the saliva of blood-feeding arthropods. Comprising four sections devoted to fibrinolysis, substances that affect platelet aggregation, procoagulant proteins and anticoagulant proteins, this publication also deals with the development of therapeutic and diagnostic agents from exogenous factors and their clinical actions.

Hematologists, biochemists, molecular biologists, protein scientists, toxicologists and toxinologists will appreciate the comprehensive overview of this important field of research provided by these proceedings.

KARGER